A Textbook of
GENERAL EMBRYOLOGY

MAVEN BOOKS

A Textbook of
GENERAL EMBRYOLOGY

WILLIAM E. KELLICOTT

Professor of Biology, College of the City of New York

MAVEN BOOKS

Chennai   Trichy   Tirunelveli   New Delhi

MAVEN BOOKS

An Imprint of **MJP Publishers**

ISBN 978-93-87867-51-2 **MAVEN Books**

All rights reserved No. 44, Nallathambi Street,
Printed and bound in India Triplicane, Chennai 600 005

MJP 714 © Publishers, 2019

Publisher : **C. Janarthanan**

Publisher's Note

The legacy of a country is in its varied cultural heritage, historical literature, developments in the field of economy and science. The top nations in the world are competing in the field of science, economy and literature. This vast legacy has to be conserved and documented so that it can be bestowed to the future generation. The knowledge of this legacy is slowly getting perished in the present generation due to lack of documentation.

Keeping this in mind, the concern with retrospective acquiring of rare books has been accented recently by the burgeoning reprint industry. MAVEN Books is gratified to retrieve the rare collections with a view to bring back those books that were landmarks in their time.

In this effort, a series of rare books would be republished under the banner, "MAVEN Books". The books in the reprint series have been carefully selected for their contemporary usefulness as well as their historical importance within the intellectual. We reconstruct the book with slight enhancements made for better presentation, without affecting the contents of the original edition.

Most of the works selected for republishing covers a huge range of subjects, from history to anthropology. We believe this reprint edition will be a service to the numerous researchers and practitioners active in this fascinating field. We allow readers to experience the wonder of peering into a scholarly work of the highest order and seminal significance.

MAVEN Books

PREFACE

General embryology should occupy an important place in the collegiate study of biology. In no other connection are the essential phenomena of life better illustrated, in no other form are they more readily appreciated. The facts of embryology lead directly to the great problems of the science of biology as it exists to-day, and many fundamental biological conceptions either are directly connected with, or are illuminated by, the study of the early phenomena of individual development.

The author's experience has clearly indicated that the subject has this value as a collegiate study. Indeed, the book is the direct outgrowth of such experience, and it has, in substance, been in use as such a text for several years. In its present form it is hoped that it will be found useful to the student who is endeavoring to comprehend the general principles of the science of life, as well as to the student preparing for the professional study of some field of biology or of medicine.

Its design as a textbook, rather than as a handbook, accounts for certain characteristics. The topics considered have throughout been approached from the standpoint of their general biological relations, and in the selection of the facts mentioned and the topics discussed, as well as in the style and method of presentation, the student has been first in mind. The arrangement of the subject matter in two sizes of type may prove useful for those undertaking a brief course. In a few instances this has involved slight repetition, but repetition is not always a pedagogic evil.

At the end of each chapter will be found a list of references to literature. Usefulness to the student has been the only criterion in determining the admission of titles to these lists. Consequently there will be found titles of works of historical importance, of recent works containing contributions of importance or representing present tendencies in research, and of papers containing extensive literature references, valuable illustrations, or general summaries. As far as possible the lists contain references to works presenting both, or several, sides of mooted questions mentioned in the text. There will also be found, in nearly every instance, the titles of papers from which illustrations may have been taken.

To a large extent the figures have been redrawn, from the original sources, for this work: it is a pleasure to notice the uniform courtesy with which authors have granted permission to make this use of their illustrations. The following special debts are gratefully acknowledged: to Prof. Edmund B. Wilson and The Macmillan Company, for clichés and for permission to copy a considerable number of illustrations in their "The Cell in Development and Inheritance"; to Prof. Gary N. Calkins, The Macmillan Company, and Lea and Febiger, for clichés and for permission to copy certain illustrations in their "The Protozoa" and "Protozoology"; to Prof. Ulric Dahlgren, Prof. William A. Kepner, and The Macmillan Company, for permission to copy certain illustrations in their "Principles of Animal Histology"; to Prof. J. W. Jenkinson and the Delegates and Secretary of the Clarendon Press, for clichés from their "Experimental Embryology"; and finally to Herr Gustav Fischer and to the several authors, for clichés and for permission to copy or otherwise make use of illustrations from Korschelt and Heider's "Lehrbuch," Oscar Hertwig's "Handbuch," Doflein's "Protozooenkunde," and Ziegler's "Lehrbuch." In every instance specific reference, both to the immediate and the ultimate sources of the figures borrowed, is made in the legends. I desire also to acknowledge my indebtedness to the authorities of The Johns Hopkins University, for the use of valuable library facilities.

W. E. K.

BALTIMORE, MD.,
March, 1913.

CONTENTS

CHAPTER I

CHAPTER II

CHAPTER III

CHAPTER IV

CHAPTER V

CHAPTER VI

CHAPTER VII

CHAPTER VIII

TEXT-BOOK OF
GENERAL EMBRYOLOGY

CHAPTER I

ONTOGENY

LIVING organisms come into existence only as the offspring of preëxisting living organisms of the same kind or species. Aristotle's belief that eels were generated from mud and slime is represented to-day, in many youthful minds, by the firm conviction that a horse hair, if only kept long enough in water, will surely "turn to a worm." From the time of Redi the belief that the living might be generated from the wholly non-living gradually became restricted, in its application, to lower and still lower groups of organisms. For a long time it remained applied only to those forms at the lower limit of the living—the bacteria. From this position the belief that "spontaneous generation" of organisms occurs nowadays, was finally driven by the brilliant demonstrations of Pasteur and Tyndall that even these simplest and smallest of organisms arise only from preëxisting living organisms of the same kind.

This property of producing new, specifically similar individuals is one of the few really distinctive characteristics of living things, and since the newly produced resemble closely the parent form, we speak of the property as *Re*production. The fact that at corresponding ages offspring resemble their parents, is the fact of heredity. But when these offspring are first distinguishable as separate and new individuals they bear little or no visible resemblance to the adult organisms producing them. This resemblance appears gradually, as the result of a long series of processes, complex and often very special, involving changes in structure, function, and form, only at the conclusion of which has the new·organism reproduced, more or less precisely, the form and other characteristics of its parents. The facts regarding all of these processes of development, of the external and internal changes in form and structure of the new organism, of the complex chain of processes leading to its first formation, and of the rôle of external factors throughout all of these, constitute the science of Embryology. We may define Embryology briefly then, as the science of the genesis of the adult organism.

As a general introduction to this subject we may first sketch in outline the broad features of reproduction among the lower animals, mentioning but a few of the almost infinitely varied forms which this process assumes. It should be understood in

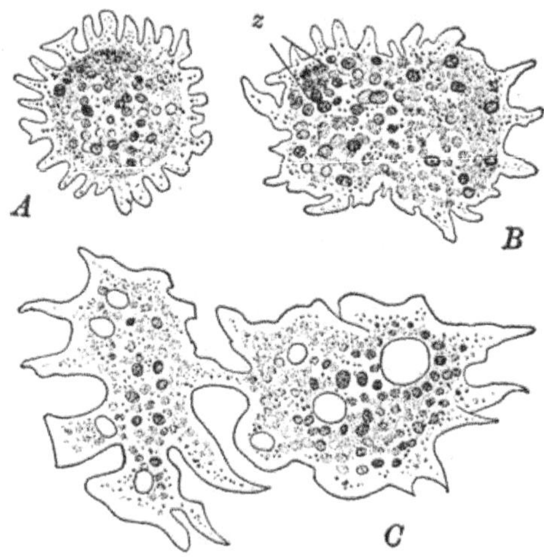

Fig. 1.—Simple fission in *Amœba vespertilio*. After Doflein. *A*. Normal vegetative form. *B*. Commencement of fission ("biscuit form"). *C*. Fission nearly completed; separation of daughter cells. *z*, cells of an Alga, *Zoochlorella*.

advance that the series of reproductive processes, of increasing complexity, to be outlined, has little if any phyletic significance; this arrangement is made for comparative purposes alone.

The simplest and apparently the most primitive mode of reproduction is that known as *fission*, characteristic of the single-celled organisms, the Protozoa and Protophyta. In the case of *simple* or *binary* fission a separation of the nuclear material of the cell into two separate masses is followed by a constriction of the cell body into two bodies, each of which may then form parts corresponding with those carried away by its sister cell, and finally develop into a creature resembling the original organism (Fig. 1). In many unicellular forms, particularly among the Sporozoa, a process of *multiple fission* or *brood formation* occurs. This is frequently preceded by growth of the organism to an unusual size; then an uninterrupted series of simple fissions, without intermediate growth or development on the part of the daughter cells, results in the formation of a

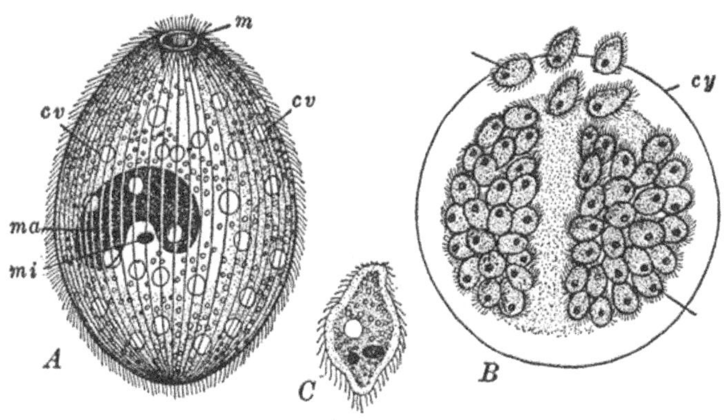

Fig. 2.—Multiple fission in a parasitic Infusorian, *Holophrya multifiliis.* After Hatschek. *A.* Normal vegetative individual. *B.* Cyst containing the products of repeated binary fission; some of the zoöspores are shown leaving the cyst. *C.* One of the zoöspores, enlarged. *cv*, contractile vacuole; *cy*, cyst; *m*, mouth; *ma*, macronucleus; *mi*, micronucleus; *t*, zoöspores.

large number of small organisms. These usually remain associated, frequently within a cyst formed by the parent cell, until the whole process of fission is completed (Fig. 2). In other cases the nucleus, or nuclei, alone may divide, either successively or simultaneously, into a large number of separate nuclei; the cytoplasm immediately surrounding each nucleus is then cut out as a separate cell, so that the organism appears to fragment simultaneously into a large number of small daughter organisms (Fig. 3). The number of new individuals formed in this way may vary from four, as in some Infusoria, up to as many as several hundred in many different forms, especially among the Sporozoa. When the number is large the process is more frequently termed *spore formation* or *sporulation* and the products are known as *swarmspores* or *zoöspores*. Technically the term sporulation or *sporogony (metagametic division)* is used only when this process of multiple fission occurs subsequently to a process of cell fusion. If no such process of cell fusion or conjugation has preceded, the process is termed *schizogony (agamogony)*, and the products of the division are called *schizonts (agamonts)*.

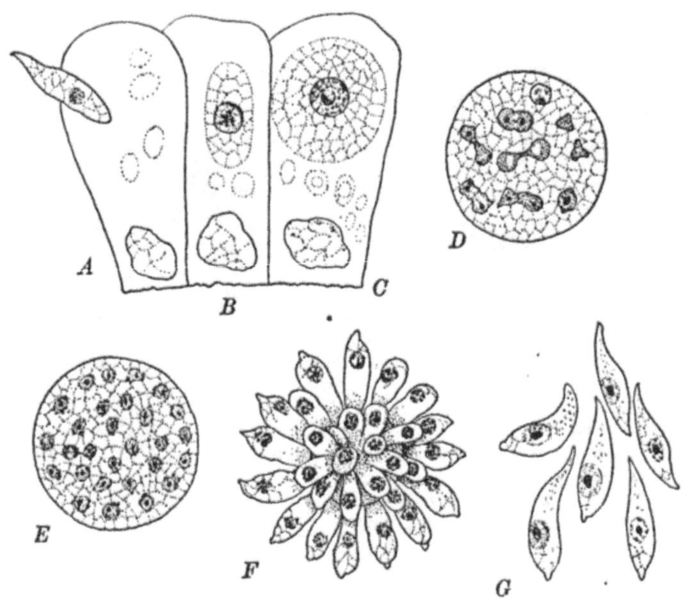

Fig. 3.—Multiple fission (schizogony) in the Sporozoan, *Coccidium schubergi*. After Schaudinn. *A*. Entrance of organism ("sporozoite") into epithelial cell of host. *B*, *C*. Two stages in growth. *D*. Multiple division of nucleus. *E*. Daughter nuclei superficial, cytoplasm still undivided. *F*. Fission of the superficial cytoplasm surrounding the nuclei, leaving a central undivided mass. *G*. Free schizonts ("merozoites").

Reproduction by analogous processes also called fission, simple or multiple, is only occasional among the many-celled animals. In a few such forms the entire body separates as a unit into two or more organisms. Organs and tissues in the plane of separation are divided, and each of the daughter organisms then regenerates parts corresponding to those carried off by the other, forming new individuals smaller than the parent form, but otherwise similar to it (Fig. 4). This fission in the Metazoa is not really comparable with the similarly named process in the Protozoa; it represents a special acquirement and is usually, though not always, associated with other more complex modes of reproduction. Normal fission is known to occur in many genera among the Cœlenterates, and less frequently among the Porifera, Platyhelminthes, Annulata, Bryozoa, and Echinoderms.

It is obvious that in reproduction by fission the individuality of the parent organism is lost in the act of giving rise to the new individuals, although none of the substance of the original organism perishes in the act.

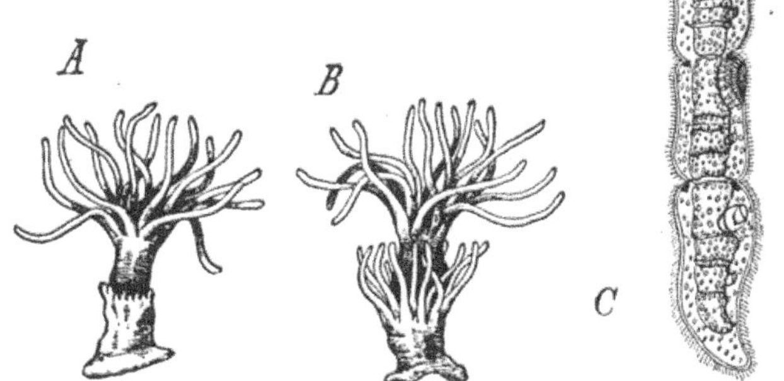

Fig. 4.—Fission in Metazoa. *A, B.* Two stages in the transverse fission of the Actinian, *Gonactinia prolifera.* From Korschelt and Heider, after Blochmann and Hilger. C. Successive transverse fissions in the Platyhelminth, *Microstomum lineare.* After L. von Graff. *I, II, III,* mark the levels of the successive fissions; a fourth fission also is indicated. *p, p,* pharynges.

In the fission of Metazoa usually many of the structures of the parent are simply transferred to the new organisms with comparatively little differentiation of parts anew, out of a visibly undifferentiated condition. In the Protozoa this may or may not be the case. In some of the highly organized Ciliata most of the structural differentiation disappears just previous to fission, after which each daughter cell differentiates a typical form and structure anew (Fig. 5). In other Protozoa there is a considerable transference of characteristics accompanied by a lesser amount of regeneration or redifferentiation. After multiple fission or sporulation, the end products of the process are commonly minute and visibly quite unlike the adult form; this they come to resemble only through growth and differentiation, that is, through processes of true development.

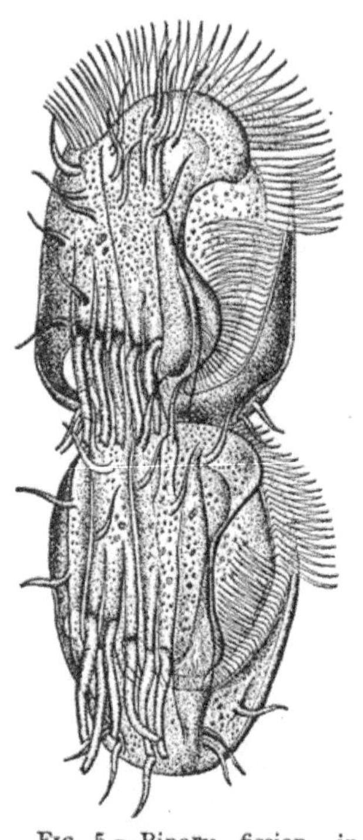

FIG. 5.—Binary fission, in the Infusorian, *Euplotes harpa.* After Wallengren. Division stage immediately before the separation of the daughter cells.

A reproductive process closely allied to fission is the familiar process of *budding.* Here one or several small outgrowths or "buds" are produced from some portion of the parent organism, and develop into forms resembling the parent, either before or after becoming detached. This process frequently appears as a sort of unequal fission and may indeed rightly be regarded as such; but usually so great is the disparity in size, as well as in extent of differentiation, between parent and buds, that the processes are properly distinguished. Budding occurs in many Protozoa (*e.g., Ephelota* (Fig. 6, *A*), some Rhizopoda), and is quite frequent among the lower Metazoa, particularly in colony forming species. It occurs among the Porifera, Cœlenterata, Platyhelminthes, Annulata, Bryozoa, and Tunicata (Fig. 6, *B*). In budding there is ordinarily very little direct transference of structures, the development of the bud occurring after it has been completely delimited as a comparatively undifferentiated mass; its development may then be almost complete before the separation of the two organisms. Nor does this process involve necessarily the loss of identity of the parent, which may continue to live and produce buds for a considerable period.

In a few Protozoa, fresh water sponges, Trematodes and

Bryozoa, *internal buds* are formed within the parent cell or body (Fig. 7). These become free and develop into new individuals ordinarily only after the death of the parent body. Although the formation of these *gemmules* (Porifera), or *statoblasts* (Bryozoa) suggests that form of reproduction characteristic of the higher Metazoa, *i.e.*, through internally formed germ cells, it will appear later that the two processes are not at all to be compared.

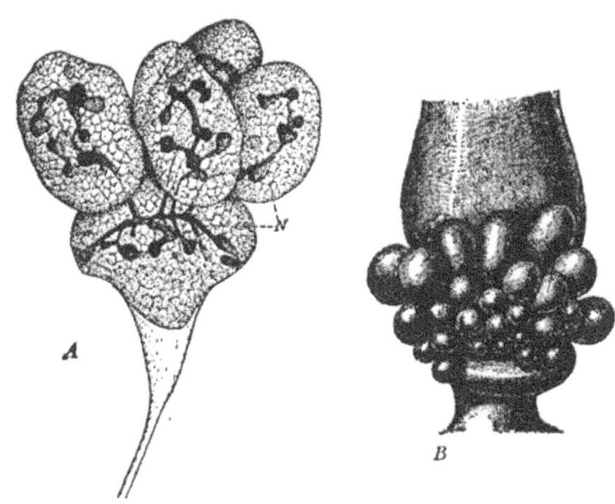

Fig. 6.—Reproduction by budding. *A.* In the Infusorian, *Ephelota.* From Calkins, "Protozoa." *N,* branched macronucleus extending from the parent cell into each bud. *B.* In the Tunicate, *Doliolum.* After Neumann. Ventral outgrowth of a phorozooid. *b,* reproductive buds of various ages; *g,* germinal knob; *s,* stalk.

In some of the shelled Rhizopods (*Euglypha, Arcella*) a combination of the processes of budding and fission occurs. In this process, called *bud-fission*, about one-half of the protoplasm flows outside the original shell, which these forms possess, and secretes a new shell like that of the parent form; or the rudiments of the new shell may be formed before the appearance of the bud. The cell body then divides and the two similar individuals may either separate completely, or they may remain in contact, forming after repeated bud-fission, an aggregate of related though distinct organisms.

In many of the Protozoa multiple fission may be incomplete,

or the daughter cells, not scattering as separate individuals, may remain associated during division and growth, forming a colony of organically related cells which as a whole is to be regarded as the individual organism. Among nearly all of these colonial or compound forms the entire colony, or *cœno-*

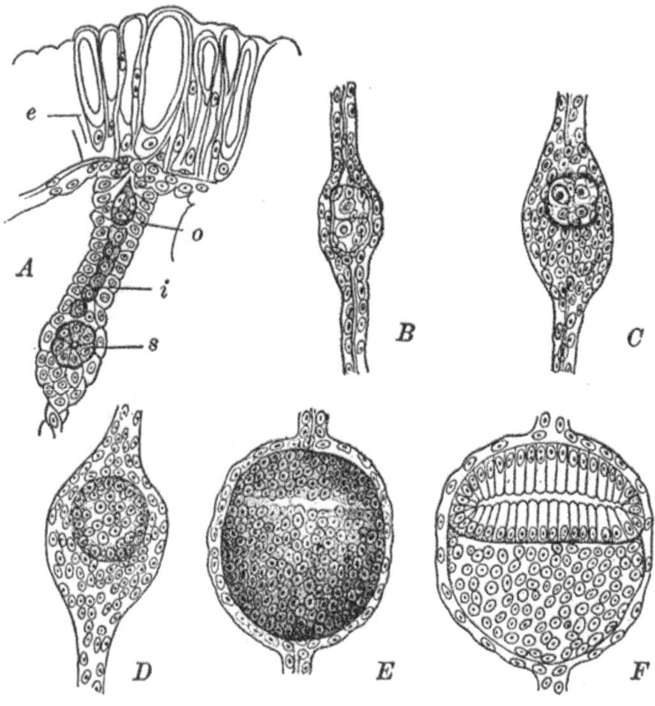

Fig. 7.—The formation and development of the statoblast, in the fresh water Bryozoan, *Cristatella*. *A*. After Braem. Others, after Verworn. *A*. Longitudinal section through the funiculus, showing the relations of the statoblast. *B–E*. Optical sections of stages in the development of the statoblast. *F*. Section showing rudiment of embryo. *e*, superficial ectoderm; *i*, inner layer of funiculus (ectoderm); *o*, outer layer of funiculus (endoderm); *s*, rudiment of statoblast.

bium as it is called, is not later involved as a unit in reproduction, but the component cells may individually undergo fission leading to the formation of new colonies. In forms like *Pandorina* or *Platydorina* all the cells are alike, and each reproduces, so that there may be as many new colonies formed as there are cells in the original colony (Fig. 8). In other forms the power of reproduction is limited to certain cells, termed

gonidia, and there results a sharp distinction between reproductive and vegetative cells (Fig. 9). Thus in *Pleodorina*, of
the thirty-two cells forming the colony, four anterior cells are
purely vegetative, the remaining twenty-eight are gonidial;

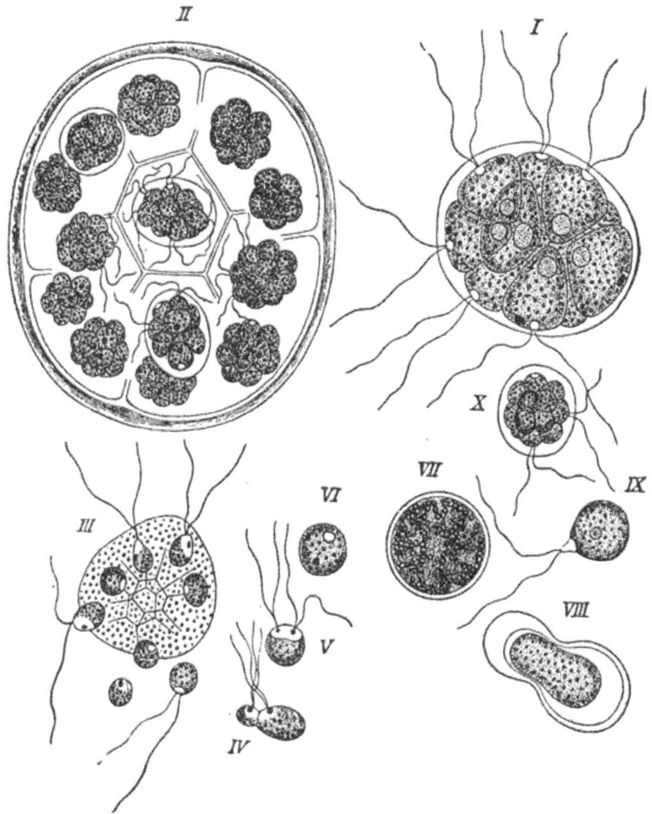

Fig. 8.—Reproduction in the colonial Flagellate, *Pandorina morum*. From
Hertwig, after Pringsheim. *I.* Free-swimming vegetative colony. *II.* Daughter colonies formed by the fissions of each individual in a colony similar to the
preceding. *III.* Gamete formation in a colony formed as above. *IV, V, VI.*
Stages in the conjugation of gametes to form a zygote (*VI*). *VII.* Zygote after
growth to full size. *VIII, IX.* Production of swarm spores by the zygote.
X. Vegetative colony formed by fissions of the swarm spore.

while in *Volvox*, where the fully developed colony may include
as many as twelve thousand or more individuals, only five to
fifty cells, scattered through the posterior half of the colony are
gonidial, the remainder being vegetative (Fig. 10).

Among most of these colonial Flagellates, and indeed in many other Protozoan groups, at irregular intervals these gonidial cells lose the property of directly giving rise to new colonies, and become very highly differentiated in structure and in behavior. These highly modified gonidia are termed *gametogonidia*, the ordinary gonidia being then given the name

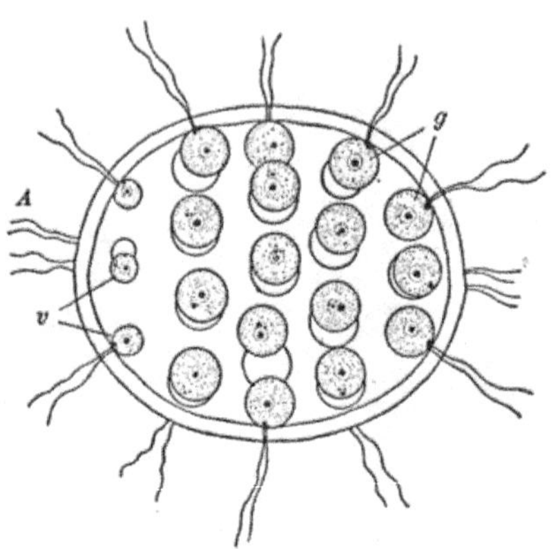

Fig. 9.—Colony of the Flagellate, *Pleodorina illinoisensis*. Lateral view. After Kofoid. *A*, anterior; *g*, gonidial (reproductive) cells. *v*, vegetative cells.

of *parthenogonidia*. The gametogonidia form specialized cells termed *gametes* which must meet and fuse in pairs, *i.e.*, *conjugate*, before reproduction may proceed. In the simplest cases these two gametes are nearly or quite alike. In most cases, however, gametes of two very unlike forms are produced by different gametogonidia. These must conjugate in pairs of unlikes before reproduction is possible. A series of forms illustrating the progressive differentiation of the gametes, or *germ cells*, is described in Chapter V. For the present we may merely notice that in such a colonial form as *Volvox*, under certain conditions, the parthenogonidia cease reproducing; certain of them (*oögonidia* or *ovaries*) enlarge, and each differentiates a large non-motile cell, the *ovum*, or *oösphere*, or *macrogamete*, while others (*spermagonidia*, or *spermaries*) after enlarging similarly, divide repeatedly, each forming a large number, often as many as one hundred and twenty-eight, small actively motile cells termed *sperm cells* or *microgametes*. A single sperm from one colony then meets and conjugates with a single ovum from another colony, forming thus a single

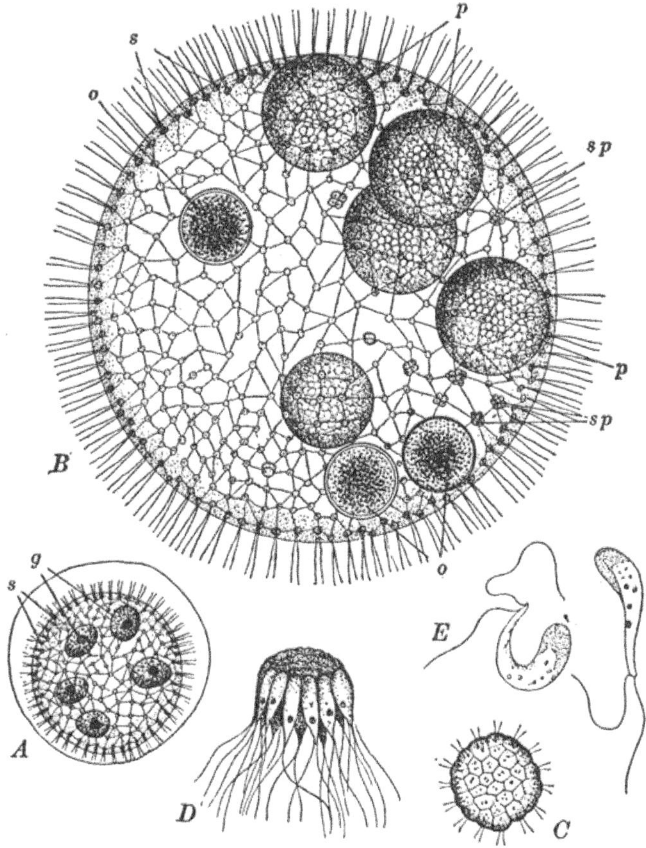

Fig. 10.—Reproduction in the colonial Flagellate, *Volvox*. *A*. After Prings-heim, *B–E*, after Klein and Schenck. *A*. Young colony showing distinction between, *s*, somatic cells, and *g*, reproductive cells (parthenogonidia). *B*. Older colony showing, *p*, parthenogonidia, *o*, oögonidia, *sp*, spermagonidia in various stages of formation. *C*. Spermagonidium consisting of thirty-two spermagametes, seen from the side in *D*. *E*. Spermagametes more highly magnified.

cell or *zygote*, which through continued fission gives rise to the individuals of a new colony.[1]

[1] Useful diagrams of these forms of reproduction and gamete formation are given in HEGNER, "An Introduction to Zoology," New York, 1910, pp. 112–115.

Reproduction following gamete formation and fusion (*syngamy*) is commonly known as "*sexual*" reproduction, while the term "*asexual*" is applied to modes of reproduction which do not involve the fusion of gametes. It is now clear, however, that several different and unrelated processes are included under each of these heads. Thus as "asexual" must be included such diverse modes of propagation as budding, fission, brood formation, parthenogenesis, development from spores, and so forth; and "sexual" reproduction would also follow many forms of syngamic fusion. Two further conditions tend to rob these terms of their precise significance; these are, the existence of transitional conditions, some of which have been mentioned above, and the doubtfully essential character of the primary relation of syngamy to reproduction. These useful terms are to be retained only as convenient though inexact expressions, and are to be used in much the same way that we still employ the convenient words "vertebrate" and "invertebrate." Even so, we may avoid any unintentional implications by substituting the more exact terms *amphigony* and *monogony* for sexual and asexual respectively.

Without suggesting the idea of direct relationship among any existing forms, we may say that it is but a short step from the processes of gamete formation and reproduction among the colonial Protozoa, to the mode of reproduction characteristic of the Metazoa, which after all may be considered highly organized cell colonies. One of the more distinctive as well as more obvious characteristics of the Metazoa is the structural and functional differentiation of large groups of cells as *tissues,* each variety of tissue performing, in the animal economy, chiefly one function, such as conduction, support, or excretion. Among these various tissues is the *reproductive* or *germinal* tissue, which, in all save a few of the lower Metazoa, is always in the form of definite organs, the *gonads.* The distinction, suggested by some of the colonial Protozoa, between the reproductive and the vegetative tissues was probably the earliest of those "physiological divisions of labor" which involved tissue differentiations in the Metazoa. The essential cells of the gonads, which correspond functionally to gametogonidia, divide repeatedly, the products of their multiplication being, with few exceptions, ultimately thrown outside the body as the highly modified *germ cells* or *gametes*. In all the Metazoa these germ cells are of two quite unlike forms; the *ova* or "*eggs*," are large and ordinarily non-motile, while the *spermatozoa* or *sperm cells*, are small and actively motile. These Metazoan germ cells clearly correspond with the ova or macrogametes and the sperms or microgametes of certain of the Protozoa. The gonads forming the ova and spermatozoa are known as the *ovaries* and *testes* respectively. In most of the Metazoa germ cells of only a single kind are formed by a single organism, a condition which leads to the primary distinction of sex; individuals forming ova are called females, those forming spermatozoa, males.

In comparatively few kinds of animals does a single individual normally possess gonads of both types, and thus become capable of forming both kinds of germ cells, either simultaneously or successively. Such a condition is known as *hermaphroditism;* it occurs chiefly among the lower Metazoa, such as the Platyhelminthes, Nemertea, some Annulates and Tunicates, and less frequently among the Molluscs, Echinoderms, Bryozoa, Brachiopoda, and Crustacea. Among the Chordata normal hermaphroditism is found only rarely (some Teleosts), though it may occur as an abnormality in any group. In a few animals a special form of hermaphroditism occurs, where a single gonad may produce first sperm and later ova, a condition known as *protandry* and found in some Nematode worms and in the Cyclostome, *Myxine*, for example. The process of forming first ova and later spermatozoa, known as *protogony*, is very rare among animals.

In the reproduction of all except a very few of the Metazoa, the initial phase is the union of an egg cell and a sperm cell; this process is known as *fertilization* or *syngamy*, and the double cell thus formed, which is called the *zygote* or *oösperm*, then gives rise directly to the new individual. Not all of the germ cells formed by an organism actually happen to give origin to new organisms, although any may do so. But with infrequent exceptions, where unusual methods of reproduction occur at times, a few of which have been noted above, new Metazoan individuals arise only from the union of two germ cells.

The substance which forms the reproductive cells or gametes of an organism is called the *germinal substance*, or briefly, the *germ*. This is visibly distinguishable at a very early period in the existence of the new organism, from that material which is to form all of the remainder of the organism, in distinction known as the *somatic tissue* or *soma*, or simply as the *body*. Among the higher forms these two kinds of substance—germ and soma—have very different histories and fates. According to the theory of Germinal Continuity, elaborated by Weismann, the germ represents or contains an organic substance which has been in a living state since the beginning of life, and which must continue in this state, in some form, as long as living things shall be produced. The soma, on the contrary, is thought to be built up around the germ, anew and under its influence, in each generation of organisms. Upon the death of the individual it is destroyed completely as living substance; somatic cells finally leave no descendants. Thus in species which reproduce by this method the soma or body is wholly temporary, while the germ may properly be said to be potentially ever enduring. For while actually the greater part of the germ substance formed in an organism is destined to perish, either before the body or with it, or at any rate with the race, some germ must always remain, producing the generations of the future. The essentials of this idea are expressed in the accompanying diagram (Fig. 11).

On account of its usefulness the value and significance of the distinction between germ and soma are frequently over-emphasized. In many organisms the distinction can scarcely be drawn at all, for under certain conditions, either normal or unusual, cells which are evidently "somatic" may take on reproductive characteristics and function as germ cells. Many such cases are known among animals, and among the plants reproduction from somatic tissues and cells is very common, indeed in some it appears to be the normal method of reproduction.

Among the Protozoa this familiar distinction between germ and soma cannot be drawn at all. In these simple forms the

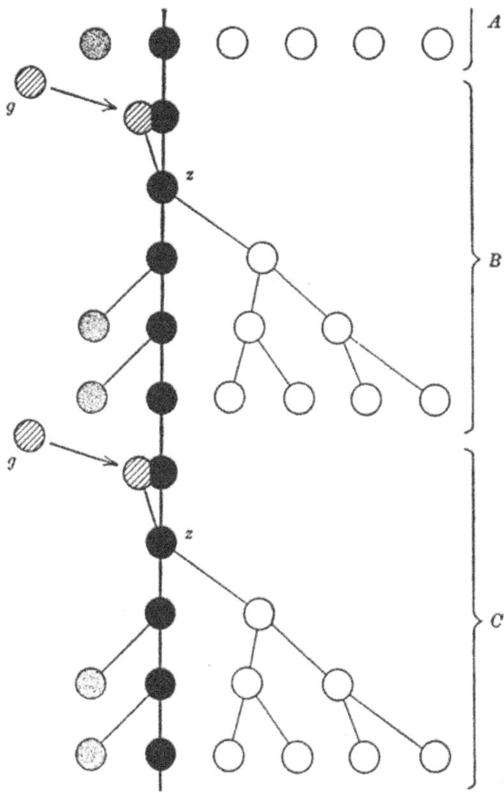

Fig. 11.—Diagram illustrating the theory of Germinal Continuity. *A, B, C,* represent successive generations; *g,* gamete (sperm cell) produced by another organism; *z,* zygote. White circles indicate successive cell divisions within the somatic tissues, the existence of which terminates with the organism of the given generation. Solid black indicates the germ. Dotted circles indicate gametes which may perish, or may unite with those of another organism.

process of reproduction is not so directly associated with the processes of gamete formation and fusion. Nearly all non-colonial Protozoa, while in the so-called vegetative state, have the power to reproduce by fission, so that the plasm of these cells is both germinal and somatic in the Metazoan sense. It is only at irregular intervals that the individuals ordinarily multiplying by fission, lose this property as well as their vegetative characteristics, become specialized as gametes, and require to undergo syngamy, later resuming their duplex vegetative and reproductive character.

While it might be misleading to say that the reproductive cells of Metazoa are, like the Protozoa, both germ and soma, yet it is quite true that in these germ cells we have a substance which produces both germinal and somatic tissues. In a sense we are hardly justified in saying that the soma is built up *anew* in each generation, while only the germ has a continuous existence. The germ cell is potentially soma as well as germ, and for a time during the early development of the organism there is no visible distinction; this distinction occurs very early in the development of a few forms, but in most organisms not until a considerable number of cells has been formed. In development the germ cells give rise to other cells like themselves (germ) and to cells unlike themselves (soma) and we may regard the "unlike" as "new."

The common conception of the life of a species as a succession of generations of individuals linked together by the germ, while superficially true, leads to a fundamentally erroneous point of view. The fertilized germ cell is just as much the individual organism as the matured individual is. The species is no more a succession of somas than it is a continuous germ. It is not the function of the germ to provide links between successive generations of "organisms" or somas, any more than it is the function of the soma to insure the continuity of the germ, and to provide materials for its increase and means of its dispersal. We should recognize that the essential continuity between successive generations is, after all, not continuity of plasm but "continuity of organization."

The term reproduction, strictly speaking, does not mean quite the same thing among Metazoa and simple Protozoa. Among the Protozoa the formation of free daughter cells, by fissions of the zygote or its descendants, constitutes reproduction. Among the Metazoa the corresponding fissions of the zygote and its daughter cells are not considered in themselves reproductive processes, but as steps (cell divisions) in the building up of the whole new individual, as but one phase in the general process of reproduction.

Some physiological reorganization of substance, such as ordinarily results from the intermingling of the plasmas of two individuals or lines, seems a necessity for the continued existence and reproductive activity of most organisms. In some Insects (Aphids, *etc.*), Rotifers, Crustacea, and other forms, reproduction occurs normally, through long periods, without any such syngamic fusion, the new organisms developing from single unfertilized ova (*parthenogenesis*). While this condition is, in these cases, clearly derived from the normal, yet it seems to illustrate the non-essential relation of syngamy and reproduction. In such forms syngamy does occur under certain conditions or during certain periods in the life cycle of the organism. For example, the difficult conditions of winter or drought may be successfully withstood by the organism while in the form of the zygote.

So too in many, perhaps most, Protozoa, reproduction or fission proceeds normally and for long periods without fertilization or conjugation. The process of conjugation is opposed to reproduction and may actually inhibit it for a time. Here it appears frequently to be associated with the onset of conditions unfavorable to the existence of the organism in its vegetative condition.

It seems, therefore, that the processes of fertilization and reproduction may not be essentially related, and that the intermingling of the plasmas of two individuals is related directly to phenomena other than reproduction. Such a modification of substance as results from fertilization, however, may be essential to continued existence, and it is certainly true of most Metazoa that such a plasmic fusion is an organic necessity. In the simple Protozoa this may be accomplished at any time by the fusion of two individuals in the form of gametes. In the Metazoa, however, it is obvious that this necessary intermingling of substance occurs only when the organisms are in the form of single cells, *i.e.*, gametes. And thus it comes about that in the many-celled animals, reproduction and syngamy are so uniformly associated, and while these processes may not have been related primarily, in some instances are not even at present, yet now they have come to be so related in the vast majority of Metazoa, that fertilization actually appears as the first and most important step in the whole chain of reproductive events.

The actual processes involved in the formation, from the zygote, of the mature Metazoan individual are extremely complicated and diverse, but they are for the most part reducible to three fundamental general processes. We must leave aside, for the present, the causal or directive processes which, though probably the essentials of development, are still obscure and little known. X The grosser external phenomena of development are, essentially, growth, cell division, and differentiation. The living germ is contained within the limits of a single cell, often of minute dimensions and only slightly differentiated visibly; the mature organism consists of an enormous number of cells, comprising a considerable mass, and exhibiting various degrees of differentiation in diverse directions. The transition from one of these states to the other is a gradual process, proceeding by minute steps; yet it is convenient to consider the whole life history of an organism as a succession of phases, each with some chief characteristic.

First, complex processes occur within the gametes or germ cells themselves, concerning chiefly their nuclei, as a result of which they come to have a constitution quite unlike that of the somatic nuclei. These preliminary events we group under the term *gametogenesis*, or *oögenesis* and *spermatogenesis* in the ova and sperm cells respectively. Then normally follows *fertilization* or *syngamy*, the fusion of the two gametes, derived from two different organisms, into a single cell which is the "new" organism. Through fertilization a typical nucleus is reconstituted in the zygote, and there follows a period of rapid cell multiplication which is called the period of *segmentation* or *cleavage*. During cleavage are formed the cellular elements which are to be built into the structures of the simple embryo, and various differentiated substances of the egg are segregated among different groups of cells. Following this are the phases of *blastula* formation, when the cells become arranged in a definite layer, and then *gastrula* formation when the cells are rearranged into two definite layers. Then comes the period of *embryo formation*, when the cells of the layers are moulded into the earliest beginnings of the chief systems and organs, blocking these out in the simplest manner. During this last phase growth becomes very rapid, accompanied by continued cell division, no longer termed cleavage, and as the formation of organs becomes more complete and more particular, the embryo increases in bulk and dimensions. This period of embryonic development may occupy a long time, and usually leads to the formation of an organism which is capable of leading an independent life, either as a larva or as a form closely resembling the adult, except in size. Finally, accompanied by continued growth, the last phases of development appear as cellular differentiation becomes more complete, and the organism begins to assume more fully the characteristics of its parents. When the reproductive tissues become functional as such, the animal is considered mature and its development complete, although in a true sense development is never entirely completed, for the form of the organism never becomes definitely fixed, and cellular differentiation seems never to cease during the life of the organism.

It is important to remember that all of these phases of development are continuous and more or less overlapping, and in all of them, excepting perhaps the earlier, where the important changes concern chiefly the structure and the composition of the germ nuclei, the three processes—growth, cell division, and differentiation—are going on together. Yet in general it is clear that in the early stages of development, after the gametic nuclei are differentiated and fused, cell division is the process of greatest activity; then follow stages during which development is characterized chiefly by growth; and lastly the final aspects are chiefly the result of cellular or tissue differentiation, processes often described separately under the term *histogenesis*.

This brief outline reflects the fundamental character of the relation of Embryology, as of all biological science, to the Cell Theory. The recognition of the ovum (Schwann, 1839; Gegenbaur, 1861) and spermatozoön (Schweigger-Seidel, 1865) as modified cells, of the basic importance of cell continuity in development (Virchow), and of the processes of fertilization, cleavage, growth, and differentiation as essentially cell processes, marked noteworthy and fundamental steps in the history of the science of development. But our recognition of the importance of this relation, and of the especial importance of the cell as the *descriptive unit* in development, should not obscure the fact that in many developmental processes we cannot recognize the cell as the actual unit of physiological activity. Many important steps in development concern elements which are distinct and individual intra-cellular elements. And later, during the cleavage period, the boundaries of specific materials behaving as units in development do not always coincide with cell boundaries or distributions. We must regard the view that the cells are the ultimate units in development as a stage in the history of opinion, and for the present recognize certain intra-cellular elements as the "ultimate" structures in development.

But the province of Embryology is not merely thus to describe the upbuilding and unfolding of the structure and form of the new organism through these successive stages of development; it is, further, to describe the more fundamental *processes* involved in this development, and still further, to summarize these descriptions of both kinds in the form of simple general statements or laws. In the historical development of the science of Embryology, as of any natural science, the description and comparison of visible forms and conditions came first. This morphological account of development, concerned chiefly with the description of *what* happens, *what* is produced in development, has now been accomplished to such an extent as to furnish a basis of this kind sufficient for immediate necessity. Next comes the study of the real processes leading to the production of one condition out of another, processes which underlie the externally visible form changes. This physiological aspect of Embryology is concerned more with *how* development occurs, *how*, and through the operation of what factors or mechanisms, one condition leads to another. In a way this is also the *why* of development—not "why" in the philosophical sense of course, but in the sense of "how does it happen that" these things occur in development. Here the two methods of observation and experiment are combined and by the artificial modification or the elimination of one condition of development after another, the essential factors are discovered and their modes of operation determined. The science of Embryology has now fairly entered upon this stage and the dominant note of the subject to-day is this search for underlying processes and modes of action. But as yet it is impossible to say that we have reached the final period of the formulation of the broad fundamental generalizations which give unity to the infinitely diverse phenomena of development, and which are expressed in the form of laws. While something has been accomplished in this direction, the basis of fact is not as yet sufficiently broad, and the necessity of frequent restatement of such "laws" shows their formulations to be premature, save as guides in investigation.

These steps in the development of the science of Embryology do not so nearly represent the course of thought and hypothesis as that of actual knowledge and achievement. For even in the eighteenth century the earliest embryologists had their hypotheses as to the causes of this mysterious process of development. They offered first what seemed to them an explanation of the facts of development which came to be termed the idea of "evolution" or *preformation*. This idea was that within the germ, either in the egg ("ovists") or in the spermatozoön ("spermists," "animalculists") there was contained a miniature organism resembling, in a general, or even in a precise way, the adult form (Fig. 12). And this miniature had merely to expand, or to unfold and grow, to produce the individual of the next generation. The relation between the germ and the adult seemed much like that between the bud and the branch—all the parts present in minute rudiments, ready to come forth and expand. We may recognize in this idea a·morphological conception of development such as we should expect to appear first. This conception, with which are associated such great names as Malpighi, Bonnet, and Haller, proves in reality an attempt to explain development by denying its occurrence. For the assumed formation of the original individuals of a species by the Creator involved at the same time the creation, within them, of the preformed germs of all the other later individuals of the species. The belief that the germ cells of an organism contained in miniature the members of the second generation necessitated the further belief that in these latter must be contained, within still smaller limits, the individuals of the third generation, and thus *ad infinitum*. And so it was estimated that some two hundred millions of human beings were actually contained in this preformed condition within the ovaries of Eve. This conception of infinite *encasement* or "*embôitement*" proved to be the *reductio ad absurdum* of the theory of preformation in this its first and crudest form.

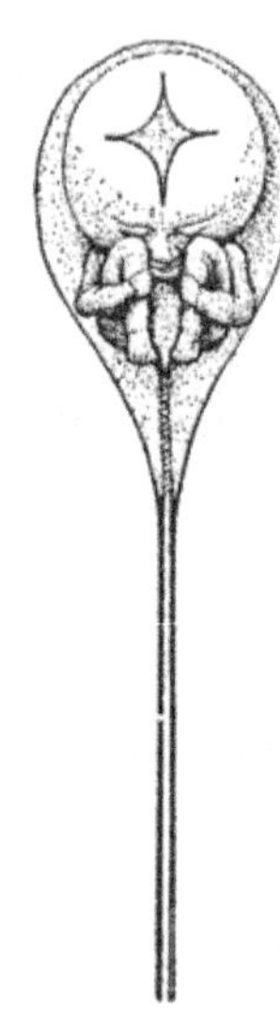

Fig. 12.—Drawing of a human sperm cell containing a miniature organism enclosed in a thin membrane. After O. Hertwig, from Hartsoeker (1694).

Those who actually observed the chick appear within the egg could not accept this naïve explanation of development, but believed that there occurred a true formation of parts anew out of unformed material not possessing at all the characters of the adult organism. This was Wolff's idea of *epigenesis*, clearly a physiological conception of development, following quite naturally the earlier morphological conception. In its original form epigenesis was chiefly a dissent from the idea of preformation rather than an explanation of develop-

ment. Indeed it seems now to have been merely a restatement of the fact that development occurs, leaving this fact to
be explained through the operation of some supernatural or
miraculous process, for the spontaneous generation of the
embryo within the egg was at first definitely assumed.

Thus we have almost from the beginning of embryological
study, two opposing explanations of the visible phenomena
of development, preformation explaining development by denying it, epigenesis explaining development by reaffirming it.
Since this early conflict of opinions, the crudity of which we understand when we think of the means then at hand for observing
such minute objects as are many eggs and embryos, there has
been constant opposition of morphological and physiological
interpretations of development. The modern understanding
of preformation is better termed *predelineation*, or better still,
predetermination, less crude, less complete and particular than
preformation. What is preformed or predetermined in the
germ in some way represents the embryo without being at
all like it. The idea of epigenesis, too, is to-day less complete;
a certain structural organization is admittedly present in the
germ as a heritage from previous generations, and real development occurs as a physiological process directed by this rudimentary structure already present. The history of these opinions indicates that neither conception is exclusively true, but
that development must involve both predetermination and
epigenesis; and the present endeavor is to find out not which,
but to what extent each, is true.

The present understanding of development seems to be an
extremely refined predetermination strongly tinged with epigenesis, using these words in their modern sense. A more
extended statement of this modern view and the facts upon
which it is based is reserved for a later chapter (Chapter VII).
Briefly stated, we believe that while the embryo, not to say
adult, is by no means preformed nor even fully predelineated
in the germ, yet there is a certain degree of protoplasmic
structure or regional differentiation in the germ cells. This is
spoken of now as the *organization* of the germ, and it may be

both material and dynamic (*i.e.*, energetic). And further this organization is definitely related to the structure of the future embryo and adult, having *reference*, but *not resemblance*, to the adult. The organization of the cytoplasmic part of the germ is itself a condition which develops (epigenesis) under the influence of the primary structure, or organization, of the nucleus. At present this inherited organization or predelineation of the nucleus seems primary and fixed, and to represent the only strictly predelineated portion of the germ, controlling and directing the later and epigenetic developmental processes, which may be said often to have commenced in the germ cells even before syngamy has occurred. But history warns us against believing that this organization of the nucleus will prove the ultimate organization. As knowledge becomes more complete this will be thrown farther back to restricted elements of the nucleus; indeed it seems probable now that the primary organization concerns, not the entire nucleus nor perhaps its chromosomal elements alone, but some, as yet invisible, problematic, chemical and physical configurations of its structure.

But we must not search for an explanation of the whole process of development, alone in the structure of the germ cells. We must look upon development as upon other forms of activity in living things, as a succession of reactions on the part of the organism to the normal stimuli of its surroundings. The things that an adult organism does are obviously *reactions;* it reacts to the conditions of its environment by making certain movements, forming certain substances, undergoing certain structural modifications; in short, by doing certain things collectively termed its *behavior*. The precise character of an animal's behavior is determined not alone by its structure, by the organs it has to react with, nor alone by its physiological condition at the time, nor alone by the nature of the external conditions acting; but by all of these combined. What the adult organism does at any particular moment is therefore determined by two interacting sets of conditions, one within the organism—its organization, the other without the organism its environment. Either set of conditions alone can lead to no action; for organismal activity is *reaction*.

Just so the developing organism, at whatever stage it be considered, reacts to the stimuli of its environment in a manner determined for the moment, on the one hand by its own state or "organization," both morphological and physiological, and on the other by the character of the stimuli acting. The ovum is not to be regarded as a mechanism wound up, ready upon receipt of a single stimulus, to go through its development into an adult organism. It is rather to be regarded as an organism which reacts to its surroundings by undergoing certain changes. This changed organism then reacts further by undergoing certain other changes. One reaction of the fertilized ovum is to cleave, of the blastula to gastrulate, and so on. Step by step, one condition succeeding another and leading to still another, the organism gradually alters its morphological and physiological characteristics. Throughout its whole existence the organism shows transformations of substance, energy, and form; we agree to set apart certain of these transformations occurring at a very early period, and to refer to them as processes of "development." The normal "behavior" of the egg or of the embryo is to develop. The processes of development are neither easier nor more difficult to explain than the phenomena of adult behavior, and they have just the same basis in the relation between the internal conditions, within the organism, and the external conditions, without the organism, at the time.

From this point of view the question why the egg develops is a problem not different, in its essentials, from why the organism grows, or why it seeks or avoids the light. None of these is to be solved by consideration of the organism alone, whether egg or adult, apart from the conditions acting upon the organism; both must be studied together.

With this conception of development in mind, we should here mention briefly one of the great generalizations that has come from the study of organic development, namely, the Biogenetic Law or the Theory of Recapitulation. Briefly stated this familiar theory is, that the organism, in its individual developmental history, tends to repeat in outline the evolutionary history of its species. This repetition is seldom particular, or detailed, never complete, yet so many of the phenomena of development can be satisfactorily interpreted from this historical point of view, seeming to have this historical significance rather than an immediately adaptive relation, that as a *general* statement the law remains fundamentally true.

This law is not so much an attempt to resume the facts of embryology as to apply these facts in the interpretation of racial history (Evolution). This application is in many instances difficult because of the fact that there has been an evolution of the egg, the embryo, and the larva, just as of the adult. The fact that the organism is specific at all stages of its existence, includes the parallel evolution of ova, and of all succeeding

Species	Zygote		Stages in Development				Adult
D	A_D		B_D		C_D		D
E	A_E	B_E		C_E		D_E	E
F	A_F	B_F	C_F	D_F		E_F	F
G	A_G	B_G	C_G	D_G	E_G	F_G	G
H	A_H	B_H	C_H	D_H	E_H	F_H G_H	H

Fig. 13.—Diagram to illustrate the essentials of the Biogenetic Law. Modified from O. Hertwig.

developmental stages, if there is to be any evolution of adult structures, else diversity of adult organization would depend upon external conditions of development, rather than upon egg organization. But we know that the eggs of any species of sea-urchin and star-fish will develop, respectively, into adult sea-urchins and star-fish of those species, although in the same dish, with identical environing stimuli.

Many important points concerning the relation between ontogeny and phylogeny may be represented schematically, as in the accompanying diagram (Fig. 13). Here we compare the ontogenies of five related species, the adults of which represent an evolutionary series; species E has evolved beyond D, F beyond E, and so forth. The first stage (*i.e.*, the fertilized ovum or zygote) of D is not merely a zygote (A), but it is the zygote of species D, and consequently indicated in our diagram by A_D; in its development this passes through the specific stages B_D, and C_D, to the adult D. Species E is more highly evolved than species D, but it begins its existence as a fertilized ovum which again is specific, this time A_E. In its development to the adult form this may pass through stages B and C similar to those of species D, but merely similar, not identical, else the result would be D and not E. Therefore, we call these intermediate stages B_E and C_E. Further, species E may pass through a stage in some particulars resembling D; this, however, does not exactly resemble D and is therefore designated D_E. Similarly for species F, G, and H; each is more highly evolved than the preceding. Each passes through stages which resemble stages in the development of the less highly evolved species, yet each stage is really specific.

Conditions of life change for the embryo as well as for the adult, and if these younger organisms are to remain in existence they must evolve to meet the changed conditions. The process of evolution concerns not merely the adult, but the organism at every stage of its existence. Stages such as B_G or C_H may finally become so highly modified that they are no longer recognizable as related to B and C, and might as well be termed X_G and Y_H. It is then said that these traits are "cœnogenetic modifications" in distinction from "palingenetic characteristics," which are obvious similarities to previous racial conditions. But recognition of the idea that the entire life history is undergoing evolution, at every point, very largely minimizes the value of this very common distinction between cœnogenetic and palingenetic traits in development, for in a very true sense all the traits of the developing organisms are in varying degrees both cœnogenetic and palingenetic.

Finally, we see that the problem why the egg develops into a form resembling its progenitors, rather than organisms of another kind, that is to say, the problem of heredity, may be more clearly understood by recognizing that the characteristics of the organism are specific at all stages of its existence. *The egg of the star-fish is just as much a star-fish as the adult is.* The germinal substance of successive generations of star-fish is directly continuous. This continuity of specific organization through the germ, combined with essentially uniform conditions of development, determines the essential uniformity of each series of interactions leading to the formation of a new adult organism. In a real sense the problem of heredity thus becomes the same as the problem of development. And the problem why the egg of the star-fish develops into a star-fish and not into a sea-urchin, is fundamentally the same as the problem why the star-fish is not a sea-urchin; it is the general problem of the evolution of organic diversity.

REFERENCES TO LITERATURE

In the "references to literature," given at the end of each chapter, the author's name and the title of the work, are followed by the reference to the journal in which the work appeared, or to the place of publication, in case the work is a separate publication. The number of the volume (Band, tome, *etc.*) is printed in black-face Arabic numerals, followed by the year of appearance. References to pages, parts, *etc.*, are omitted except in a few necessary instances.

The abbreviations of the more common references are as follows:

Amer. Jour. Anat. *American Journal of Anatomy.* Baltimore and Philadelphia.

Amer. Jour. Physiol. *American Journal of Physiology.* Boston.

Amer. Nat. *American Naturalist.* Boston and New York.

Anat. Anz. *Anatomischer Anzeiger.* Jena.

Anat. Hefte. *Anatomische Hefte.* Wiesbaden.

Arch. Anat. u. Entw. *Archiv für Anatomie und Entwicklungsgeschichte.*

Arch. Anat. u. Phys. *Archiv für Anatomie und Physiologie.* Leipzig.

Arch. Biol. *Archives de Biologie.* Leipzig and Paris.

Arch. Entw.-Mech. *Archiv für Entwickelungsmechanik der Organismen.* Leipzig.

Arch. gesamte Physiol. *Archiv für die gesamte Physiologie des Menschen und der Tiere.* Bonn.

Arch. mikr. Anat. *Archiv für mikroscopische Anatomie und Entwicke-lungsgeschichte.* Bonn.

Arch. Protist. *Archiv für Protistenkunde.* Jena.

Arch. Zellf. *Archiv für Zellforschung.* Leipzig.

Arch. Zool. Exp. *Archives de zoologie experimentale et général.* Paris.

Biochem. Bull. *Biochemical Bulletin.* New York.

Biol. Bull. *Biological Bulletin.* Woods Hole, Mass.

Biol. Centr. *Biologisches Centralblatt.* Leipzig.

Botan. Gaz. *Botanical Gazette.* Chicago.

Bull. Mus. Comp. Zool. Harvard Coll. *Bulletin of the Museum of Comparative Zoology at Harvard College.* Cambridge, Mass.

Ergebnisse und Fortschr. Zool. *Ergebnisse und Fortschritte der Zoologie.* Jena.

Ergebnisse Anat. u. Entw. *Ergebnisse der Anatomie und Entwickelungsgeschichte.* Wiesbaden.

Jena. Zeit. *Jenaische Zeitschrift für Naturwissenschaft.* Jena.

Jour. Anat. Phys. Paris. *Journal de l'anatomie et de la physiologie normales et pathologiques de l'homme et des animaux.* Paris.

Jour. Coll. Sci. Imp. Univ. Tokyo. *Journal of the College of Science, Imperial University of Tokyo.*

Jour. Exp. Zool. *Journal of Experimental Zoology.* Baltimore and Philadelphia.

Jour. Morph. *Journal of Morphology.* Boston and Philadelphia.

Mitt. Stat. Neapel. *Mitteilungen aus der zoologischen Station zu Neapel.* Berlin.

Morph. Jahrb. *Morphologisches Jahrbuch.* Leipzig.

Phil. Trans. Roy. Soc. *Philosophical Transactions of the Royal Society of London.*

Pop. Sci. Mo. *Popular Science Monthly.* New York.

Proc. Am. Phil. Soc. *Proceedings of the American Philosophical Society.* Philadelphia.

Q. J. M. S. *Quarterly Journal of Microscopical Science.* London.

Sitz.-Ber. Acad. Wiss. Berlin. *Sitzungsberichte der königlich preussischen Akademie der Wissenschaften zu Berlin.*

Sitz.-Ber. Phys.-Med. Ges. Würzburg. *Sitzungsberichte der Physicalisch-medizinisch Gesellschaft zu Würzburg.*

Sitz.-Ber. Ges. Morph. Phys. *Sitzungsberichte der Gesellschaft für Morphologie und Physiologie in München.*

Trans. Am. Phil. Soc. *Transactions of the American Philosophical Society.* Philadelphia.

Zeit. Indukt. Abstamm. Vererbungslehre. *Zeitschrift für induktive Abstammungs- und Vererbungslehre.* Berlin.

Zeit. wiss. Zool. *Zeitschrift für wissenschaftliche Zoologie.* Leipzig.

Zool. Jahrb. *Zoologische Jahrbücher.* (Abteilung für Anatomie und Ontogenie der Tiere, unless otherwise specified.) Jena.

REFERENCES TO LITERATURE—Chapter I

CALKINS, G. N., The Protozoa. Columbia Univ. Biol. Ser. VI. New York. 1901.
Protozoology. New York. 1909.

DOFLEIN, F., Lehrbuch der Protozoenkunde. Jena. (3 Aufl.) 1911.

HARTSOEKER, N., Essay de dioptrique. Paris. 1694.

HERTWIG, O., Zeit- und Streitfragen der Biologie. I. Präformation oder Epigenese. Grundzüge einer Entwickelungstheorie der Organismen. Jena. 1894. English translation by P. C. Mitchell, "The Biological Problem of To-day: Preformation or Epigenesis," *etc.* London. 1896. Die Entwickelungslehre im 16. bis 18. Jahrhundert. Handbuch der vergl. u. exp. Entwick. d. Wirbelt. I, 1, 1. Jena. 1906. (1901.) Ueber die Stellung der vergleichenden Entwickelungslehre zur vergleichenden Anatomie, zur Systematik und Descendenztheorie. (Das biogenetische Grundgesetz, Palingenese und Cenogenese.) Id. III, 3. 1906.

HERTWIG, R., Mit welchem Rechte unterscheidet man geschlechtliche und ungeschlechtliche Fortpflanzung? Sitz.-Ber. Ges. Morph. Phys. **15.** 1899. English translation by Curtis, W. C., Science. **12.** 1900.

KOFOID, C. A., On *Pleodorina illinoisensis*, a new Species from the Plankton of the Illinois River. Bull. Ill. State Lab. Nat. Hist. **5.** 1898.

KORSCHELT und HEIDER, Lehrbuch der vergl. Entwickelungsgeschichte der wirbellosen Thiere. IV Abschnitt. Ungeschlechtliche Fortpflanzung and Regeneration. Jena. 1910.

LILLIE, F. R., The Theory of Individual Development. Pop. Sci. Mo. **75.** 1909.

LOCY, W. A., Biology and its Makers. New York. 1908.

MORGAN, T. H., The Problem of Development. International Monthly. Burlington, Vt. 1901.

SCHULTZ, E., Prinzipien der rationellen vergleichenden Embryologie. Leipzig. 1910.

WEISMANN, A., Essays on Heredity. English Translation. I and II Series. Oxford. 1891, 1892. The Germ Plasm. English Translation. New York. 1893. The Evolution Theory. English Translation. New York. 1904.

WHITMAN, C. O., The Inadequacy of the Cell-theory of Development. Woods Holl Biol. Lect. 1893. Bonnet's Theory of Evolution. Id. 1894. Evolution and Epigenesis. Id. 1894.

WILSON, E. B., The Problem of Development. Science. **21.** 1905.

CHAPTER II

THE CELL AND CELL DIVISION

Two universal characteristics of living things are the possession of protoplasm and a cellular composition. Recognition of these fundamental facts was dependent upon the use of the compound microscope, and so we find them comparatively late acquirements in the history of biology. The cell-unit structure of an organic tissue was first described in plants (cork tissue), by Robert Hooke in 1665, and quite naturally, therefore, emphasis was laid upon what we now know as the "cell walls." As a consequence the term "cell" was applied to these small box-like units which seemed to resemble the cells of a honey comb. When, about the middle of the nineteenth century, it became apparent that the cell content, and not the cell wall, was the important thing, the word "cell" had become so definitely fixed that it could not but be retained, although its utter inaptness was fully recognized.

This is not the place to discuss the general importance and significance of the Cell Theory of Schleiden and Schwann (1839) and their successors. It will become clear as we proceed that the cell, in structure and in action, is the basis of modern Embryology. Most of the early processes of organic development are strictly cell processes and must be studied from the standpoints of both Cytology and Embryology, from neither alone, and throughout development constant reference must be had to cellular phenomena.

As known to-day cells of different organisms and different tissues exhibit an unending variety in size, form, structure, and function (Figs. 14, 15), but throughout there are two essentials of structure expressed by the definition of a cell given by Leydig (1852) and by Schultze (1861) as " a mass of protoplasm con-

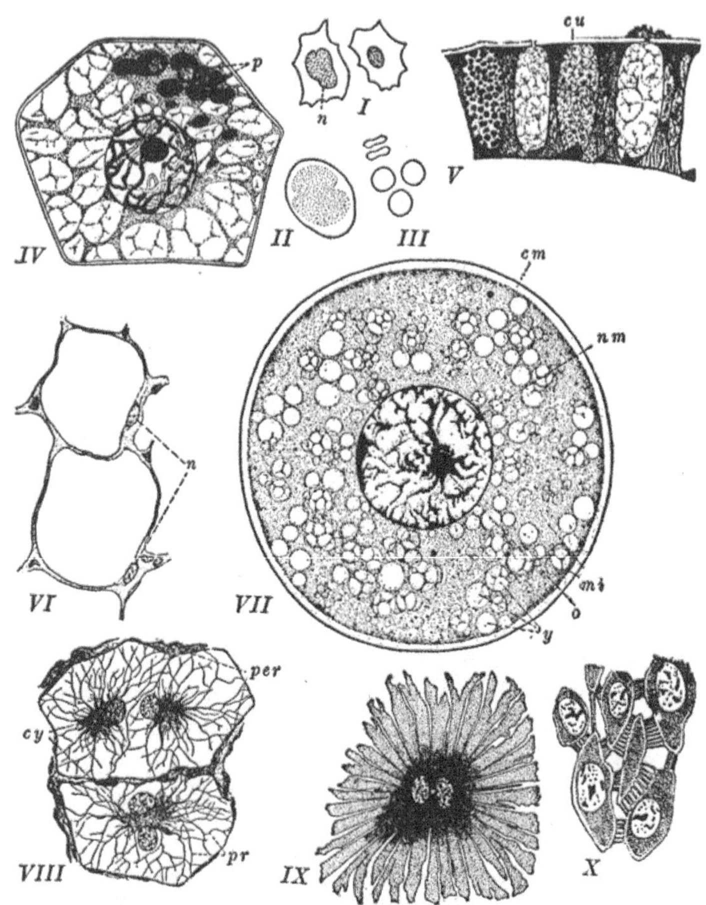

Fig. 14.—Various forms of cells. *IV–IX*, from Dahlgren and Kepner, *X*, after Prenant and Bouin. *I, II.* Human leucocyte, × 350. *III.* Human red blood-corpuscle, × 350. *IV.* Cell from root cap of calla lily, × 350; *p*, plastids. *V.* Epidermis of earthworm, showing four mucous cells in various stages of secretion, and *cu*, cuticle, × 550. *VI.* Fat cells in skin of chicken; *n*, nucleus. × 435. *VII.* Ovarian ovum (oöcyte) of cat; *cm*, cell membrane; *mi*, microsomes; *nm*, nuclear membrane; *o*, nucleolus; *y*, yolk alveoli. *VIII.* Connective tissue cells from the lobster; *cy*, cytoplasmic mass; *pr*, cytoplasmic processes; *per*, peripheral layer of cytoplasm upon which the rigid material of the tissue is laid down. *IX.* Pigment cell from the peritoneum of the fish, *Ammodytes.* Fully extended; the processes can be completely retracted. Two nuclei, × 90. *X.* Stratified epithelium from human pharynx, showing intercellular connections or bridges, × 375.

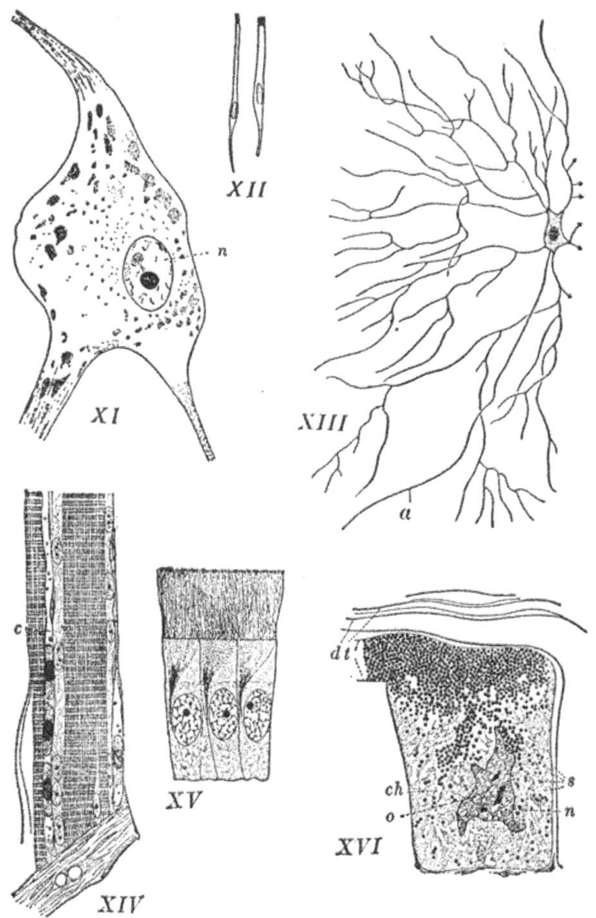

Fig. 15.—Various forms of cells, continued. *XI–XIII*, after Prenant and Bouin, *XIV–XVI*, from Dahlgren and Kepner. *XI*. Ganglion cell from human spinal cord; *n*, nucleus, × 250. *XII*. Sensory (receptor) cells from human olfactory epithelium, × 175. *XIII*. Multipolar ganglion cell from optic ganglion of horse. (The processes of the right side are cut off.) *a*. Axon, × 63. *XIV*. Part of muscle cell from the fish, *Catostomus*; *c*, capillary containing blood cells and platelets, × 500. *XV*. Ciliated cells from the digestive tract of the Mollusc, *Cyclas*. *XVI*. Gland cell from the leech, *Piscicola*; *ch*, cytoplasmic channels containing the secretion granules; *dt*, discharging tubes; *n*, nucleus; *o*, nucleolus; *s*, secreted materials in various stages of elaboration.

taining a nucleus." To-day we modify this but slightly and define a cell as *a limited mass of protoplasm containing nuclear material.*

The mass of the cell is usually very small. The smallest cells known are the Bacteria, some of which may be only 0.001 mm. (1 *micron,* or 1/25000 inch) in length (Streptococci), or even less (Staphylococci). But among the Metazoa, cells are never so minute. The human white blood corpuscle or leucocyte (Fig. 14, *I*), is perhaps of average size, measuring something less than 0.01 mm. (8–10 *micra*) in diameter. Tissue cells having a diameter of 0.05 mm. (50 *micra*) are considered large, although a few specialized cells may far exceed this (*e.g.,* muscle or nerve cells). The egg cells of animals are usually larger than tissue cells, but this is a special condition, and is frequently due to the accumulation of stored food substance, rather than to the possession of a larger amount of protoplasm. Within the species the sizes of specific varieties of cells are very constant. The size of an organ or of an individual is related to the number of its component cells rather than to their size (Amelung, Conklin).

We may proceed now to describe the essentials of structure exhibited by a typical cell—an imaginary thing which has no more real existence than the "average man." Such a cell would consist of a spheroidal or irregular mass of protoplasm, limited by a definite *cell membrane* or *cell wall.* The wall may be a surface condensation of the protoplasm or, more frequently, a true secretion of the cell body, either membranous as in most animal tissues, or thick and rigid, like the cellulose walls of most plants. In many cells the viscid transparent protoplasm just within and in contact with the cell wall forms a thin layer, the *ectosarc* or *ectoplasm* (Fig. 16), clearer than the granular and more refractive central *endosarc* or *endoplasm,* which contains, besides the granules, many cell organs and inclusions.

The protoplasm itself is made up of a combination of two forms, perhaps two kinds, of material plainly differing in density and arrangement (Fig. 17). The denser material called the *mitome, spongioplasm, reticulum,* or *filar substance,* forms a sort of complex framework or fine network of irregularly woven paths along which are scattered minute granules called *microsomes.* The spaces or meshes of this spongioplasmic network are filled with the less dense *ground substance* or *cell sap,* called also the *hyaloplasm, paraplasm,* or *interfilar substance.* The

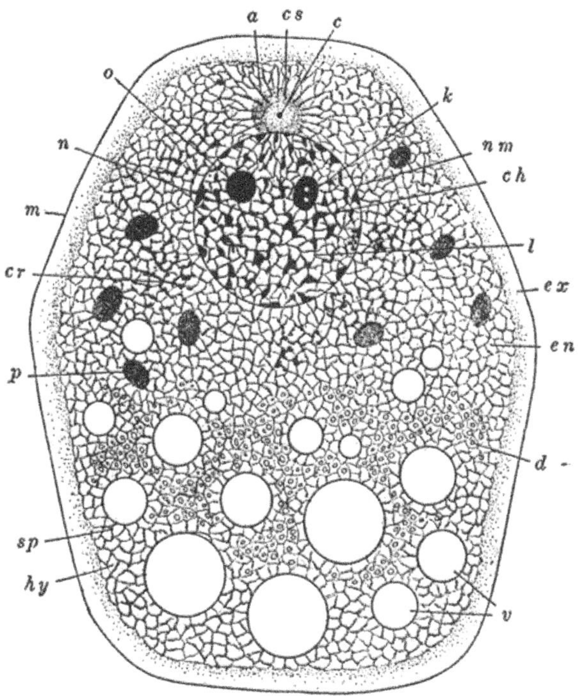

FIG. 16.—Diagram of a typical cell. *a,* aster; *c,* centrosome (centriole); *ch,* chromatin; *cr,* chromidia; *cs,* centrosphere; *d,* deutoplasmic granules; *en,* endoplasm; *ex,* exoplasm (cortical plasm); *hy,* hyaloplasm; *k,* karyosome; *l,* linin network; *m,* cell membrane; *n,* nucleus; *nm,* nuclear membrane; *o,* nucleolus; *p,* plastids; *sp,* spongioplasm; *v,* fluid vacuoles (metaplasm).

actual relation of these two kinds of substances varies in different kinds of cells or even at different times in the same cell. A frequent arrangement is that of a *reticulum* just described, in which the spongioplasm is definitely fibrous, forming a felt-work holding the more fluid hyaloplasm. In other cells, or at other times, protoplasm has a distinctly alveolar structure resembling a fine emulsion. Here the hyaloplasm is

in the form of minute drops or *alveoli*, while their walls or the
irregular interalveolar spaces are of the denser material.
Occasionally other structural relations are seen, such as the
granular, where the fibrous reticulum is represented by $_r{}^o{}_{ws}$ of
excessively minute granules, and the *fibrillar*, where the fibers

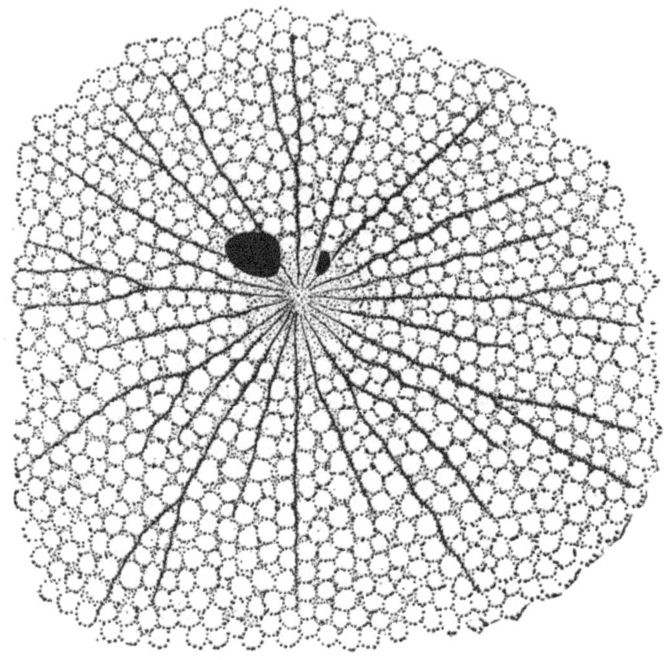

Fig. 17.—Alveolar protoplasmic structure in the egg of the sea-urchin, *Toxopneustes*, one and one-half minutes after the entrance of the spermatozoön.
From Wilson, "Cell," × about 2000. The protoplasm consists of alveoli surrounded by microsomes. In the middle is the centriole, surrounded by the
centrosphere, while radiating from it are the rays of the aster. The large and
small black masses are the sperm head and middle-piece.

of the reticulum are larger, longer, and less branched than in
the ordinary reticulum. It is still uncertain how exactly the
real structure of living protoplasm is represented by its appearance after it has been killed, in preparing it for examination.
It should be remembered that in living protoplasm these
fibrils, reticula, *etc.*, are in all probability fluid structures of
greater density than the ground substance.
    The cell is far from being a simple unit, for it contains a

variety of structures and materials differing chemically and functionally; these may not all be directly visible as organized structures. The only constantly differentiated substance is the nuclear material which is usually contained in a definitely formed body, the *nucleus*, though it may be scattered through the protoplasm. There are many reasons for believing that primitively the nuclear substance was not thus organized into a definite nucleus, but that it was distributed through the cytoplasm in the form of small granules, as it is still in many of the simplest organisms, and that gradually these became aggregated into fewer larger masses. / The Protozoa show many stages in the gradual enlargement and numerical reduction of the nuclear elements, but in the Metazoa the nuclear material is nearly always collected into a single body. The nucleus is to be regarded as a specialized portion of the protoplasm of the cell, highly differentiated in structure, chemical composition, function, and behavior. All cell activities seem to involve mutual interaction between the nucleus and the remainder of the cell and neither is able long to function normally without the other. But the action of the nucleus is primary and directive, to a large extent controlling and regulating cell activities and cell life as a whole. In most cases the nucleus is a spherical or ovoid body of fixed form; in some very active cells it may be elongated or of irregular form, or even branched, ramifying all through the cell; in a few rare instances the nucleus may be amoeboid (Figs. 14, 15).

Typically this complex center of cell activity shows much the same fundamental structure as the remainder of the protoplasm, which in distinction from the nucleus is called the *cytoplasm*. The nucleus is limited by a definite nuclear wall or membrane formed either from the cytoplasm or from the nucleus itself. The substance of the nucleus as a whole is termed the *karyoplasm*. The equivalent of the spongioplasmic reticulum of the cytoplasm is here termed the *linin* network, and the hyaloplasmic ground substance is known as the *nuclear sap*, or *karyolymph*, or *paralinin*. The chief distinction of the nucleus is the presence of a very special nucleo-protein substance called *chromatin*, a name given it on account of the ease with which it is colored with such dyes as hæmatoxylin (logwood) or carmine. This chromatin is ordinarily in the form of small flakes or granules of variable size, termed *chromioles* (Eisen). These are always distributed along the linin fibers, which stain less readily and are therefore described as *achromatic*. Sometimes a few large masses of chromatin called *karyosomes* may be seen within the nucleus. Apparently the chromatin may be rapidly dissolved or condensed, or even formed anew, so that its visible amount and condition vary considerably from time to time.

The structural differentiation of the nucleus is frequently complicated further by the presence of one or more bodies called *plasmosomes* or *nucleoli*, which often superficially resemble the karyosomes, at least morphologically, although chemically and physiologically they are quite unlike (Fig. 16). The nucleoli are spheroidal bodies, staining densely, though not of the same material as the true chromatin. The karyosomes are sometimes called *chromatin nucleoli*. The nucleoli vary considerably in number, size, and form, and their significance in the nucleus is not altogether understood; probably several unlike bodies having different functions have been included under this single term.

Another organ of the cell is the *centrosome*. This is not always to be seen for it may be lost from the cell at certain times and reappear later. While commonly a cytoplasmic structure lying just outside the nucleus, in some forms it is an intra-nuclear organ and it is quite possible that primitively it was an essential part of the nucleus, as it still is in many Protozoa (Figs. 29, 30). The centrosome is a minute densely staining granule or pair of granules, sometimes hardly larger than the granules or microsomes of the cytoplasmic reticulum. Occasionally it consists of several separate granules closely associated. The cytoplasm in the neighborhood of the centrosome and directly under its influence is ordinarily differentiated as the *archoplasm*. This substance consists typically of two portions. A medullary region forming a small spheroidal mass called the *centrosphere* or *attraction sphere* immediately surrounds the centrosome, and peripherally, radiating from this out into the cytoplasm, there is at times a collection of diverging rays or fibers called collectively the *aster*. In the ordinary vegetative cell the centrosome and archoplasmic structures are usually reduced in size and perhaps even absent in some cells, but in the dividing cell they may be very large and prominent organs extending nearly throughout the cell, for their chief function is in connection with the process of cell division. Similar dense granules and fibers are found associated with organs of the cell which are motile, for example, the granules at the bases of cilia and flagella, the axial filaments of some flagella, undulating membranes, and some pseudopodia, the fibrillæ of muscle cells, *etc.* All such modifications of protoplasm connected particularly with the production or regulation of motion are included under the general term *kinoplasm* (Strasburger). It has been suggested that the kinoplasm of the cell is of the nature of a definite and permanent cell organ. In many cells the centrosome is such a permanent organ, but many other kinoplasmic structures seem to be more or less temporary and may disappear or be formed anew at the time of cell division or at other times. The rays of the aster, for example, in many cases apparently are formed from the enlargement and rearrangement of the cytoplasmic reticulum resulting from the activity (probably chemical) of the centrosome.

In addition to these nearly constant cell structures of comparatively uniform characteristics, there are various other organs or bodies which are peculiar to certain special kinds of cells. Among these are the bodies called in general *plastids* (Fig. 14, *IV*), of which the more familiar are the pigment bodies, such as *chloroplastids* or *chlorophyl bodies*, the *chromoplastids* or colored bodies not containing chlorophyl, *amyloplastids* or starch-forming bodies, protein-forming plastids, and others. *Vacuoles* of various sizes and kinds are very common, such as the digestive, excretory, food, water, and food-storage vacuoles; occasionally one or more are specialized as *contractile* or *pul-*

sating vacuoles. Nutritive substances are often stored temporarily in cells. These materials are collectively known as *deutoplasm, metaplasm, or paraplasm,* and may be starch or protein granules, yolk plates, oil drops, *etc.* There are also

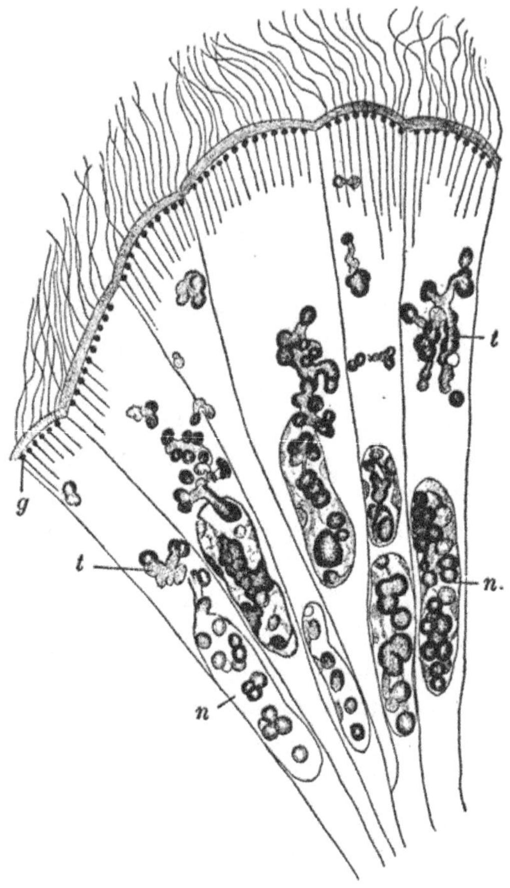

Fig. 18.—"Trophospongien" in cells of the hepatic duct of the snail, *Helix vomatia.* After Erhard. *g,* basal granules; *n,* nucleus; *t,* trophospongien.

granules of various other kinds—of materials being secreted or excreted, of food in various stages of ingestion or digestion, pigment granules, crystals, and formed substances of many kinds, usually specific for the organism or tissue.

In many cells true chromatic structures are found in the

cytoplasm outside the definite nucleus. These are usually small granules and bits of chromatin of varied form and significance in different cells. Collectively they are termed *chromidia;* for the most part they are not formed *in situ*, but are derived directly from the nucleus (Fig. 18). They are most frequent in very active cells such as gland cells, the rapidly growing germ cells, and many others. Many forms of chromidia, functionally as well as structurally distinct, have been given special names such as *chondriosomes,* "*Chondromiten,*" *idiochromidia, mitochondria, pseudochromosomes,* "*Nebenkern,*" "*Trophospongien,*" etc.

It has been said already that no single cell shows typically all the structures described above and illustrated in Fig. 16. Without stopping for further description many of the details of structure truly characteristic of a few varied types of cells are shown for comparison in Figs. 14, 15. (For descriptions and further details the student should consult the standard texts of Cytology and Histology, *e.g.*, Dahlgren and Kepner's "Principles of Animal Histology," New York, 1908.)

One further aspect of cell structure remains to be mentioned. This is the important fact that there is in most tissues a fairly definite cell form combined with a nearly constant arrangement of the cell organs and contents (Van Beneden, Rabl, Heidenhain). In any epithelial cell the different physiological conditions at the free and the attached surfaces lead to this definite relation of cell structures which is called *polarity*. In such cells the nucleus lies toward the basal end of the cell, the centrosome either toward the free end or on that side of the nucleus. Thus an imaginary axis may be passed through the centrosome and nucleus perpendicular to the free surface, about which the cell organs are arranged symmetrically, either bilaterally or radially (rotatorially). This polarity extends also to other less constant structures and even occasionally to non-epithelial cells. It may be seen for example in the arrangement of cilia, cuticle, conductile and contractile fibers, granules or drops of substances being excreted or secreted, and yolk or other stored materials (Figs. 14, 15, 16, *etc.*). In the germ cells, as we shall see later, this fact of polarity becomes of the greatest importance, for it is frequently related closely to the symmetry of the mature organism.

While we may thus describe the Metazoan tissue cell as a separate morphological unit, complete in itself, it is also true that in a great many tissues the cells are in direct material continuity with one another through minute protoplasmic connections or *bridges* which pass through fine perforations in the cell walls. These have been observed in a great variety of tissues in many different forms (Fig. 14, *X*); whether or not they are present in all tissues it is yet impossible to say definitely, and we must recognize clearly that in the physiology of the organism the cells do not behave as completely autonomic units. While each represents a localized field of activity, the life of the cell is subordinated to the life of the organism as a whole—a fact that comes out with especial clearness in the *development* of the organism. In some way not yet understood the cell, or groups of cells, influence the activities of other cells, and are in turn influenced by them. The activities of the Metazoan organism of course equal the sum of the activities of its component cells; but these combined activities are organized and unified into a whole in such a way that this represents more than the unit activities when considered separately, just as the action of a community represents something beyond the sum of the actions of its members taken individually.

From the embryological point of view one of the most important phases of cell activity is cell reproduction or cell division. For cells arise only from preëxisting cells. The history of opinion regarding the genesis of cells parallels roughly that regarding the genesis of organisms. It was Virchow who finally demonstrated convincingly (1855, 1858) the universality of the fact of cell continuity in the tissues of a single organism, and further the fact that in a succession of generations of organisms the process of the formation of cells from preëxisting cells is not interrupted. We know now that in this process of cell division all the essential organs of the cell take an active part—the cytoplasm, nucleus, the centrosome, and even many of the plastids. So that the final result of the process is typically the formation of two daughter cells similar to each other and also to the parent cell in all essential respects save in size.

The division of cells occurs in two quite dissimilar ways. The simpler method, and the less frequent, is termed *direct division* or *amitosis* (Flemming). Here the first step is sometimes the elongation and constriction of the nucleolus, when this is present, into two separate daughter nucleoli, or in other cases the appearance of a new second nucleolus (Fig. 19). Next the whole nucleus divides into two, sometimes by simple constriction into two separate elements, sometimes by the ingrowth of

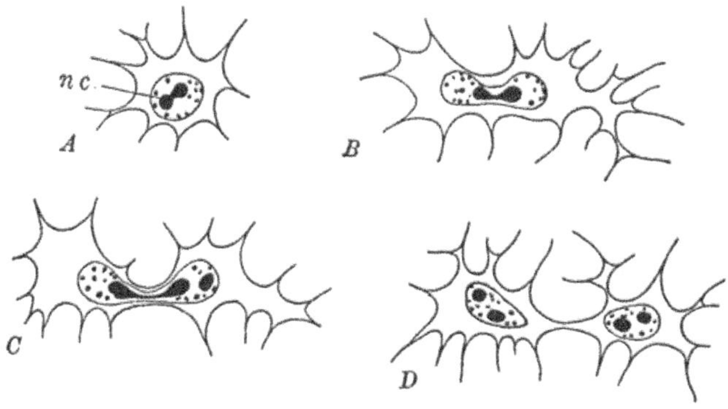

Fig. 19.—Amitosis in tendon cells of a new-born mouse. After Nowikoff, × 800.
nc, nucleolus.

a partition wall, or in still other cases, by the formation of two new nuclear membranes within the original membrane, the disappearance of the latter freeing the two daughter nuclei. This division of a nucleus is typically followed by the division of the cytoplasmic portion of the cell which is ordinarily accomplished by the development of a cell wall between the two daughter nuclei. Very frequently, however, division of the cell body does not follow and the cell remains binucleate; or this process of nuclear fission may be repeated, a multinucleate cell resulting, such as a striated muscle cell. In such cases of incomplete cell division the essence of the process seems to be the rapid increase of nuclear surface and then volume; it is usually associated with special forms of cell activity. Conditions in some of the Protozoa suggest that primitively division of the nucleus or the multiplication of the nuclear bodies might not have been associated with a corresponding division of the cytoplasmic body, but that these originally independent divisions have gradually come to be uniformly associated.

It has commonly been supposed that the direct form of cell division occurs but rarely and then usually in cells which are moribund. It is becoming clear, however, that amitosis is in reality not particularly infrequent. It seems to occur normally in many tissues (mesenchyme), and often where there is an unusual or a sudden increase in nuclear activity and energy expenditure on the part of the whole cell, as in ovarian follicle cells and tapetal cells, muscle cells, rapidly growing or redifferentiating cells in regenerating tissues; it is also true that amitosis is frequent in such tissues as stratified epithelia whose cells are nearing the end of their life or activity.

The second and more usual method of cell fission is that termed *indirect cell division*, or *mitosis* (Flemming), or *karyokinesis* (Schleicher). This is a complicated process involving the establishment and operation of an intricate mechanism within the cell, shared in by nearly all its living parts. The essential result of the division of the cell by the action of this complex mechanism concerns in particular the chromatic substance of the nucleus, for in nearly all known instances the chromatin sharing in the process is very precisely divided into two equal portions, each of which goes to one of the daughter cells. We may give here only a brief outline of the essentials of this process of mitosis, again describing an imaginary schema with which to compare later some of the variations in detail shown by actual cells.

As the first step in mitosis we should consider the division of the centrosome into two, which remain lying together within the undivided kinoplasmic centrosphere. This division of the centrosome is usually quite removed in point of time from the other phenomena of mitosis, for it occurs normally during the reconstruction of a daughter cell immediately after its formation, and so is separated by a considerable intervening vegetative period from the other events of mitosis or the doubling of other parts (Fig. 20, *A*). This vegetative phase of cell life is frequently referred to as the *"resting"* period or *interkinesis;* a state of inaction is not implied by the term "resting," for during this period the cell is performing its normal and characteristic functions as a tissue cell; the word merely indicates that the cell is not undergoing any active phase of division. The termination of the vegetative phase and the immediate inauguration of mitosis is ordinarily first distinguishable in the structure of the nucleus. The chromatin granules become more distinct, enlarge rapidly, and undergo some change in chemical constitution indicated by an increase in staining capacity (Fig. 20, *A*; 22, *A*). As the chromatin increases some of the granules or flakes come to be arranged in a linear, or sometimes bilinear series, still upon some of the linin threads which share in this arrangement. Thus the chromatin and linin form a tangled thread or ribbon called the *skein* or *spireme* (Figs. 20, *B*; 21, *B*; 22, *B*).

We should note here that at this time the chromatin of the nucleus which is not included in the spireme, often indeed the greater part of the whole amount of this material, is thrown out into the cytoplasm and dissolves (Fig. 32); the more fluid parts of the nucleus are also thrown into the cytoplasm by the dissolution of the nuclear membrane. It may thus be only a comparatively small part of the whole nuclear structure that is formed into the spireme proper.

The spireme may be quite *continuous* throughout the nucleus, or it may appear from the first as a fragmented thread composed of several short pieces; when in this latter condition it is spoken of as a *segmented* spireme. In a few cases the spireme stage is largely suppressed and the chromatin granules collect immediately into compact groups without indication of a skein stage. The linin network in part becomes a sort of fine core throughout the spireme and in the extra-chromatic region remains as a network of naked fibers. The latter portion soon becomes polarized so that its fibers converge, more or less

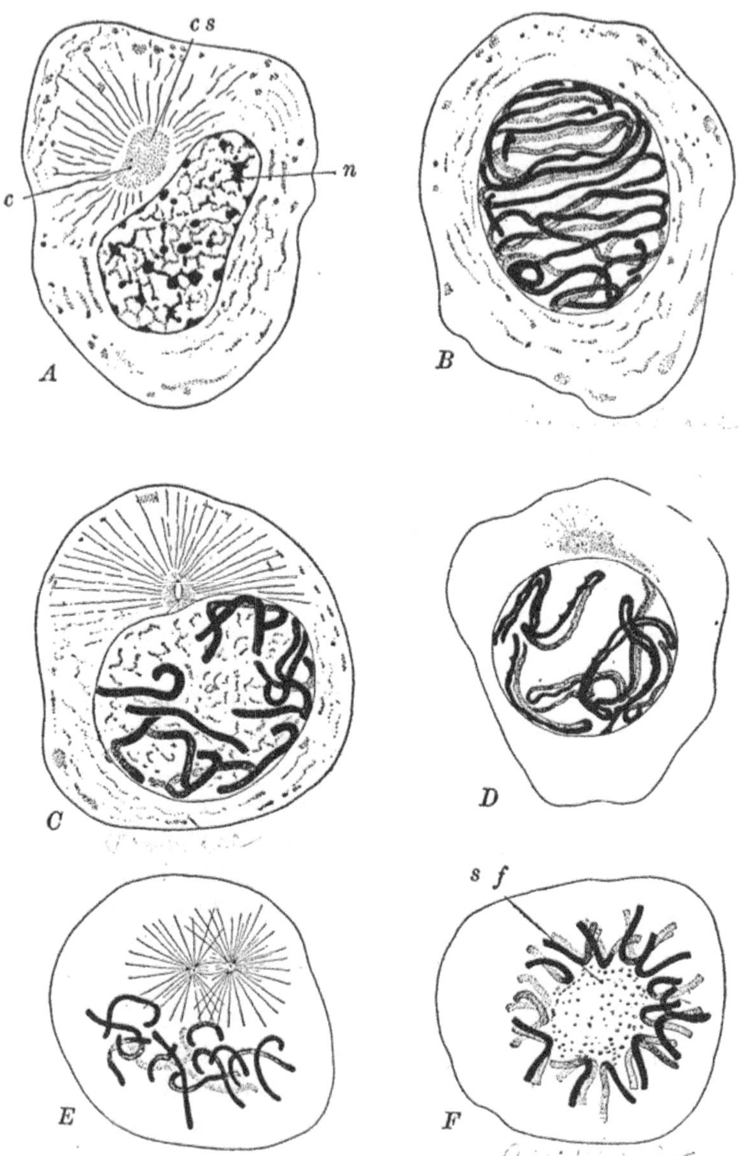

Fig. 20.—Mitosis in cells of *Salamandra maculosa*. After Prenant and Bouin. *D, H.* Primary spermatocytes, others, spermatogonia. *A, B, C,* × 1000, others, × 800. *A.* Interkinesis or resting stage. *B.* Early prophase; spireme continu-

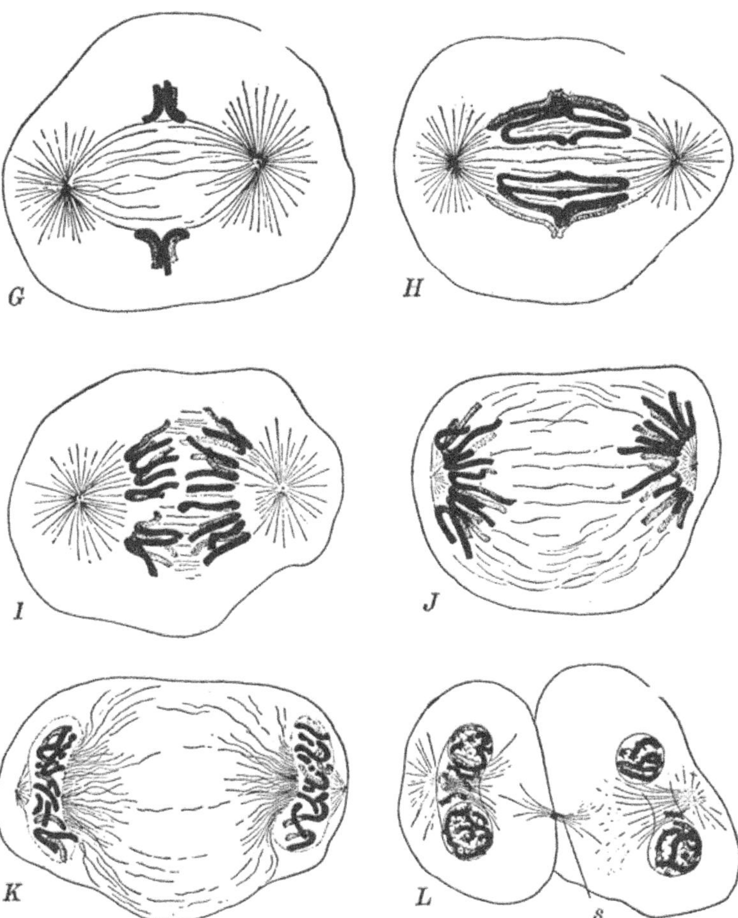

ous. Centrosomes omitted. *C*. Prophase; spireme segmented into chromosomes. Centrosomes commencing to diverge; spindle forming. *D*. Longitudinal splitting of chromosomes. *E*. Disappearance of nuclear membrane; continued divergence of centrosomes and asters. *F*. Mesophase; formation of equatorial plate. Polar view. Chromosomes V-shaped. G. Same in side view. Only a few of the chromosomes are shown. *H*. Anaphase; daughter chromosomes diverging, still united at ends. *I*. Anaphase; continued divergence of chromosomes, now entirely separated. *J*. Late anaphase; complete divergence of chromosomes. Spindle breaking down, asters disappearing. K. Telophase; beginning of reconstruction of daughter nuclei. Chromosomes disintegrating. *L* Late telophase; division completed. Nuclei reconstructed; centriole divided; cell walls completed. Nuclear membrane forming. *c*, centrioles; *cs*, centrosphere; *n*, nucleus; *s*, spindle remains; *sf*, spindle fibers cut across.

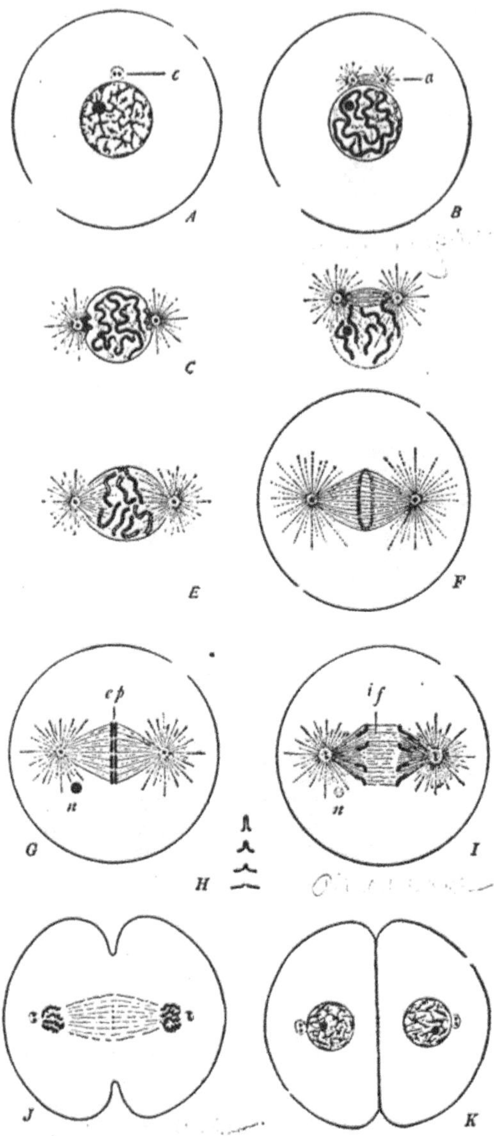

Fig. 21.—Diagrams of the process of mitosis. From Wilson, "Cell," slightly modified. *A.* Resting-cell with reticular nucleus and true nucleolus; at *c* the attraction-sphere containing two centrosomes. *B.* Early prophase; the chromatin forming a continuous *spireme*, nucleolus still present; above, the amphiaster (*a*). *C.D.* Two different types of later prophases; *C.* Disappearance of the primary spindle, divergence of the centrosomes to opposite poles of the nucleus (examples, many plant-cells, cleavage-stages of many eggs). *D.* Persistence of the primary spindle (to form in some cases the "central spindle"), fading of the

distinctly, toward the region of the centrosphere; in many cells such a polarization of the linin toward the centrosome exists throughout the vegetative phase. The two centrosomes now begin to diverge and the surrounding centrosphere pulls in two, one portion accompanying each centrosome (Figs. 20, *E;* 21, *D*). As the centrospheres diverge they enlarge, and within each appear fibers radiating from the centrosome as a center and producing the *asters.* While the chromatic and achromatic parts of the nucleus have been passing through these early stages of mitosis, the nucleolus when present becomes vacuolated, commences to dissolve and finally disappears. Soon the nuclear membrane also commences to break down and dissolve, first in the region of the asters, leaving the nuclear substance free in the cytoplasm. Next the chromatin thread shortens and thickens, breaking in the case of the continuous spireme, into a number of separate segments or rods; or if the spireme itself is of the segmented type, its elements now shorten and thicken. When the spireme segments, the linin thread upon which the chromatin granules are strung may remain continuous between as well as through the chromatic rods. These chromatic segments now become quite homogeneous, clearly differentiated structures called the *chromosomes* (Figs. 20 *C, E;* 21, *D;* 22, *C*). Strictly speaking, each chromosome consists of a dense mass of fused chromatin granules with a portion of linin embedded.

In practically all organisms in which the nucleus is a definitely formed structure, the number of chromosomes appearing during mitosis is fixed, and is constant throughout all divisions of

---

nuclear membrane, ingrowth of the astral rays, segmentation of the spireme-thread to form the chromosomes (examples, epidermal cells of salamander, formation of the polar bodies). *E.* Later prophase of type *C;* fading of the nuclear membrane at the poles, formation of a new spindle inside the nucleus; precocious splitting of the chromosomes (the latter not characteristic of this type alone). *F.* The mitotic figure established. *G.* Metaphase; splitting of the chromosomes (*e.p.*); *n,* the cast-off nucleolus. *H.* Four stages in the divergence of the two halves of a chromosome. *I.* Anaphase; the daughter-chromosomes diverging, between them the interzonal fibers (*i.f.*), or central spindle; centrosomes already doubled in anticipation of the ensuing division. *J.* Late anaphase or telophase, showing division of the cell-body, mid-body at the equator of the spindle and beginning reconstruction of the daughter-nuclei. *K.* Division completed.

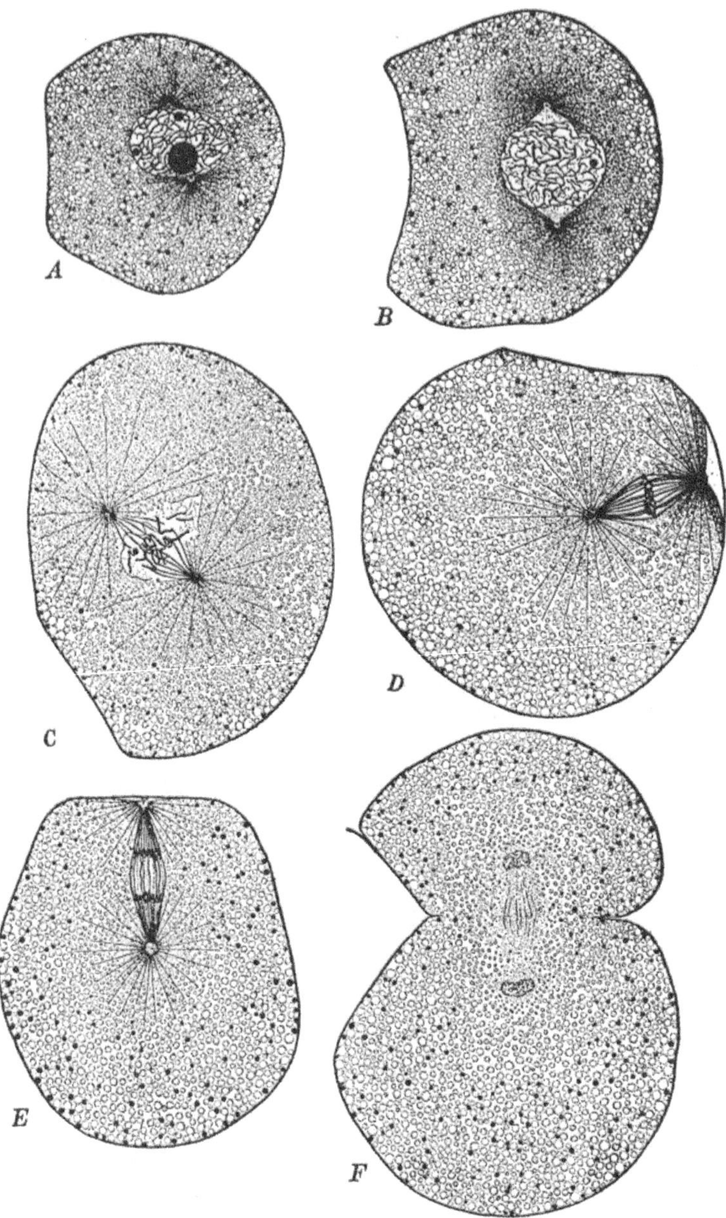

Fig. 22.—Mitosis in the segmenting egg of the clam, *Unio*. From Dahlgren and Kepner. *A*. Prophase of the fourth cleavage. Chromatin reticulum; centrosomes on opposite sides of nucleus. *B*. Prophase. Spireme beginning to segment into chromosomes; nuclear membrane disappearing; spindle forming. *C*. Late prophase. Chromosomes formed; spindle becoming completed; nucleolus nearly disappeared. *D*. Mesophase. Chromosomes in equatorial plate. *E*. Early anaphase. Divergence of the daughter chromosome groups. *F*. Telophase. Nuclear division completed and daughter nuclei reformed; cytoplasmic division commencing.

somatic cells. In all but a few groups the chromosomes appear in an even number, ordinarily between twelve and thirty-six, although these limits are frequently passed. The chromosomes are at present considered the most important elements in the cell, and interest in the whole process of mitosis centers in their behavior. The details of the process of mitosis seem directed toward the exactly equal division and distribution of these elements, the importance of which justifies the more detailed consideration which we give them after the general description of mitosis is completed.

While the chromosomes are forming, the centrosomes and asters continue to diverge, passing around toward opposite sides of the nucleus. The linin fibers of the nucleus tend throughout to remain polarized toward the centrosomes and the separation of these bodies from one another draws out the linin fibers into an elongated bundle converging at each end toward the centrosome. Finally the centrosomes come to lie on exactly opposite sides of the nuclear structures and as the nuclear membrane disappears completely we find the rays of the asters penetrating into the nuclear region and forming, together with the linin, a spindle-shaped structure lying between the centrosomes, its component fibers passing among the chromosomes (Figs. 20, *G*; 21, *F*; 22, *D*). In many cells the centrosomes do not thus migrate to opposite sides of the nucleus, but separate directly; the nucleus in this case is simply drawn up to lie between them (Fig. 21, *D*). The result is the same, but the difference in relative behavior of the centrosomes and nucleus is real and must be taken into account in some cases. The spindle and asters now form a figure resembling a diagram of the lines of force within a simple bipolar magnetic field. This figure is called the *amphiaster*, sometimes the *achromatic figure*, emphasizing its distinctness from the chromatic portion of the nucleus, now all or largely included in the chromosomes. The parts of the linin network directly continuous with those upon which the chromosomes were formed originally, seem to have a somewhat different history from the remainder of the linin. They remain attached to the chromosomes and extend thence toward the centrosomes, forming in this stage a superficial sheath around a central portion of the spindle, and are hence termed the *mantle fibers*. The central core of the spindle seems in many cases to be formed largely from the remainder of the nuclear linin, though in other cases this is formed in the same way that the asters are, and from cytoplasmic materials. Thus the amphiaster is usually of mixed origin, nuclear and cytoplasmic, though in some cases the spindle at least seems to be wholly nuclear (linin); the asters are always cytoplasmic in origin. All these fibers form definite threads, enlarged as compared with the original linin reticulum.

The definite formation of the chromatic portion of the nucleus into chromosomes and the achromatic substance into the amphiaster marks the termination of the first phase of mitosis which is known as the *prophase*. During the prophase there has occurred the actual division of only the centrosome and centrosphere; the other important changes have been preparatory to further divisions—the dissolution of the nuclear membrane, the enlargement and rearrangement of the chromatin granules, the formation of definite chromosomes, and the establishment of the achromatic figure. We should remember that the nucleolus meanwhile has fragmented and, together with a large or small amount of chromatin which is not formed into chromosomes, has passed out into the cytoplasm and disappeared.

The arrangement of the materials forming the achromatic figure is evidently the result of certain tensions within the cell, the effect of which is first to draw the chromosomes, until now distributed irregularly, into a circle about the equator of the spindle. When in this position the chromosomes are said to form the *equatorial plate* (Figs. 20, *F, G*; 21, *F*). This phase of mitosis is also in general preparatory to actual division but it is carried on after the division mechanism is completely established. This period of division is known as the *mesophase* (Lillie).

Following the mesophase is the *metaphase*. The chief event of this phase is the longitudinal splitting or division of each chromosome into two parallel halves. This forms two equal and similar groups of daughter chromosomes, each group similar to the original group except in size. As a matter of fact this longitudinal splitting of the chromosomes is by no means always deferred until this time, for frequently it occurs during the prophase of division, even in the spireme stage; or rarely the chromatin granules may divide even before a definite spireme is constituted (Fig. 23). In such cases the chromosomes are in the form of double rods throughout the entire prophase and mesophase; the metaphase is then present only virtually.

The mantle fibers from the opposite poles of the spindle are now attached to the daughter chromosomes, usually in their middles. Next the mantle fibers begin to shorten as the result of some process centering in or about the centrosomes at the poles of the spindle. As the poles are relatively fixed, perhaps by the anchoring asters or through the rigidity of the central part of the spindle, the result is the separation of the two groups of daughter chromosomes which move along the central spindle fibers toward the regions of the centrosomes.

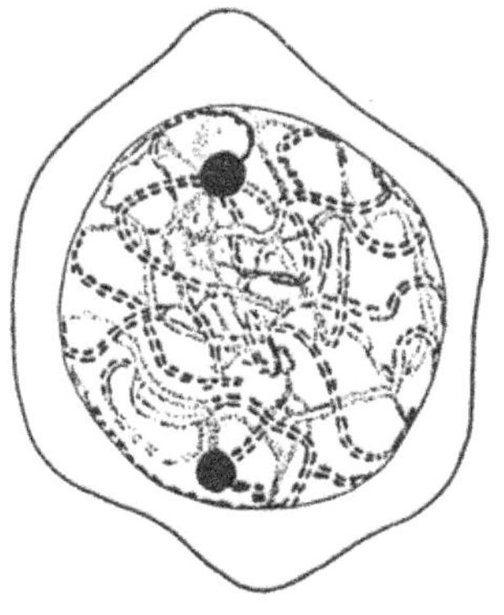

Fig. 23.—Longitudinal fission of the spireme in the division of spore-mother-cell in *Lilium candidum*. After Farmer and Moore.

If the mantle fibers are attached to the middle of the chromosome it is first drawn out into a ⊃- or >-shape; frequently this form is assumed during the mesophase, the apex of the > pointing centrally in the equatorial plate (Figs. 20, *F, H;* 21). The period of mitosis occupied by the divergence of the two chromosome groups is called the *anaphase;* this is usually very brief as the chromosomes diverge rapidly. Their divergence exposes the central fibers of the spindle which are then called the *interzonal* or *connecting fibers,* and which frequently come to have an important share in the formation of later structures (Figs. 21, *H;* 22, *E*). In their divergence the chromosomes themselves seem to be entirely passive, except in a very few isolated instances where they are definitely amœboid (*Opalina*) (Fig. 31). The two chromosome groups are finally drawn completely to the opposite poles of the spindle and the process of mitosis then enters upon its final period, the *telophase*.

During this phase the cytoplasmic portion of the cell becomes divided into two parts, usually equal, though occasionally extremely unequal. Sometimes, as in many animal cells, this division of the cytoplasm results from its peripheral constriction in a plane corresponding with that of the equatorial plate, the constriction deepening until the cell body is completely severed and the two daughter cells formed (Figs. 20, *L;* 21, *I, J;* 22, *F*). More frequently, in some animal and in practically all plant cells, the division of the cytoplasm results from the formation of a partitioning cell wall, in the formation of which the interzonal fibers of the spindle usually take an important part. These seem to increase in number and to thicken in the middle, ultimately fusing and forming a transverse plate which is the rudiment of the future cell wall; the remainder of the wall forms as a distinct secretion of the cytoplasm in that region.

As the diverging chromosomes approach the poles of the spindle they lose their distinct outlines, become vesicular, and gradually lose their visible identity and separateness to a large extent; with a few exceptions they finally seem to disintegrate completely and form scattered granules, and a new typical nucleus is constituted in each daughter cell. Meanwhile the centrospheres and asters have diminished in extent and clearness and have returned again to the condition which is characteristic of the interkinesis. About the new nucleus a membrane is formed, either from the nucleus or the cytoplasm, and the mitosis is accomplished (Figs. 20, *L;* 21, *H–J*). In most cases just about this time the centrosome divides in preparation for the next mitosis. During the interkinesis the nucleus and cytoplasm increase in size and soon the process of division is repeated.

The length of time occupied by the whole process of mitosis varies greatly. In the division of some egg cells it may be completed in fifteen minutes, or it may occupy one or even two hours, and in some special cases a much longer period.

This account of mitosis, although brief and including only some of the essentials, brings out clearly the unity of the nucleus as an organ; it behaves as a more or less separate unit of cell organization throughout all this intricate process. And we see clearly this extremely important fact, that the nucleus of a cell is formed from a preëxisting nucleus of the same constitution. Nuclei arise only from preëxisting nuclei; there is a nuclear continuity quite parallel with cell continuity. And going one step farther, it is probable that chromosomes are derived only from preëxisting chromosomes. This idea of *genetic continuity* is not completely applicable to all cell organs, however, for occasionally the centrosomes are not derived from preëxisting centrosomes, but may arise *de novo*, and in the development of the new organism the centrosome is typically derived from the sperm cell alone. Among the plants many of the plastids seem to be genetically related and to be formed by the division of preëxisting plastids. The other less living cell structures are usually distributed passively to the daughter cells, and such structures may be formed anew in the new cells.

The relation between the direction of the plane of cell division and the general morphology of the cell body demands a word. From the preceding account it is obvious that the plane of division is at right angles to the longitudinal axis of the spindle, but the position which the spindle assumes is itself predetermined. The position of the spindle axis is fixed by the location of the centrosomes. When the single centrosome divides and the daughter centrosomes pass to opposite sides of the nucleus, they usually migrate equal distances from the original position of the centrosome; it follows, therefore, that this region falls within the plane of the equator of the spindle and consequently in the plane of the new division. When the centrosome does not alter materially its relative position in the cell, the next division, again being through the plane occupied by the centrosome and the center of the nucleus, will be in

general at right angles to the preceding division (Fig. 24). Thus
the planes of succeeding divisions tend to alternate, each perpendicular to the preceding. Any other relation involves the
migration of the centrosome from the position originally
occupied by it in the daughter cell; or the spindle may change
its position during or after its formation, and this regular
relation thus be disturbed. Typically the spindle takes such
a position that its long axis lies in the direction of the greatest
protoplasmic extent of the cell (O. Hertwig), a position which
would result from the tensions between a comparatively
elongated body and a fluid medium in which it is suspended

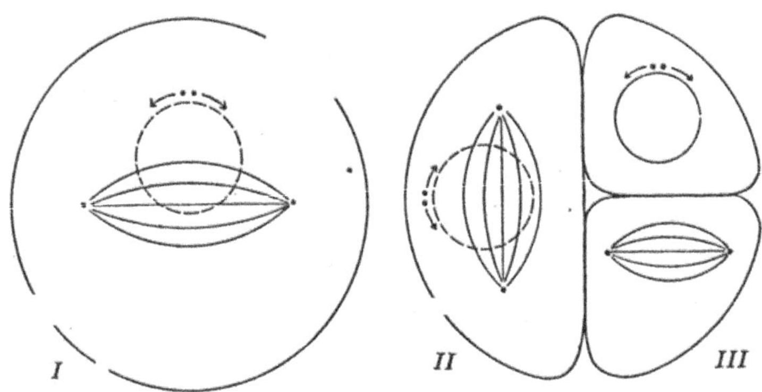

Fig. 24.—Diagram illustrating the relation between the position of the centrosome and the plane of cell division. Symmetrical motion of the daughter centrosomes results in the regular alternation, at right angles, of successive division
planes.

and free to move in any direction. There are many exceptions to these two general rules in special conditions, such
as simple columnar epithelia, stratified epithelia, the maturing
germ cells, *etc.;* these indicate perhaps that a more fundamental cause of the direction of cell division remains to be
discovered.

Some important modifications of this schema of cell division given
above will be noted in other connections, but a few special conditions
are conveniently mentioned here. The most important and fundamental
modifications are doubtless those which occur during the forming and
maturing of the germ cells; these are to be described in detail in Chapter
IV, and here we should only note that the behavior of the chromosomes

in these divisions is very complex; that there is here only one-half the number of these bodies formed in the cells of the somatic tissues; that mitoses may occur without an intervening interkinesis; that in the case of the egg cell the division may be of extreme inequality, so that one of the daughter cells is like an extremely small bud, although with respect to nuclear structure the two cells are alike; and that in certain special mitoses one or more of the chromosomes may fail to divide and therefore may be unrepresented in one of the daughter cells.

Among the higher plants an important characteristic is the absence of definite centrosomes and asters, although these structures are normally present among the lower plants. The absence of centrosomes results in the formation of a characteristically blunt or truncated

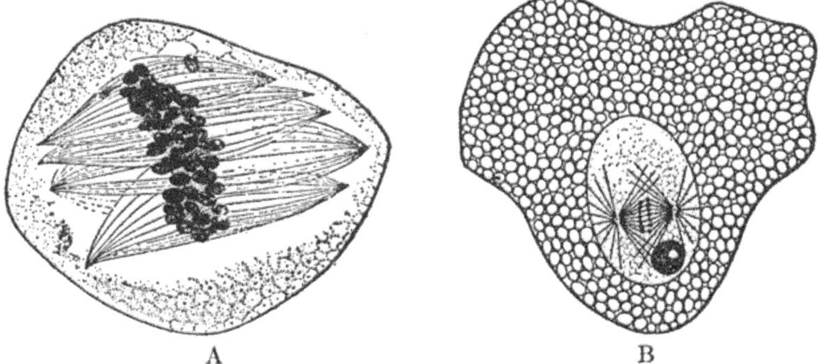

FIG. 25.—*A*. Multipolar spindle in spore-mother-cell of *Equisetum*. From Wilson, "Cell," after Osterhout. *B*. Intranuclear spindle in the oöcyte of the Copepod, *Canthocamptus staphylinus*. From Hegner. × 850.

spindle in most of the higher plants. In some animals the spindle is rather truncated also, but this is usually found to be in reality multipolar, composed of many small bundles of spindle fibers terminating in a row of centrosomes or centrosome-like bodies (Fig. 25, *A*). In the tissue cells of most animals the asters are relatively small, though the spindle remains large and distinct; in a few cases it seems that even in animal cells division may be effected in the absence of centrosomes.

Many significant modifications of mitosis occur among the Protozoa, where we find certain conditions which seem to offer suggestions as to the evolution of the mitotic figure and process, as well as of some of the chief cell organs themselves. Among these forms the process of division is often complicated through its being at the same time the essential step in reproduction, rather than merely a step in or condition of differentiation, as in the Metazoan. Thus the process of budding or gemmation is essentially an unequal cell division; and in brood ("spore") formation we see a form of cell division in which the nucleus divides many times without any corresponding cytoplasmic divisions, until finally, all at once, the cytoplasm is cut into a large number of small cells. In these forms of division no mitotic figure is formed ordinarily, and one of the modifications due to association with reproduction is seen in the fact that the resulting bud, or brood-cell, may have a form and structure entirely unlike that of the mother cell. In budding, especially in that form called bud-fission, the nucleus does not ordinarily divide until

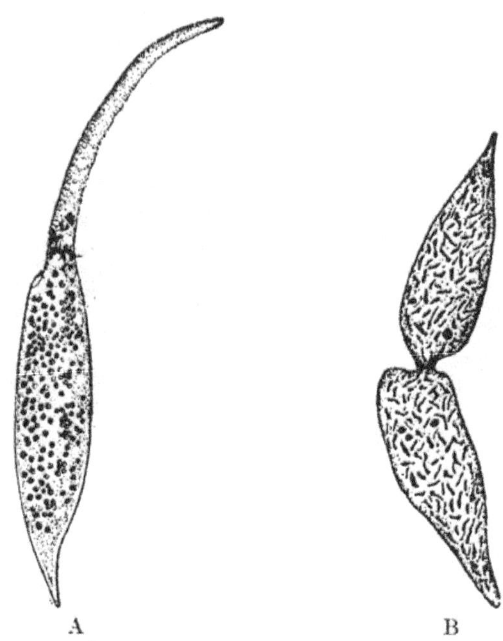

Fig. 26.—Nuclear division in the Ciliate, *Dileptus.* From Calkins, "Proto-zoology." *A.* Vegetative form. Nucleus in the form of chromatin granules scattered through the greater part of the cell ("distributed nucleus"). *B.* During division. Each chromatin granule elongates and divides into two.

after the bud is practically completed and ready to be cut off, *i.e.*, cytoplasmic division tends to precede nuclear division. Another complication due to the same association is the frequent differentiation of two forms of chromatic substance in the nucleus. These are the reproductive chromatin, or *idiochromatin*, and the vegetative, or *trophochromatin* (Dobell); in some Protozoa these may come to be organized into two separate nuclei which are sometimes equivalent to what are called the micro- and macronucleus respectively. We have already mentioned the distributed nucleus of many unicellular organisms in which the chromatin is not organized into a definite nuclear organ, but is in the form of scattered granules or collections of granules throughout the cell (Fig. 26, *A*). Such a condition indicates strongly that the nuclear and cytoplasmic parts of the cell have arisen through the gradual differentiation of a common protoplasmic basis. In cell division each of these chromatic bodies may first divide into two (*Dileptus*), though the members of the resulting pair are not distributed to different daughter cells, for the accompanying division of the cell is completed without any rearrangement of the chromatin granules (Fig. 26, *B*). In other forms with distributed nuclei (*Tetramitus*), the scattered granules collect about an active kinoplasmic organ termed the *division center* (Fig. 27); this divides, and the two products separate,

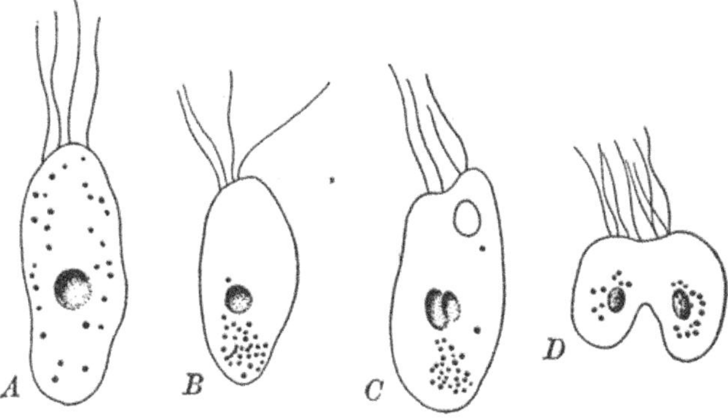

Fig. 27.—Cell division in the Flagellate, *Tetramitus*. After Calkins. *A* Vegetative condition showing scattered chromatin granules (distributed nucleus) and division center. *B*. Collection of chromatin granules preparatory to division. *C*. Fission of the division center. *D*. Separation of the division centers accompanied by the daughter groups of granules (nuclei).

each accompanied by a group of chromatic granules which are then redistributed equally to the daughter cells, although no definite mitotic figure is formed. Even when a definite nucleus does come to be established, much of the chromatin of the cell may not be contained within it but may remain distributed (Heliozoa, Radiolaria). Finally, of course, all of the chromatin becomes contained within one or more definite nuclear structures which may be simple spherical bodies, or they may show considerable complication and variation in form. A definite nuclear membrane may be absent at first, as in *Chilomonas* and *Trachelomonas*, though it is formed in practically all cases where the chromatin granules form definite nuclear groups. Within the nucleus the chromatic substance may not be definitely organized into chromosomes, or these bodies may appear only in certain divisions associated with gametic reproduction (*Paramœcium*); in some Protozoa, however, definite chromosomes are typically established and become clearly marked during each mitosis (Fig. 28).

Probably the most interesting modifications of mitosis among the Protozoa are those connected with the formation and behavior of the centrosome and mitotic spindle, upon the origin of which they may

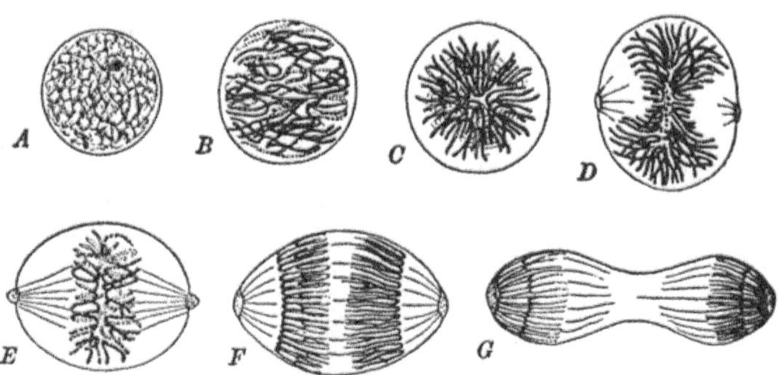

Fig. 28.—Nuclear division in the Rhizopod, *Euglypha*. After Schewiakoff. × 800. *A.* Coarse network of chromatin. *B.* Contraction of chromatin fibers and beginning of formation of loops (chromosomes). *C.* Arrangement of chromosomes in equatorial plate. Polar view. *D.* Equatorial plate. Lateral view. Spindle forming. *E.* Splitting of chromosomes and beginning of divergence. *F.* Continued divergence of chromosomes. *G.* Division nearly completed.

perhaps throw some light. In some species there is no indication of a specialized organ concerned particularly with division. In a few forms, as mentioned above, a division center may be formed, although

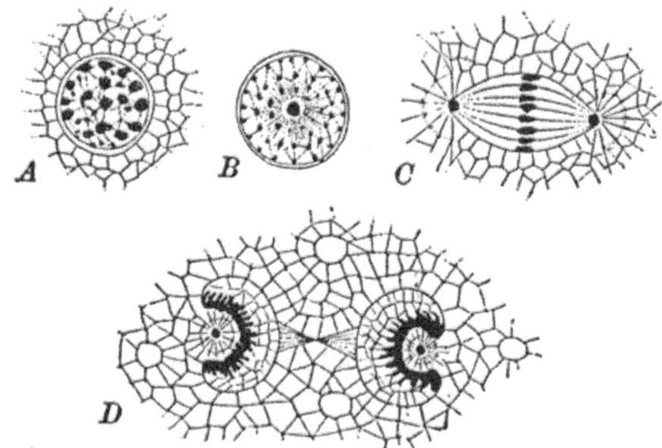

Fig. 29.—Nuclear division in the Rhizopod, *Centropyxis aculeata*. After Schaudinn (Doflein). *A.* Nucleus in vegetative stage. *B.* Appearance of centriole. *C.* Equatorial plate. Spindle with centrosomes; plasma radiations. *D.* Beginning of reconstruction of daughter nuclei.

no definite nucleus is present, the distributed chromatin granules collecting into a single group only at the time of cell fission. In some species of *Amœba*, and in a few other forms, one of the large chromatin bodies, or karyosomes, *within the nucleus* is specialized as an organ of division, called the *central body* and functionally equivalent to a centrosome (Fig. 29). This body does not lose its chromatic character, may be surrounded by a definite membrane, and appears to have functions other than those of a centrosome which it exercises during the intervals between divisions. In cell division this central body remains wholly or in part intranuclear. Apparently this represents a very early stage

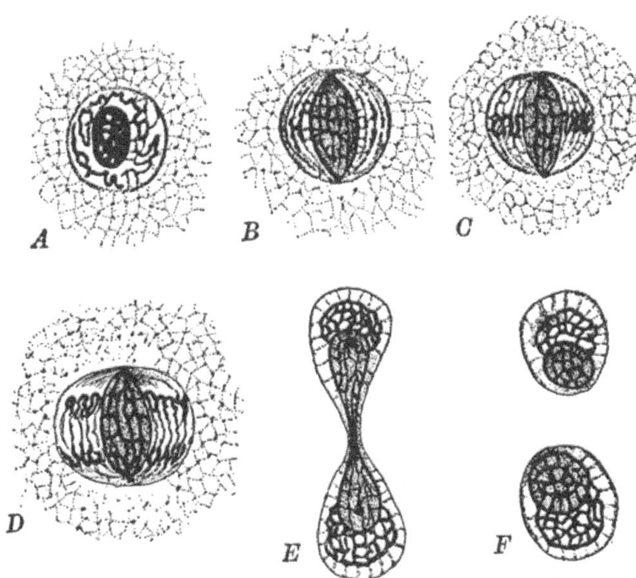

Fig. 30.—Nuclear division in the Rhizopod, *Chlamydophrys stercorea.* After Schaudinn (Doflein). *A.* Nucleus with central body and chromatin threads. *B.* Elongation of central body and beginning of formation of equatorial plate. *C.* Equatorial plate. Central body spindle-shaped with polar centrosome-like thickenings. *D.* Equatorial plate divided into two. Plasma radiations from the "centrosomes." *E.* Fission of central body and chromatin masses. *F.* Division completed. Daughter nuclei being reconstructed.

in the differentiation from a chromatic body of the centrosome which later becomes wholly achromatic and typically extranuclear. Several other forms have a dynamic *division center* equivalent to the centrosome but intranuclear; a nucleus of this type is known as a *centronucleus* (Fig. 30; see also Fig. 25, *B*). In the division of such nuclei the nuclear membrane may remain entirely intact (*Euglena*) or, as in *Noctiluca*, the nuclear membrane may partly break down during mitosis. In several forms possessing a definite extranuclear centrosome this body remains undivided in cell fission, passing to one daughter cell alone, while a new centrosome develops in the other cell; but this forms first as an intranuclear structure which later moves out into the cytoplasm alongside the nucleus. These conditions indicate strongly the nuclear origin of the centrosome. And there is some reason for believing the spindle also originally a nuclear structure, as it still is, in part at least. The spindle is a less constant organ than the centrosome, compared with which it is of secondary importance. In several Protozoa and simple plants the spindle is entirely absent, usually where the centrosome is intranuclear, so that no definite mitotic figure is formed. In other forms the spindle is intranuclear, and then the centrosomes or their equivalents may be absent, as in *Opalina* (Fig. 31). This form is further remarkable for the chromatic character of some of its spindle fibers, and in

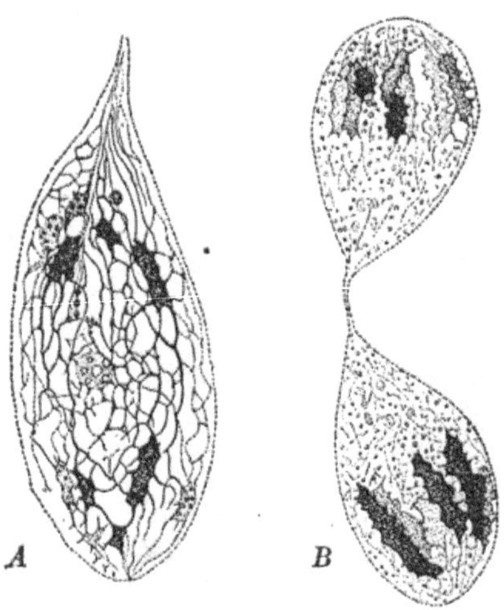

Fig. 31.—Nuclear division in the Infusorian, *Opalina intestinalis*. After Metcalf. × 1200. *A*. Nucleus in anaphase showing chromatic reticulum, which may be regarded as equivalent to the spindle, and branched, amœboid chromosomes. *B*. Late telophase.

that, in the absence of centrosomes, the chromosomes separate by an active amœboid movement, suggesting a possibly primitive behavior of the chromosomes in cell division (Metcalf).

It is of course impossible now to get definite information regarding the evolution of the cell organs and the process of mitosis, and in these connections the conditions found at present among the Protozoa are only suggestive. Many of these conditions do suggest strongly, however, that the nucleus has been developed from the gradual aggregation of scattered, chemically differentiated particles; that within the nucleus certain chromatic elements became specialized into division centers which finally became extranuclear, achromatic bodies—the centrosomes; and that certain achromatic elements became specialized into a spindle, also at first intranuclear (linin), and at present in nearly all forms both intra- and extranuclear in origin. The chromosomes we must now consider more particularly.

Interest in the process of mitosis centers in the chromosomes and their behavior, for, as we have said, this whole process seems to be directed toward the equal distribution of the daughter chromosomes to the daughter cells, while the cytoplasm may or may not be equally divided at the same time. We do not know how the constituents of the nucleus other than the chromosomes are distributed in cell division. There is little reason for supposing that these are distributed with exact equality.

We may recognize two chief aspects of chromosomal behavior in mitosis. The first is the division of the chromatin granules or chromioles composing the chromosomes; these granules all divide in the same direction so that the total result appears as an exactly equal longitudinal division of each chromosome, or of the spireme in those cases where the division of the granules occurs very early. This is the essential act of chromosome reproduction and it is obviously a process concerning the chromatin alone, independent of the remainder of the mitotic mechanism. The second important fact is the distribution of the two chromosome halves or daughter chromosomes to the two daughter cells. This is accomplished by the extra-chromosomal elements of the mitotic figure, the chromosomes apparently taking a purely passive part in the process. We see in the mitotic figure, not a mechanism for cell division merely, for this is frequently accomplished in the absence of mitosis, nor for chromosome division, for this frequently precedes the formation of the mitotic figure; but essentially the mitotic figure is a mechanism effecting the equal distribution to the daughter cells of the products of chromosomal division within the nucleus of the parent cell, so that each new cell has a complete group of chromosomes similar to those of the parent cell. The precision and wide occurrence of this equal distribution, through mitosis, of the chromosomes and of these cell organs alone, leads to the assumption that the chromosomes constitute the essential physiological elements of the nucleus and therefore of the cell. There are many subsidiary facts indicating the great importance of these bodies. Consideration of the relation of the chromosomes to the special problems of heredity and sex will be deferred to a later chapter (Chapter VII) for fuller consideration, but we should mention here a few of the important facts and hypotheses regarding these bodies.

In the nuclei of many of the Protozoa definite chromosomes are already present, but in some unicellular organsims there are conditions suggesting a possible mode of evolution of these structures. We have mentioned the collection of the distributed chromatin granules into small groups through the cell; these groups have been regarded as the rudiments of chromosomes. After a definite nucleus is established the chromatin granules remain as definite bodies, and each divides in cell fission. Even when the chromatin granules merely become rearranged about a division center, as in *Chilomonas* and *Trachelomonas*, although definite chromosomes may not be formed, the division of the granules occurs as the essential step in fission, just as later when the granules collect and fuse into chromosomes. In *Paramœcium* definite chromosomes are formed only in certain divisions, namely, those immediately preceding conjugation, that is during gametogenesis (Calkins and Cull, Fig. 82), and it is but a short step from this condition to the regular formation of chromosomes in all mitoses.

In the reconstruction of the daughter nucleus the chromosomes become vacuolated and finally break up into scattered granules whose distribution through the nucleus is so irregular that in nearly all cases no trace of chromosomal structure is apparent. The nucleus then grows rapidly, the chromatin content often increasing to many times that in the original chromosome group, until it soon reaches a quite definite size, varying widely in different cells, but fairly constant for cells of a single kind in a given species. The nucleus and chromatin then remain without much further change, quantitative at any rate, until toward the close of the vegetative or normally functional period of cell life. At the time of the next division much of the chromatin is usually eliminated from the nucleus, is cast out into the cytoplasm and disappears along with the nucleolus (Fig. 32); the chromosomes which then appear

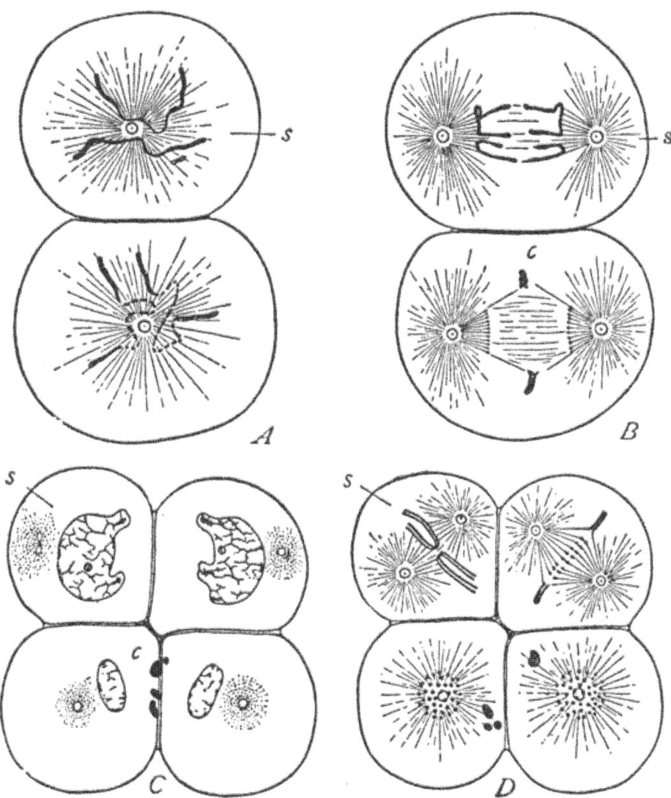

Fig. 32.—Early cleavages of the egg of the Nematode, *Ascaris*. Origin of primordial germ cells and casting out of chromatin in the somatic cells. From Wilson, "Cell," after Boveri. *A*. Two-cell stage dividing; polar view. *s*, stem-cell, from which the germ cells are derived. *B*. Later stage of same division; lateral view. *c*, chromatin in somatic cell being thrown out into the cytoplasm. *C*. Completion of four-cell stage showing eliminated chromatin. *D*. Division of four-cell stage showing continued chromatin elimination in the third somatic cell.

or reappear, are similar in every respect to those in the preceding division (Van Beneden). The chromosomes, therefore, show the same specific constancy as any other characteristic of the organism.

The most obvious characteristic of the chromosomes is that of numerical constancy. In different species of organisms the number varies greatly but there is in general little if any relation between the grade or relative complexity of the organism and the number of chromosomes in its nuclei; closely related species of a single genus may differ widely, *e.g.*, *Ascaris megalocephala* has four, *A. lumbricoides* forty-eight. In general the number seems highest in some of the Protozoa. Where there are very many minute chromosomes the difficulty of counting them exactly is very great and it cannot then be said precisely what or how constant the number is. In *Mastigella* there are about forty, in *Actinosphærium* one hundred and thirty to one hundred and fifty, in *Paramœcium* about two hundred. Among the Metazoa the smallest number is two, in a variety of *Ascaris*, the largest known is one hundred and sixty-eight, in *Artemia*. Frequent numbers are twelve, sixteen, and twenty-four, but any number may be found within these known limits. The number is practically always an even one in somatic cells, even or odd in the germ cells or their immediate predecessors. The numbers found in the tissue cells of some of the familiar organisms are the following: rat, guinea-pig, ox, sixteen; Amphioxus, salmon, salamander, frog, mouse, man (female), twenty-four; earthworm, thirty-two; shark, thirty-six; sea-urchin, eighteen in one species, thirty-six or thirty-eight, in another; pine, onion, sixteen; lily, peony, twenty-four.

While the number of chromosomes is thus practically constant it is not absolutely invariable and deviations from the normal are now known to occur in several forms. Of course the most frequent variation is the typical reduction of the number to one-half the somatic $\left(\dfrac{s}{2}\right)$, during certain phases in the formation of the germ cells (Van Beneden), or throughout the gametophytic generation of many, perhaps most, plants. But as we shall see later, this should hardly be called a variation from normal. A deviation of an entirely different kind is seen in a few cases where the chromosome number is $\dfrac{s}{2}$ in cleavage or tissue cells of certain individuals; thus in the cleavage cells of *Ascaris megalocephala* the number is two or four, in the tissues of *Helix pomatia*, twenty-four or forty-eight, in *Strongylocentrotus*, eighteen or thirty-six. In such cases each of the lesser number is said to be *bivalent*, and it is supposed, not without reason, that each is actually composed of two ordinary or *univalent* chromosomes. In a few instances the number may be even less than one-half the normal and each is then said to be *pluri-valent;* thus in the formation of the embryo-sac in the lily a variation in certain nuclei has been found, the number varying by fours from twelve to twenty-four ($s = 24$). The significance of these unusual cases is varied and sometimes doubtful. Again, constant differences in chromosome number, in both somatic and germ cells, are associated with sex differences in a large number of species of several widely divergent phyla. In such cases the females have a somatic group from one to five, or even more, in excess of that of the male; in such cases the specific number is fixed, though some variable species are known.

None of these unusual deviations from the normal is of the character of a "normal variability." Indeed very few in stances of this kind of variability in chromosome number are known. One instance is that of the salamander larva, in certain tissue cells of which the number is said to vary (Della Valle) in different individuals from nineteen to forty, the normal being twenty-four, and in a single specimen limits of nineteen and twenty-seven have been described.

This very high degree of specific numerical constancy of the chromosomes indicates strongly that the appearance of a chromosome in mitosis is not determined by a large number of causes, but that it is the result of the operation of a single and simple factor; what this may be is a matter of conjecture only.

Another important characteristic of the specific chromosome group is the constancy of the form and size differentiations among the members of the group. In many organisms it can be seen clearly that the chromosomes of a single nucleus are not all of the same size and form; they may differ in shape, dimensions, and proportions (Fig. 33) (Montgomery). Moreover these differences are constant from one cell generation to the next, so that similar chromosomes may be identified in successive mitoses. The form is somewhat more variable than the volume, which is remarkably uniform. It should be said that the size of a given chromosome is not fixed throughout the entire cell history, for at certain periods, particularly in the

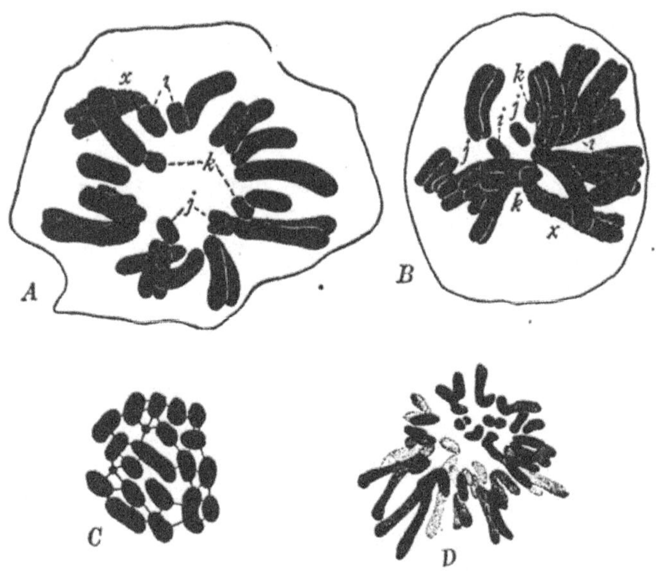

FIG. 33.—Various chromosome groups illustrating variation in size and form, and coupling of chromosomes. *A,B,* from Sutton, *C,* after Wilson, *D,* after Agar. *A,B.* Spermatogonia of the grasshopper, *Brachystola magna. C.* Spermatogonium of the squash-bug, *Anasa tristis. D.* Spermatogonium of the lung-fish, *Lepidosiren.*

germinal tissues, the chromosomes may be many times larger than at other periods (Fig. 34). But at corresponding cell ages the corresponding chromosomes are practically equal in volume, and in somatic cells such volume changes of single chromosomes are relatively infrequent.

One most significant and very important fact in this connection is that in the somatic cell the chromosomes are present in couples of similar elements; there are two of each size or form (Montgomery) (Figs. 33, 72, *A;* 142, *A*). The exceptions are found chiefly in those forms where sex differences are found; in such cases one or more chromosomes are unpaired, or the members of the pair may be dissimilar in size. And further in the germ cells with $\frac{s}{2}$ chromosomes, none is paired—all are single, each somatic pair is represented, and the groups in eggs and sperm are alike.

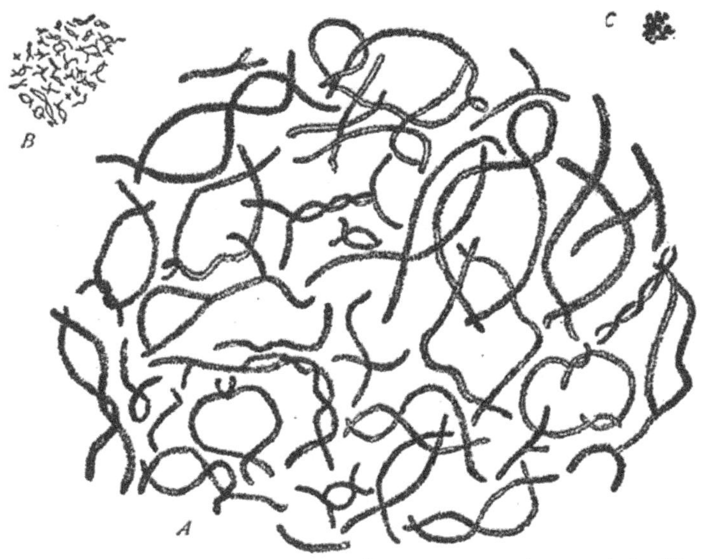

Fig. 34.—Changes in the volume of chromosomes in the egg of the Elasmobranch, *Pristiurus*. All drawn to same scale. From Wilson, "Cell," after Rückert. *A*. In egg of 3 mm. diameter. Chromosomes at maximal size and minimal staining capacity. *B*. In egg of 13 mm. diameter. *C*. In fully grown ovarian egg. Minimal size and maximal staining capacity.

Such facts as those given above taken in connection with the precision with which each chromosome is halved in mitosis, lead almost irresistibly to the supposition that the chromosomes must be qualitatively unlike. Such qualitative differences cannot be observed directly, and can only be inferred, but as we shall see in connection with the relation of the chromosomes to heredity, this inference seems to be justified from the results of the experimental or accidental modification of the chromosomal content of the nuclei and the character of the resulting cells or cell groups (Chapter VII).

In connection with the chromosomes there remain to be mentioned two important hypotheses. The first is the hypothesis of the *specificity* of the chromosomes. Stated in its barest form the essence of this idea is that each chromosome functions, in cell life, in its own particular way, representing a center for reactions of a specific kind only; that the chromosomes are cell "organs," functionally differentiated and representing a division of labor roughly analogous to the functional differentiations of the whole animal body. It is impossible to discuss this hypothesis satisfactorily here and it is deferred to Chapter VII, where it occupies a natural place in our account of the mechanism of differentiation.

The second hypothesis is that of the *genetic continuity* of the chromosomes. The essential of this idea is that the chromosomes which appear during the preparation for a mitosis, are definitely related in a precise way, to the chromosomes entering that nucleus at the close of the preceding division. In its first form this hypothesis was called that of the *individuality* of the chromosomes (Rabl, Boveri), and it was held that the chromosomes actually, though not visibly, preserve their structural identity during the period of interkinesis, that the chromosomes of one mitosis are not related to those of the preceding division, but are actually the same chromosomes. The fact that little or no direct evidence of chromosomal identity during the interkinesis, is to be had, has led to the remodeling of the idea of individuality, into the hypothesis of genetic continuity.

The nature of the evidence bearing upon this hypothesis, while not scanty, is largely circumstantial, and hardly affords definite proof, either affirmatively or negatively. We may suggest some of this evidence without pretending to give a detailed account of the facts.

At the very beginning we must recognize that the chromosomes are actually visible, as differentiated structures, only during that comparatively brief period of cell life occupied by the process of mitosis. Considering first some of the facts opposed to this hypothesis, we should say that those who deny the fact of continuity, maintain that during the vegetative period of cell life the dissolution and disappearance of the chromosomes is not only apparent but real. There is truly little visible and direct evidence of chromosomes during interkinesis. And further, much new chromatin is formed during this period which cannot be distinguished from the chromosomal chromatin; in the early stages of mitosis much chromatin is again thrown out of the nucleus and takes no part in the formation of chromosomes. It is impossible to say whether the chromatin forming the chromosomes is or is not then, the same as that previously derived from them, or that resulting from growth. Many instances are known where the chromatin of the vegetative nucleus is nearly all contained within a single homogeneous chromatin nucleolus or karyosome (*e.g.*, *Asterias*), and in preparation for mitosis some granules pass out of the karyosome and form into distinct chromosomes while the greater part of the karyosome dissolves (Fig. 35); it is difficult to understand how the chromosomes could have preserved their integrity through such a history. In amitosis no chromosomes are visible yet the nuclei of cells thus formed seem to function normally and such cells are capable of typical differentiations and in a few instances are said, with some question,- however, to be subsequently capable of division by mitosis, and of normal chromosome formation.

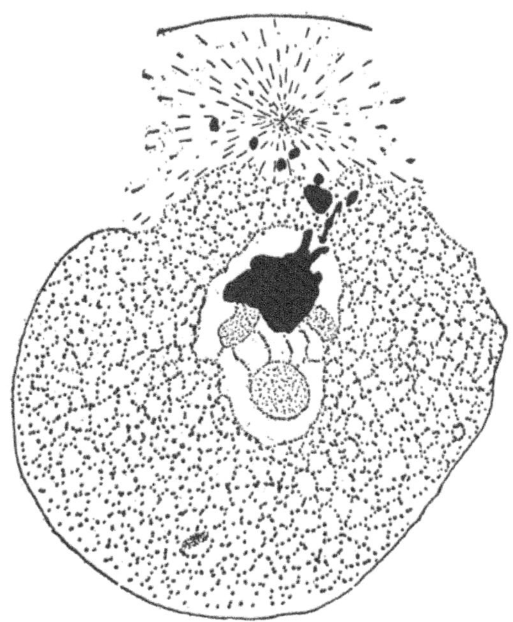

Fig. 35.—Primary oöcyte of the star-fish, *Asterias forbesii*, at beginning of division. From Dahlgren and Kepner (Jordan). Chromatin leaving the "chromatin reservoir" or chromatin nucleolus, and being added to the chromosomes.

Such an interruption of a series of mitotic divisions by a period of amitotic division would seem to exclude the possibility of both genetic relation and specificity of the chromosomes. In those forms where chromosomes are not normally formed, the chromatin granules are the units in nuclear division; and even when these are formed into chromosomes the essential step in nuclear division is the fission of these granules, which thus seem to be the real units of the chromatic substance. Yet one could hardly maintain the genetic continuity of these granules upon other than logical grounds, and to many there seems no stopping place

short of this, if the fact of continuity of organized chromatic structure
be accepted to begin with.

On the other hand, those who adhere to the continuity hypothesis
find many supporting facts in the phenomena of fertilization and
maturation, the importance of which can be appreciated more fully
after our consideration of these subjects. They assume, from the consideration of the behavior of the chromosomes, that they only apparently
lose their structural and functional identity during the interkinesis,
and that something directly representative of the chromosomal structure

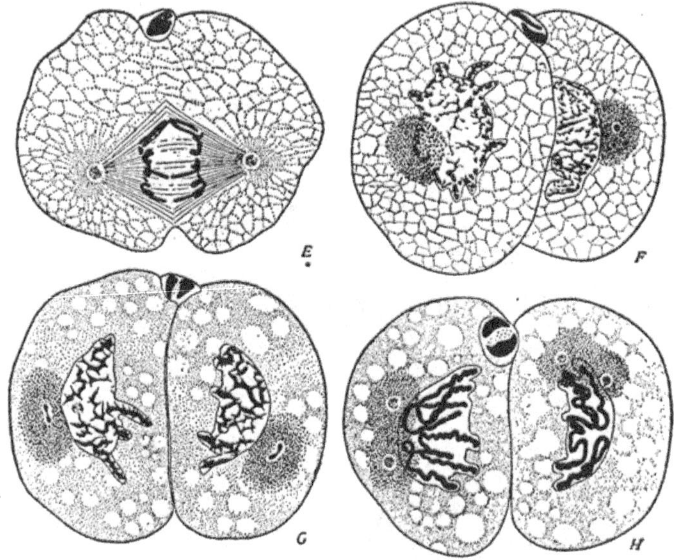

Fig. 36.—Indications of the individuality of the chromosomes in the cleavage
of the egg of *Ascaris*. From Wilson, "Cell," after Boveri. *E*. Anaphase of first
cleavage. *F*. Two-cell stage with lobed nuclei, the lobes formed by the ends of the
chromosomes. *G*. Early prophase of next division. Chromosomes reforming,
centrosomes dividing. *H*. Late prophases of same division, the chromosomes
lying with their ends in the same position as at the close of the preceding division.

is present in the resting nucleus, invisible directly and known only from
its consequences. There are, it is true, a few instances in which the
chromosomal arrangement is said really to be visible in the interkinesis
(Figs. 36, 37), but these cases are not very clear, except in the maturation
divisions leading directly to the formation of the specialized germ cells,
where the chromosomes are definitely known to be directly continuous.
The remarkable constancy of form, volume, and number of chromosomes
throughout the cells of a given organism and species, is important evidence favoring the hypothesis under consideration. The probability
is so high as to amount almost to certainty, that if the chromosomes

were new and independent formations in each mitosis, they would show a normal variability in number throughout long series of mitoses. However, numerical variability is so rare as to be practically absent, although the few exceptions known are emphasized by those who do not accept the idea of continuity. Constancy of form seems to be less precise than constancy of volume, but both are sufficiently marked to be noteworthy. It is difficult to get precise observations here on account of the liability to shrinkage or deformation of the chromosomes in the preparation of the material for study. There is often undoubtedly a definite pairing of chromosomes in the somatic nuclei (Fig. 33), while in the germ nuclei, with $\frac{s}{2}$ chromosomes, the same categories of chromosomal form and size found in somatic nuclei are distinguishable, but these are no longer paired—there are only single representatives of each. This is one of the strongest points favoring this hypothesis. And no less significant is the fact that the odd or unpaired chromosomes associated with sex differentiation, mentioned above, remain constant in size and form and number, throughout the tissue cells of certain individuals, and are easily recognizable, not only through their peculiar morphology but on account of their peculiar behavior as well. It is important in this connection, to notice that these particular chromosomes may remain undissolved even in the resting nucleus, where they had often been described as chromatin nucleoli or other bodies, before their significance was appreciated or their history known.

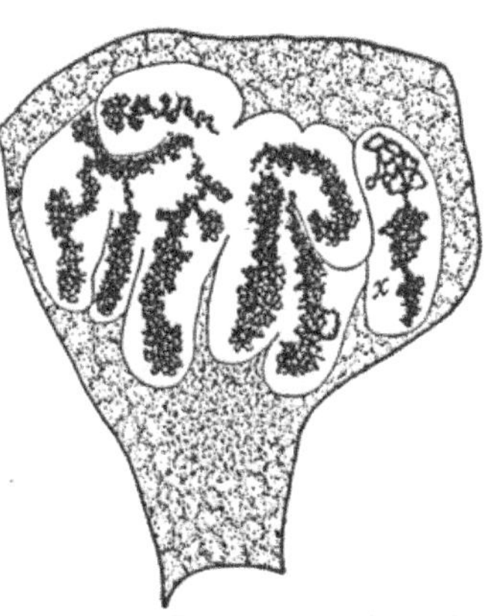

Fig. 37.—Indications of the individuality of the chromosomes in early prophase of division of a spermatogonium of *Brachystola magna*. From Sutton. Spiremes forming in lobes of the nucleus corresponding with the chromosomes which entered the nucleus at the close of the preceding division.

Another group of facts of quite a different character has an important bearing upon these hypotheses. It sometimes happens in mitosis that one or more chromosomes belonging to one daughter group, accidentally become included with the other group so that one of the daughter nuclei has fewer, the other more, than the normal somatic number. In subsequent divisions of these cells the number of chromosomes appearing is not the normal, but the increased or diminished number, the sum of the two, however, always being 2s. Or in fertilization of the egg by the sperm, each of which has $\frac{s}{2}$ chromosomes, various abnormalities occur in the distribution of the chromosomes, and it is always the case that *the number of chromosomes forming out of a nucleus is the same as the number passing into it,* no matter how that deviates from the normal (Boveri) (Fig. 38). There seems to be no regulation within the nucleus in this respect, such as would result were it a unified structure tending always to maintain its own normal. In many instances of

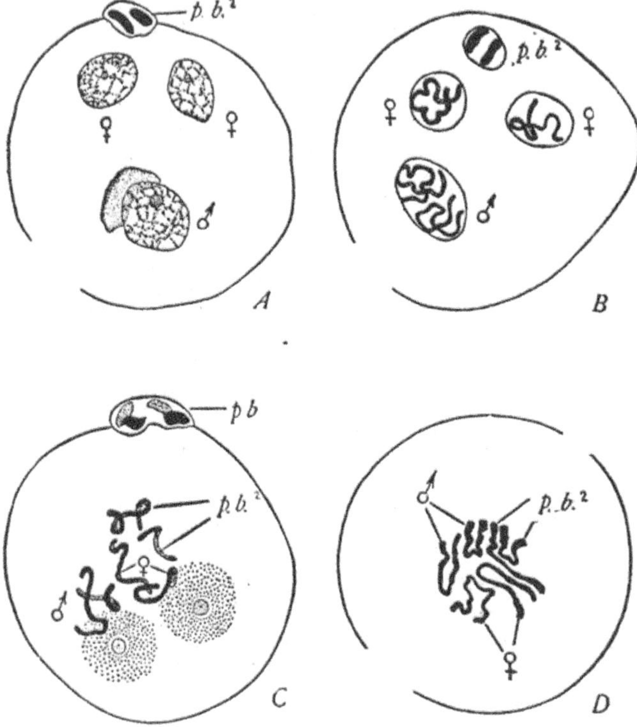

Fig. 38.—Indications of the individuality of the chromosomes in the fertilization of *Ascaris.* From Wilson, "Cell," after Boveri. *A.* The two chromosomes of the egg-nucleus, accidentally separated, have given rise each to a reticular nucleus (♀, ♀); the sperm-nucleus below (♂). *B.* Later stage of the same, a single chromosome in each egg-nucleus, two in the sperm-nucleus. *C.* An egg in which the second polar body has been retained; *p.b.²* the two chromosomes arising from it, ♀ the egg-chromosomes, ♂ the sperm-chromosomes. *D.* Resulting equatorial plate with six chromosomes.

alternation of generations, the agametically produced generation is formed from a cell with $\frac{s}{2}$ chromosomes; in all the cells of such an organism, the nuclei show the same reduced number, even in so complex an organism as the fern prothallus. In some forms, the chromosome groups derived from the egg and sperm nuclei, each of $\frac{s}{2}$ chromosomes, remain distinguishable through a considerable series of the early divisions of the zygote; two separate spiremes, each forming $\frac{s}{2}$ chromosomes, may even be distinguished sometimes (Rückert). And in some hybrids where the chromosomes are unlike in number, or

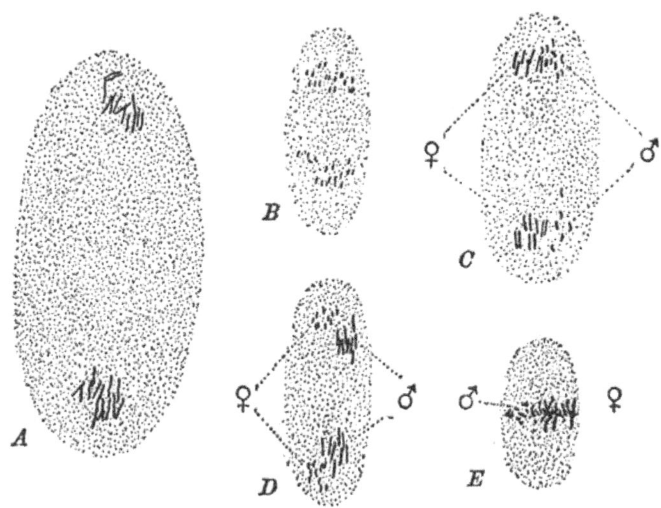

Fig. 39.—The chromosome group in the hybrids of the Teleosts, *Fundulus* and *Menidia*, showing the distinctness of the paternal and maternal elements. From Moenkhaus. *A.* Late anaphase of first cleavage of normally fertilized *Fundulus*. All chromosomes of the long type. *B.* Anaphase of first cleavage of normally fertilized egg of *Menidia*. All chromosomes of the short type. *C.* Anaphase of first cleavage of *Fundulus* egg fertilized with *Menidia* sperm. To the left long (*Fundulus*) chromosomes only, to the right short (*Menidia*) chromosomes only. *D.* Anaphase of first cleavage of *Menidia* egg fertilized with *Fundulus* sperm. *E.* Metaphase of first cleavage of *Fundulus* egg fertilized with *Menidia* sperm. ♂, chromosomes of paternal origin. ♀, chromosomes of maternal origin.

form, or size, the two groups derived from the male and female parents remain distinct (Fig. 39), often for a very long time, perhaps even throughout life (Moenkhaus, Herbst, Baltzer).

While evidence of the kind suggested above does not constitute definite proof of the genetic continuity of the chromosomes, it is very difficult to explain the facts upon any other basis. In the absence of any other satisfactory hypothesis in this field it seems wise to accept a certain modification of the essential idea, as a working basis, while admitting the difficulties of demonstration and the existence of some apparent contradictions. We may recognize in this difference of opinion regarding the continuity of the chromosomes, an outcropping of the opposed preformational and epigenetic conceptions, which pervade all descriptions of developmental phenomena. The hypothesis of chromosomal continuity is essentially a preformational view; those who deny continuity assume the epigenetic view.

Perhaps the immediate solution of the difficulty here may not be unlike the solution of the greater problem and more fundamental difference of opinion. It seems likely that what is directly continuous from nucleus to nucleus, *i.e.*, what is preformed, is some sort of fundamental organization determining the chromosomal structure of the nucleus, just as the "organization" of the egg determines the structure of the embryo developed from it. Yet what appears is a new structure, formed epigenetically under or through the influence of the "organizational" factor, by the material present, and subject to the modifying influences of changing conditions external to the nucleus. That is to say, the formation of the chromosomes out of the chromatin of the vegetative nucleus is to be regarded as a true process of development. The reaction between the fundamental organized structure of the nucleus, and the stimuli acting upon it, consists in the formation of the chromosomes and other structures, not to be seen in the nucleus previous to this reaction. The chromosomes are thus no more genetically continuous than the organs of adults, and yet there is a real continuity of organization underlying.

In many respects the nucleus is analogous to an organism, the chromosomes and other nuclear structures representing the organismal organs; both are functionally specific, are constant in number, form, and size throughout the species; both reproduce and exhibit development as a form of response. As Wilson points out, the analogy is far from complete—no complete analogy is known, but that there is an underlying organization of some kind, continuous and specific, seems clear although we remain entirely ignorant of what it really is, and just how it operates and is affected by new conditions.

Before leaving the subjects of the cell and cell division we must consider briefly two other questions which are of great importance, but which are also still in a hypothetical state. These are, the nature of the causes leading to cell division, and the nature of the fundamental mechanism of the process. The interactions between the nucleus and cytoplasm, and between these and the external medium, which constitute the life of the cell, are largely, probably wholly, of a physico-chemical nature. As such their normal procedure is dependent upon certain mass relations of the interacting substances, and upon the maintenance of adequate pathways of interaction between them. For these reasons we look quite naturally, in searching for a possible cause of cell division, to the volumetric relations of the nucleus and cytoplasm, and to the extent of the surface of these parts of the cells in relation to the two masses.

Immediately after mitosis is completed the nucleus grows very rapidly for a brief period, and then much more slowly or not at all. The cytoplasm, however, does not show this rapid initial growth, but maintains a fairly uniform and continuous growth rate. As a result, in a newly formed cell the ratio of nuclear mass to cytoplasmic mass becomes quite high, but soon diminishes, and in an older cell diminishes rapidly, since the cytoplasm continues to grow after the nucleus nearly ceases. Considering first the relation of the surface to the mass, and assuming for illustration that the cell and nucleus are both spherical, we can see how this increase in size alters the relation of mass to path of interchange, since the area of the surface of an enlarging sphere increases relatively slower than the volume. Should the cell double its diameter through growth, it increases its volume eight times and its surface only four times. A cubical cell doubling the length of its sides reduces its ratio of area to volume as 2 : 1. Or expressing a similar relation somewhat differently, a spherical cell doubling its volume, lessens its ratio of surface to volume approximately as 5 : 4, while a cubical cell doubling its volume reduces the same relation approximately as 6 : 4.75.

These relations are probably important for both of the chief interactions of cell life, those between nucleus and cytoplasm, and between cytoplasm and surrounding medium. The only pathway between cytoplasm and nucleus is the nuclear membrane, while the surface of the cytoplasm or cell wall forms the pathway between cell and medium. There is of course a limit to the capacity, so to speak, of these surfaces, and as the cell increases in volume this limit tends to be reached. The ratios of these surfaces to the masses are raised by the division of the cell, which reduces volume more than surface, and thus restores the efficiency of the surfaces as pathways of interaction. The tendency for the cytoplasmic mass to increase more rapidly than the mass of the nucleus in older cells seems to be of even greater importance than these surface to mass relations. There seems to be a fairly definite specific limit to the ratio of nuclear mass to cytoplasmic mass, although it is difficult to say whether after all the nuclear surface relation is not even here an equally important factor. This mass relation is called the "*kern-plasma*" or *nucleo-cytoplasmic* relation, the importance of which is emphasized particularly by Richard Hertwig. There is a definite average cell size in a given tissue and species; a large or a small organ or organism does not possess respectively larger or smaller cells, but larger or smaller numbers of cells of the same average dimensions. The limiting factor, however, seems to be not the actual bulk of the cell, but the proportion between the volume of the nucleus and that of the cytoplasm, *i.e.*, $\left(\dfrac{V_n}{V_c}\right)$. In many cells during, and immediately after, division the nucleus grows at the expense of the cytoplasm, and the ratio $\left(\dfrac{V_n}{V_c}\right)$ is raised (Fig. 40). The cell then enters upon its vegetative phase, during which the cytoplasm grows more rapidly than the nucleus, and the ratio diminishes toward the lower functional limit; as the ratio approaches this limit the functional activities of the cell change, and the normal vegetative processes give place to that form of action which we call cell division, during which the ratio again rises to a value permitting normal vegetative functioning. It is not yet possible to state in precise quantitative terms what the limits of these ratios of volume

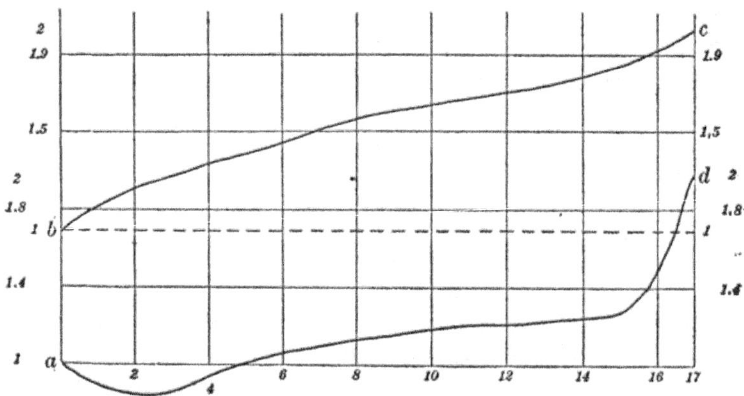

Fig. 40.—Curve showing the increase in volume of nucleus (*a–d*) and cytoplasm (*b–c*) during interkinesis, in the Infusorian; *Frontonia leucas*, at temperature of 25° C. After Popoff. Ordinates, volume; abcissas, time in hours. Each curve shows a doubling of volume (1:2) during the seventeen hour period of the interkinesis. Each curve is based upon its own units of measurement, which are different for the two curves. The nucleo-cytoplasmic relation is identical at the beginning and end of the period. The size of the nucleus is *relatively* smallest at fifteen hours; then the nucleus begins to grow very rapidly, so that at the time of the division of the whole cell, the original relation is restored.

and surface are in specific instances, nor even to say whether the volume or surface, or volume and surface relations are those essentially involved. But so far this nucleo-cytoplasmic hypothesis is the most plausible explanation of the nature of the immediate conditions of cell division. It should be said, however, that the applicability of Hertwig's "Kern-plasma Relation" is still chiefly limited to Protozoan cells, and that even here there are many contradictions. Some of the more obvious exceptions to the definition of such a limiting ratio are to be explained as special adaptations. Such are the very great cytoplasmic bulk of many egg cells or the relatively large size of the nucleus in the sperm cell. Many other exceptions of special character are to be found, usually associated with reproductive processes, *e.g.*, brood formation.

Regarding the real nature of the mechanism of mitosis, even less can be said to be known than with respect to its causes. The arrangement of the achromatic amphiaster and the behavior of the chromosomes show that in the mitotically dividing cell the forces or tensions are

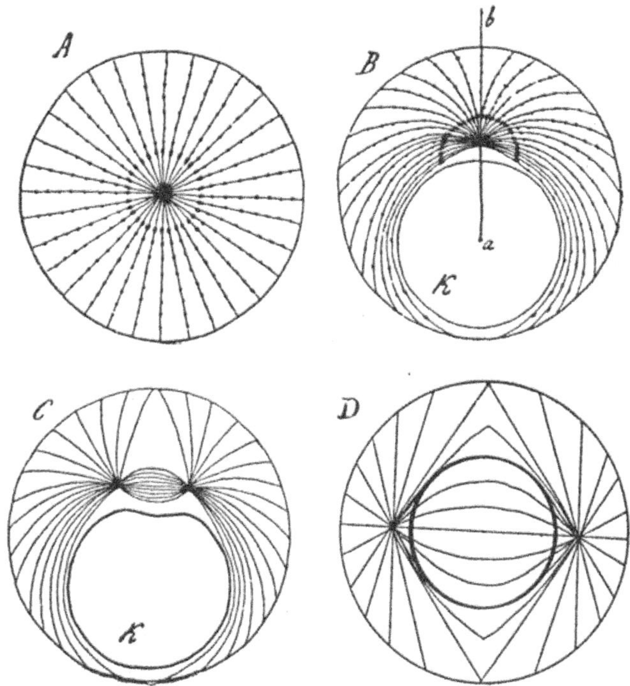

Fig. 41.—Diagrams of the arrangement of the spongioplasmic network (mitome) in the cell. *A, B, C,* from Korschelt and Heider, after Heidenhain. *A.* Schema of the arrangement of the spongioplasm as it would appear in the absence of a nucleus. Symmetrical monocentric system. *B.* Arrangement in the presence of a nucleus. Asymmetrical monocentric system. *C.* Arrangement at the beginning of mitosis. Commencement of dicentric system. *D.* Arrangement during the process of mitosis. Symmetrical dicentric system. *a,b,* cell axis; *k,* nucleus.

arranged in a *dicentric* system, whereas in the vegetative cell the system is *monocentric* (Fig. 41). As the result of the action of the forces in this dicentric system the chromosome halves are separated and the cytoplasm divided. The problem here is to discover the nature of the forces and the cause of the formation of the amphiaster in the form which it has. In explanation of the first part of the problem, *i.e.*, the divergence of the chromosomes, it was formerly believed that the fibers in the amphiaster were the active elements and that different groups of these fibers had different functions. Thus the mantle fibers attached to the chromosomes were supposed to be contractile and by shortening to draw the chromosomes toward the ends of the spindle, the central spindle remaining rigid and resisting any tendency for the ends of the spindle to approach as the result of the contraction of the mantle fibers, and at the same time serving as a sort of track upon which the chromosomes would slide along. The asters then, either as anchors helped to fix the ends of the spindle, or by contraction served to draw the fully diverged chromosomes further into the daughter cells (Fol, Van Beneden, Heidenhain). This naïve explanation cannot be applied *in toto* to any known instance of division, although certain features may be correct descriptions of the events in certain forms. And there are many facts opposing such an account of the action of the forces of mitosis, such as the absence of mantle fibers or of asters in many mitoses. Another hypothesis, in greater favor at present, and apparently well founded, is that the centrosomes and centrospheres rather than the achromatic fibers are the active elements, and further, that their activity is primarily of a chemical nature. Thus the chromosomes are believed to be chemically attracted toward the centrosomes, the achromatic fibers being passive and formed merely as the result of the rearrangement of cytoplasmic granules along the paths of the chemical transformations which have their seat in the centrosomes and extend thence through the cytoplasm (Strasburger). This hypothesis goes farther and makes it possible to explain the division of the cytoplasm on the same basis. For the chemical transformations centering in the centrosomes might easily influence the tensions of the comparatively impermeable surface film of the cell so that in the region near the centrosomes the tensions would be increased while that region farthest from the centrosomes and symmetrically related to them both, namely, the plane of the equatorial plate, would be a region of lower tension; the result of this would, of course, be a constriction in this plane. It is quite likely too that differences in electric tension accompany these chemical transformations, and these might assist in the alteration of the surface tensions in such a way as to contribute to the same end (R. S. Lillie). It is known that there are differences in the electric potential in different regions of the dividing cell, in some cases at least.

A further development of the chemical hypothesis attempts to explain the formation of the amphiaster itself. Thus, increase in the ratio of the volume of the cytoplasm to volume or surface of the nucleus beyond a certain point leads to a chemical alteration of the centrosomes such that they become the centers of two equivalent series of chemical reactions with the cytoplasm, the result of which is the formation of the

dicentric system. It is suggested, reasonably, that the chemical alteration is such that the centrosomes absorb or condense the more liquid
parts of the cytoplasm, leaving this considerably more dense than in the
vegetative condition (Bütschli, Rhumbler). The existence of two such
centers withdrawing the more liquid parts of the cytoplasm would
lead to the radiations seen in the asters and spindle, which would thus result from a physical alteration of the structure of the cytoplasm induced
by the chemical changes within the centrosomes. This is all extremely
hypothetical of course, but there are many inorganic phenomena as
well as processes to be seen in the cleavage of the egg which lend considerable support here. At any rate these hypotheses represent the
state of our present ignorance of the nature and origin of the mitotic
figure and process. There are many reasons for believing that the chemical differentiations within the cell are of fundamental importance here,
such as the fact that cells can be made to divide artificially by altering
their chemical structure. And that interactions of the nucleus and
cytoplasm are involved is indicated by the important observation that
while the amphiaster may be formed in the absence of a nucleus, no
real division of the cell may occur without the presence of nuclear
material (Boveri, Ziegler).

REFERENCES TO LITERATURE

von Baer, K. E., Die Metamorphose des Eies der Batrachier vor der
Erscheinung des Embryo und Folgerungen aus ihr für die Theorie
der Erzeugung. Arch. Anat. u. Phys. 1834.

Baltzer, F., Zur Kenntnis der Mechanik der Kerntheilungsfiguren.
Arch. Entw.-Mech. **32.** 1911.

Benda, C., Die Mitochondria. Ergebnisse Anat. u. Entw. **12.** 1903
(1902).

Van Beneden, E., Recherches sur la composition et la signification de
l'œuf, *etc.* Mem. Acad. roy. Belgique. **34.** 1870. Recherches
sur la maturation de l'œuf et la fecondation. Arch. Biol. **4.**
1883.

Bonnevie, K., Chromosomenstudien. Arch. Zellf. **1.** 1908.

Boveri, T., Zellenstudien II. Die Befruchtung und Teilung des Eies
von *Ascaris megalocephala.* Jena. Zeit. **22 (15).** 1888. Zellenstudien III. Ueber das Verhalten der chromatischen Kern-
substance bei der Bildung der Richtungskörper und bei der
Befruchtung. Jena. Zeit. **24 (17).** 1890. Ueber die Befruch-
tungs- und Entwickelungsfähigkeit kernloser Seeigel-Eier, *etc.*
Arch. Entw.-Mech. **2.** 1895. Zur Physiologie der Kern- und
Zelltheilung. Sitz.-Ber. Phys.-Med. Ges. Würzburg. 1897.

Bütschli, O., Untersuchungen über mikroskopische Schäume und das Protoplasma. Leipzig. 1892.

Calkins, G. N., Nuclear Division in Protozoa. Woods Holl Biol. Lect. 1899. (See also ref. Ch. I.)

Calkins, G. N., and Cull, S. W., The Conjugation of Paramæcium aurelia (caudatum). Arch. Protist. **10.** 1907.

Conklin, E. G., Karyokinesis and Cytokinesis in the Maturation, Fertilization and Cleavage of *Crepidula* and other Gasteropoda. Jour. Acad. Nat. Sci. Philadelphia. **12.** 1902. Cell Size and Nuclear Size. Jour. Exp. Zool. **12.** 1912.

Dahlgren, U., and Kepner, W. A., Text-book of the Principles of Animal Histology. New York. 1908.

Delage, Y., La structure du protoplasma et les théories sur l'hérédité et les grands problèms de la Biologie générale. Paris. 1895.

Della Valle, P., L'Organizzazione della chromatina studiata mediante il numero dei chromosomi. Archiv. Zoologico. (Napoli.) **4.** 1909.

Dobell, C. C., Chromidia and the Binuclearity Hypotheses: a Review and a Criticism. Q. J. M. S. **53.** 1909.

Doflein, F. (See ref. Ch. I.)

Duesberg, J., Nouvelles recherches sur l'appareil mitochondrial des cellules séminales. Arch. Zellf. **6.** 1910.

Erdmann, R., Kern- und Protoplasmawachsthum in ihren Beziehungen zueinander. Anat. Hefte. **18.** 1910.

Erhard, H., Studien über "Trophospongien." Zugleich ein Beitrag zur Kenntnis der Secretion. Festschr. f. R. Hertwig. Jena. 1910.

Farmer, J. B., The Structure of Animal and Vegetable Cells. Lankester's Treatise on Zoology. I, **2.** London. 1902.

Fick, R., Vererbungsfragen, Reduktions- und Chromosomenhypothesen Bastardregeln. Ergebnisse Anat. u. Entw. **16.** 1906 (1907).

Flemming, W., Zellsubstanz, Kern und Zellteilung. Leipzig. 1882.

Gurwitsch, A., Morphologie und Biologie der Zelle. Jena. 1904.

Häcker, V., Praxis und Theorie der Zellen- und Befruchtungslehre. Jena. 1899.

Hartmann, M., Die Konstitution der Protistenkerne und ihre Bedeutung für die Zellenlehre. Jena. 1911.

Heidenhain, M., Cytomechanische Studien. Arch. Entw.-Mech. **1.** 1895. Plasma und Zelle. Jena. 1907.

Hertwig, O., Die Zelle und die Gewebe. Jena. I, 1893. II, 1898.

Hertwig, R., Ueber neue Probleme der Zellenlehre. Arch. Zellf. **1.** 1908.

Jörgensen, M., Zur Entwicklungsgeschichte des Eierstockeies von

Proteus anguineus (Grottenholm). Festschr. f. R. Hertwig. Jena. 1910.

LILLIE, F. R., Karyokinetic Figures of Centrifuged Eggs. An Experimental Test of the Center of Force Hypothesis. Biol. Bull. **17.** 1909.

LILLIE, R. S., The Physiology of Cell-division. IV. The Action of Salt Solutions followed by hypertonic Sea-water on unfertilized Sea-urchin Eggs, and the Rôle of Membranes in Mitosis. Jour. Morph. **22.** 1911.

METCALF, M. M., *Opalina*. Its Anatomy and Reproduction, with a Description of Infection Experiments and a Chronological Review of the Literature. Arch. Protist. **13.** 1909.

MOENKHAUS, W. J., The Development of the Hybrids between *Fundulus heteroclitus* and *Menidia notata*, with especial Reference to the Behavior of the Maternal and Paternal Chromatin. Am. Jour. Anat. **3.** 1904.

MONTGOMERY, T. H., JR., A Study of the Chromosomes of the Germ Cells of the Metazoa. Trans. Amer. Phil. Soc. **20.** 1901. On the Dimegalous Sperm and Chromosomal Variation of *Euschistus*, with Reference to Chromosomal Continuity. Arch. Zellf. **5.** 1910.

NOWIKOFF, M., Zur Frage über die Bedeutung der Amitose. Arch. Zellf. **5.** 1910.

POPOFF, M., Experimentelle Zellstudien. Arch. Zellf. **1.** 1908.

PRENANT, A., Théories et interprétations physiques de la mitose. Jour. Anat. Phys. Paris. **46.** 1910. Les mitochondries et l'ergastoplasme. Id. **46.** 1910.

PRENANT, A., BOUIN, P., et MAILLARD, L., Traité d'Histologie. Paris. 1910.

PRZIBRAM, H., Experimental Zoology. I. Embryogeny. Cambridge. 1908. Anwendung elementarer Mathematik auf biologische Probleme. Leipzig. 1908.

SCHAXEL, J., Die Morphologie des Eiwachstums und der Follikelbildungen bei den Ascidien—Ein Beitrag zur Frage der Chromidien bei Metazoen. Arch. Zellf. **4.** 1910.

SCHEWIAKOFF, W., Ueber die karyokinetische Kerntheilung der *Euglypha alveolata*. Morph. Jahrb. **13.** 1887.

SCHNEIDER, K. C., Histologisches Praktikum der Tiere. Jena. 1908. Histologische Mitteilungen. III. Chromosomengenese. Festschr. f. R. Hertwig. Jena. 1910.

STRASBURGER, E., Zellbildung und Zellteilung. 3 Auf. Jena. 1880.

SUTTON, W. S., On the Morphology of the Chromosome Group in *Brachystola magna*. Biol. Bull. **4.** 1902.

Virchow, R., Die Cellularpathologie in ihrer Begründung auf physiologische und pathologische Geweblehre. Berlin. 1858.

Wilson, E. B., Atlas of Fertilization and Karyokinesis. New York. 1895. On Protoplasmic Structure in the Eggs of Echinoderms and some other Animals. Jour. Morph. **15,** Suppl. 1899. The Cell in Development and Inheritance. Columbia Univ. Biol. Ser. IV. 2 ed. New York. 1900.

Ziegler, H. E., Experimentelle Studien über die Zelltheilung. II. Furchung ohne Chromosomen. Arch. Entw.-Mech. **6.** 1898.

CHAPTER III

THE GERM CELLS AND THEIR FORMATION

THE reproductive elements of the Metazoa are single cells, often greatly specialized in form, and always highly differentiated in internal structure and in function. They differ from the reproductive cells of several groups of the Metaphyta, in that they cannot function, *i.e.*, develop, until two single cells, usually derived from two different individuals, shall have met and fused, or conjugated. In many plants single reproductive cells (spores) are formed which develop directly without any such conjugation, and which are therefore to be distinguished from the true germ cells, or gametes, which develop only after conjugation. We shall describe the process of conjugation, or fertilization, in a following chapter, but in order to appreciate the significance of many of the details of germ-cell form and structure, we must remember that they are adapted toward ensuring the conjugation of two unlike cells, egg and spermatozoön.

Reproductive cells are set apart from vegetative cells in many of the colonial Protozoa. In some cases they are distinguishable only at certain times, when cells usually vegetative may give up such characteristics and become reproductive; in others the reproductive and vegetative cells remain permanently distinct though only slightly differentiated structurally. Finally in a few colonial Protozoa the reproductive cells are considerably modified from the vegetative condition, and in form and composition, as well as in function and behavior, are readily distinguished as germ cells. Many of the details regarding these cells have already been mentioned, others can be more conveniently and more significantly considered later, in connection with the process of fertilization (Chapter V). In this chapter we shall first describe the form and structure of the typical germ cells of the Metazoa, as they appear when fully formed and ready to function, that is, just prior to fertilization. We shall then mention some of the more important modifications of structure shown in different groups, and finally give a brief account of the formation and history of the germ cells up to the time when they are ready actually to enter upon the real process of development. The details of certain phases in the history of the germ-cell nuclei, namely, the maturation processes, are of such importance that we shall refer to them only briefly in this chapter and devote that following to a more extended account of these events.

Among the Metazoa the fully formed germ cells are always of two very unlike types, the *ova* or eggs and the *spermatozoa* or sperm cells. These cells are alike only with respect to their nuclear structure and composition. Their form differences are associated with fundamental differences in function. The egg cell contains by far the greater share of the substance which is to form the material basis of the new individual. The sperm, on the other hand, contributes little substance, and that chiefly nuclear, to the new individual. One of its important functions seems to be that of ensuring to the comparatively passive egg, a stimulus which leads it to react, *i.e.*, to commence development; and the sperm's nuclear configuration, with that of the egg, together appear to determine the course of development to a large extent, if not wholly.

The substance of which the ovum is composed is not a homogeneous protoplasm. The cytoplasm is differentiated and organized into a definite structural and chemical (energetic) configuration. The details of this configuration are uniform in the eggs of each animal kind, *i.e.*, it is *specific*. This cytoplasmic structure of the ovum, although itself apparently determined primarily by nuclear activity, is of great importance in maintaining the continuity and uniformity of organismal characteristics through successive generations (heredity).

The ova are less modified in external form than the spermatozoa, and often approach the form of a typical cell, except that they are nearly always larger than ordinary cells (Fig. 42). The size of the ovum is not related to the size of the organism producing it, but is in general related to the amount of food substance stored in it, the actually living protoplasm showing much less variation in amount than the deutoplasm. The smallest eggs are those of the Mammals (Figs. 43, 14, *VII*), in man only 0.25 mm. (250 *micra* or 0.01 inch) in diameter, others being still smaller—0.07–0.10 mm. (70–100 *micra*) in the deer,

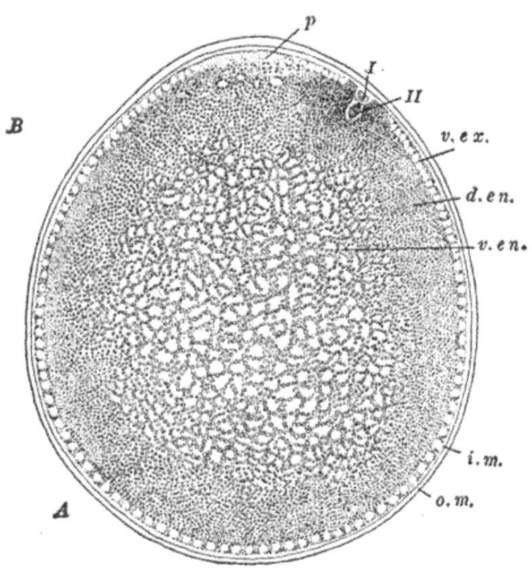

FIG. 42.   *A*. Section through the egg of the lamprey, *Petromyzon fluviatilis*. After Herfort. *B*. Spermatozoön, drawn to scale. *d.en.*, dense endoplasm; *i.m.*, inner membrane (? vitelline); *o.m.*, outer membrane (? chorion); *p*, granular "polar plasm;" *v.en.*, vacuolated endoplasm; *v.ex.*, vacuolated exoplasm; *I*, first polar body; *II*, second polar spindle.

and only 0.065 mm. (65 *micra*) in the mouse.   The largest eggs are, in volume, the largest known cells; such are the "yolks" of birds' eggs, the largest of which are several inches in diameter and equalled in other groups only by the eggs of one of the sharks (*Heterodontus*) which are nearly 2 inches (4.0 to 5.0 cm.) in diameter.   In a very few cases (some Cœlenterates and Porifera, and a few worms) the ova may be capable of loco-motion, performing amœboid movements (Fig. 44), but in nearly all cases they are quiescent, passive structures although

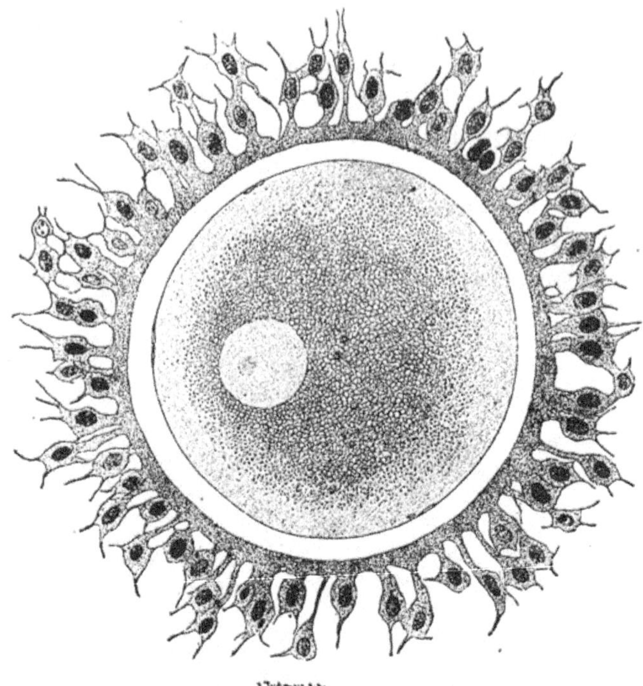

Fig. 43.—Fully grown human oöcyte freshly removed from the ovary. Outside the oöcyte are the clear zona pellucida and the follicular epithelium (corona
radiata). The central part of the oöcyte contains deutoplasmic bodies and the
eccentric nucleus (germinal vesicle); superficially is a well marked exoplasmic or
cortical layer. From Waldeyer-Hertwig.

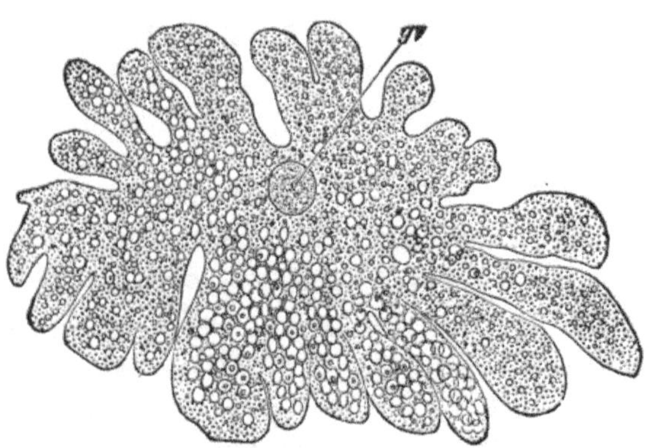

Fig. 44.—Fully grown egg of *Hydra viridis*, containing nuclei of ingested cells.
gv, nucleus of egg. The egg is amœboid. From Waldeyer-Hertwig, after
Kleinenberg.

containing a great deal of potential energy which becomes kinetic after fertilization, during the early stages of development.

Upon examining the internal structure of the egg we find that the nucleus is unusually large in most cases, spherical or ovoid, and with or without a nuclear membrane (Figs. 43, 45). It is located centrally or eccentrically, usually the latter if the egg contains an appreciable amount of food material. The chromatic network of the nucleus may be either dense, or so open as to give, even after staining, the appearance of a lighter

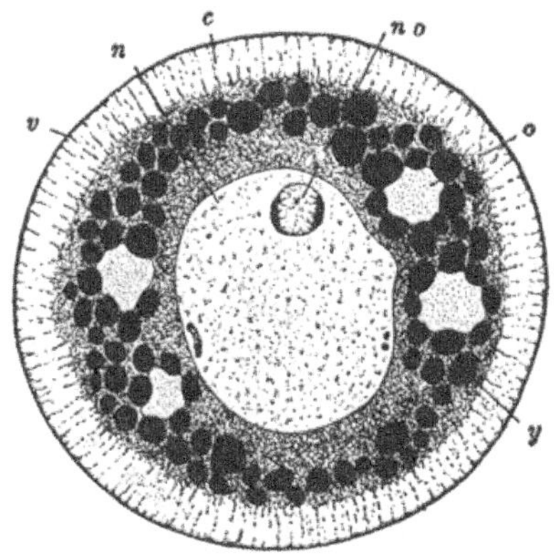

Fig. 45.—Axial section through the oöcyte of the Annulate, *Nereis*. After Lillie, slightly modified. *c*, cortical protoplasmic layer (exoplasm); *n*, nucleus; *no*, nucleolus; *o*, oil vacuoles; *v*, vitelline membrane; *y*, yolk bodies.

area in the cytoplasm (Figs. 43, 45), often known as the *germinal vesicle*. In some eggs the nucleus is in the process of division at the time of egg-deposition (Figs. 42, 46); this condition will be explained later. The actual morphological composition of the nucleus is one of the chief characteristics of the egg, and this too is reserved for consideration in Chapter IV. There is commonly a large plasmosome or nucleolus of varying form and size (Figs. 43, 45). Centrosome and centrosphere are typically present at this time, though often minute and difficult to observe; later these structures disappear entirely. Frequently the cytoplasm contains unusual bodies termed "yolk-nuclei." The term *yolk-nucleus* includes organs of several different types, in some way related apparently to the formation and deposition of the yolk and certain other substances within the egg cytoplasm. Sometimes the cytoplasm appears homogeneous, in other cases it shows considerable differentiation. Often the peripheral layer of the cytoplasm is more vacuolated and less granular than the central portion; the former is then spoken of as *exoplasm* or as the *cortical* layer, the latter as *endoplasm* (Figs. 42, 43, 45, 46). There may also be various materials in the cytoplasm which have been laid down during the formation of the egg, under the influence of the nucleus, or yolk-nucleus, and deposited in different regions (Figs. 42–48).

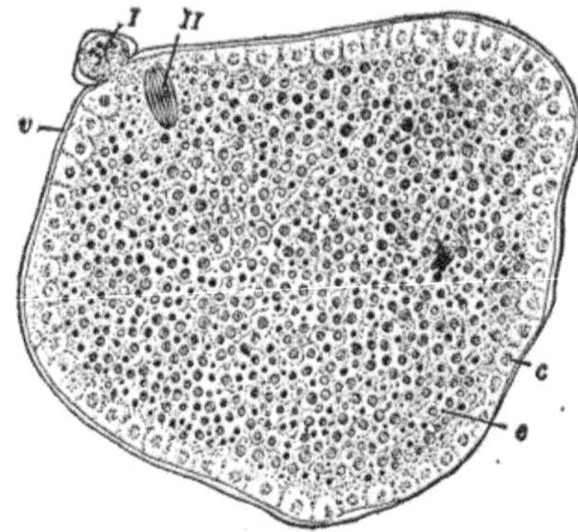

Fig. 46.—Section through the ovarian egg (oöcyte) of Amphioxus. After Sobotta. × 525. *c*, vacuolated cortical layer (exoplasm); *e*, endoplasm containing deutoplasmic bodies; *v*, vitelline membrane; *I*, first polar body; *II*, second polar spindle.

The extent of the cytoplasmic differentiation varies greatly in eggs of different species. In many forms it can hardly be demonstrated in the egg at the time it is fully formed; in such eggs this differentiation appears later, during or after the processes of fertilization, or even still later, during cleavage. We shall have to return to this subject in connection with the subject of cleavage, after we have described the fertilization process. But there are two or three fundamental aspects of this fact that should be mentioned here. There are quite commonly three, sometimes more, distinct forms of cytoplasmic material arranged as definite regions of the ovum, occasionally as zones, or layers, or as localized masses (Figs. 42, 45). These may be distinguished by the more or less vacuolated character of the protoplasm, or by the collection of various pigments and dif

ferently colored granules, or by forms of deutoplasmic materials other than yolk, or in various other ways. The disposition of these substances usually expresses, incompletely, however, an underlying *organization* or morphology of the egg substance as a whole, which is considered a fundamental structure of the egg as a specific organism. This organization is practically always polar, *i.e.*, disposed symmetrically with reference to one chief axis (Von Baer), and in the eggs of most bilateral animals examined, it is bilateral also (Roux, Van Beneden). In some way this morphology of the egg is related to the morphology of the embryo developed from the egg, and hence is called its *promorphology.*

This promorphology is better termed *organization*, for it is not only grossly material, but also dynamic, *i.e.*, energetic, depending upon chemical and physical arrangements not often visible directly. The extent and nature of this organization are often obscure, but this, and the nuclear structure of the ovum, are probably its most important characteristics, for together these determine the course of its development as a specific creature.

Polarity is one expression of this organization. The polarity of the fully formed ovum is related to the polarity of the egg cell

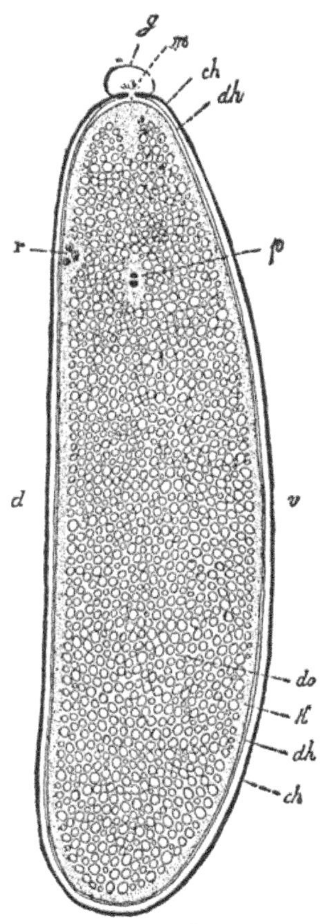

Fig. 47.—Semidiagrammatic drawing of a median section through the fertilized egg of the fly, *Musca*. From Korschelt and Heider, after Henking and Blochmann. *ch*, chorion; *d*, flattened dorsal side of the egg; *dh*, vitelline membrane; *do*, yolk bodies; *g*, jelly-like material extruded through the micropyle; *k*, cortical layer; *m*, micropyle; this also marks the anterior end of the egg; *p*, egg and sperm pronuclei in process of fusion; *r*, polar bodies; *v*, ventral side of egg.

as it was placed in the epithelium of the gonad, the chief egg axis corresponding with the axis passing through the attached and free surfaces of the epithelial cell. This polarity apparently determines the primary position of the egg nucleus and centrosomes, and thus secondarily determines also the arrangement of the cytoplasmic substances which develop through the interactions between the nucleus and cytoplasm, processes which may frequently be observed during the growth period of the ovum. The two poles of the egg are commonly unlike (Figs. 42, 48), so that we distinguish an *animal* and a *vegetal pole*, corresponding in most cases with the originally free and attached surfaces respectively, although this relation may occasionally be reversed (Echinoderms). In general the animal pole is that toward which the nucleus is eccentrically displaced, and nearer which the centrosome, or similar body, is located; it is the more protoplasmic, and therefore the more active region of the egg. The vegetal pole is frequently occupied largely by the relatively inert food substance, the materials in general related with the vegetative organs of the developing embryo.

As regards the nature of the organization, or promorphological relations, of the egg, two views have been taken and will be discussed in Chapter VII. The first is that the differentiated substances visible within the cytoplasm are genuinely "organ-forming substances," or at any rate tissue forming, in their potentiality. Thus they represent the organs of the embryo in an intracellular form. The second view is that these substances are only secondarily related to the real morphology of the embryo, and that both embryonic structure and the differentiated substances of the egg, are the result of an underlying, invisible, and as yet little-known organization of the ground substance of the egg cytoplasm. According to this view the correspondence between the "organ-forming substances" of the ovum, and the organs and tissues of the embryo, is not in itself direct, but results from their common relation to the primary underlying arrangement or organization of the substance of the egg. In some cases the arrangement and position of these substances may be considerably altered experimentally without disturbing the normal course of development. It should be added, however, that in some other cases such a disarrangement does effect a corresponding disarrangement of the organs or tissues of the embryo.

In addition to the formative substances mentioned above eggs may contain varying amounts of nutritive substance of many different kinds, collectively termed *yolk* or *deutoplasm.* The yolk may be in the form of granules, small spherical bodies, large plates, fluid drops of various sizes, or in compact masses (Figs. 45, 48).   These substances may be of different chemical compositions and staining reactions in a single egg.   They may be formed within the egg by its own activity, or they may be contributed indirectly by cells associated with the egg during its formation.   The arrangement of the food substances in the egg has an important bearing upon its later development, especially upon the form of its cleavage (Balfour).   Eggs in which the yolk is distributed quite uniformly through the cytoplasm, and in which the protoplasm is therefore more or less completely intermingled with the yolk granules, or plates, are termed *homolecithal* or *isolecithal* eggs.   Some eggs have been described as *alecithal, i.e.,* without yolk, but many of these have been found really to contain a small amount of quite uniformly distributed deutoplasm, and a truly alecithal egg is rarely if ever found.   Eggs of some species among nearly all the large groups are of this homolecithal type, for example, the star-fish, sea-urchin, and also the Mammals, which were formerly thought to be alecithal (Fig. 43).   More frequently the yolk and cytoplasm are not uniformly mingled but are chiefly accumulated in different parts of the cell.   Ordinarily these materials occupy opposite poles of the egg so that this retains a radial or rotatorial symmetry; the yolk is accumulated toward the vegetative pole, the protoplasm toward the animal pole (Fig. 48).   Such eggs are termed *telolecithal.*   They show great variation in the relative amount of yolk contained.   On the one hand it is often difficult to distinguish the telolecithal egg from the homolecithal type, for the tendency toward polar accumulation of the yolk may be very slight.   The egg of Amphioxus illustrates such a transitional condition.   At the opposite extreme we find eggs such as those of the Reptiles and Birds, which are relatively immense cells, in which it is difficult to distinguish, before development begins, any definite region which is entirely free from yolk. Between these extremes all intermediate conditions are found. This telolecithal type of egg is very common among the Invertebrates, and is characteristic of all the Craniata except the true Mammals. Among the Chordata successive stages in the accumulation of yolk are represented by Amphioxus, Lampreys, Ganoids, Dipnoans, Amphibians, Reptiles, and Birds. In the last two groups the

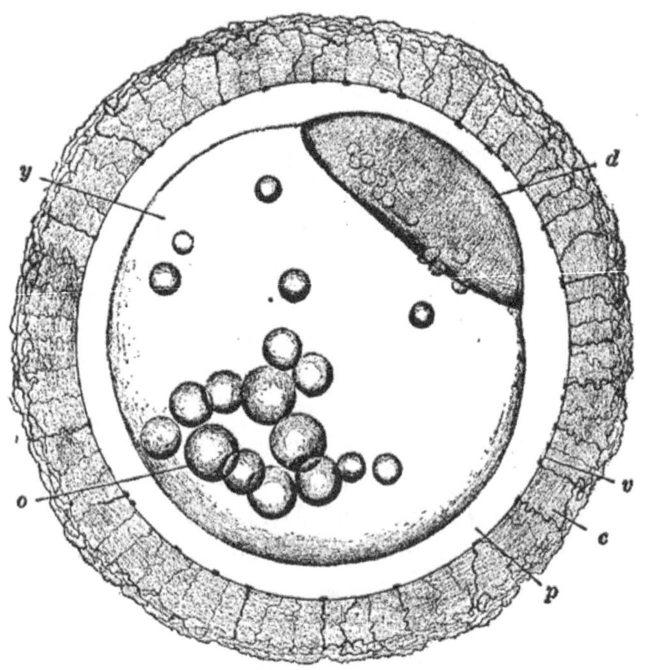

Fig. 48.—Egg of the Teleost, *Fundulus heteroclitus*. Total view, about an hour after fertilization. *c*, chorion; *d*, protoplasmic germ disc or blastodisc; *o*, oil vacuoles; *p*, perivitelline space; *v*, vitelline membrane; *y*, yolk.

protoplasm is extremely limited in amount, and is found only as a small disc or layer on the surface of the spherical yolk-mass, at the animal pole. A third and less common arrangement of yolk is that seen in the *centrolecithal* eggs of many Arthropods, chiefly Insects. Here the yolk occupies a greater or lesser portion of the center of the egg while the protoplasm forms a superficial layer all around it (Figs. 47, 117, 118).

The various constituent materials of the egg differ, often very considerably, in density, and since they are definitely disposed with reference to the chief egg axis, the eggs tend to assume a definite position with respect to gravity when they are free to move. Usually, in such cases, the yolk is heavier than the protoplasm, and the animal pole is therefore directed upward; this position is reversed occasionally, particularly when the deutoplasm is in the form of oil drops, *e.g.*, *Nereis* and most Teleosts.

In some forms the egg cells are naked, without cell coverings or membranes, as in many Cœlenterates and some Molluscs. Or the egg may be naked at first, but soon after becoming free from the parental body it may acquire a thin membrane over its surface, as in Echinoderms.   Ordinarily the egg is surrounded by definite membranes of varying nature and origin.   The *primary* egg membrane is the *vitelline membrane*, or true egg membrane. Typically this is a thin membrane secreted by the superficial protoplasm of the egg and closely applied to its surface (Figs. 42, 45, 46, 47). In most cases it is quite structureless; sometimes it is thicker and may be perforated radially by minute canals or pores, when it is termed the *zona radiata*. Occasionally the vitelline membrane may appear double, showing an inner zona radiata and an outer structureless layer (Figs. 49, 50). In such cases it is possible that the membrane is not wholly vitelline, or it may be the case more frequently that the pores originally passed completely through the membrane and disappeared from the outer portion of it upon contact with an external medium.

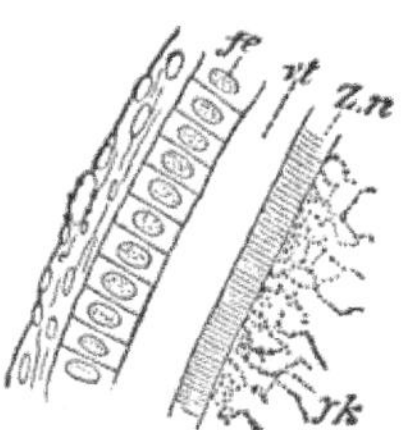

Fig. 49.—Section through the egg membranes of the Elasmobranch, *Scyllium*. (Ovarian egg.) From Ziegler, after Balfour. *fe*, Follicular epithelium; *vt*, outer portion of the vitelline membrane (zona pellucida); *yk*, surface of the ovum; *Zn*, inner portion of the vitelline membrane (zona radiata).

The vitelline membrane may envelop the egg completely, or there may be left a minute, funnel-shaped perforation through it at the point where the egg was attached or otherwise especially related to the epithelium of the ovary. This aperture

is the *micropyle* (Fig. 50), and, as we shall see, this sometimes
affords an aperture for the entrance of the sperm cell.

A *secondary* membrane called the *chorion* often surrounds the
egg outside the vitelline (Figs. 42–50). This is a secretion
formed while the egg is still contained within the ovary, from
the cells there surrounding the egg, and its presence depends
upon the arrangement of these ovarian cells in the form of a
definite layer or epithelium surrounding the egg, termed the

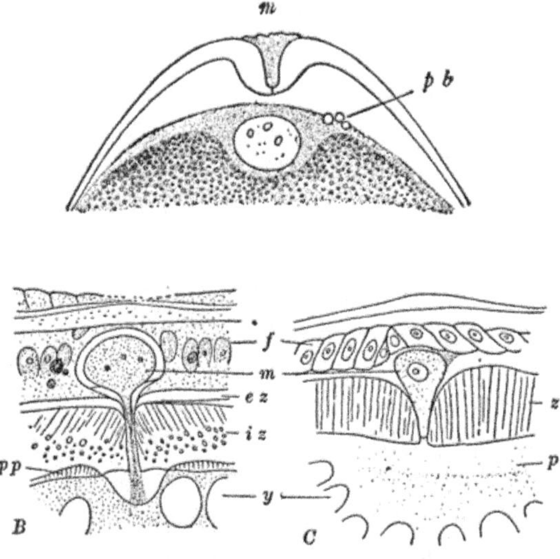

Fig. 50.—*A.* Animal pole of the egg of the Cephalopod, *Argonauta.* From
Wilson, "Cell," after Ussow. Surrounding the egg is the chorionic membrane
perforated by the funnel-shaped micropyle, *m.* Beneath the micropyle lies the
egg nucleus in the cortical protoplasmic layer. *p.b,* polar bodies. *B, C.* Sections through the egg membranes and micropyle of the egg of the Teleosts *Esox*
(*B*) (ovarian egg) and *Pygosteus* (*C*) (ovarian egg, 0.4 mm. in diameter). After
Eigenmann. × 375. *ez,* zona radiata externa; *f,* egg follicle; *iz,* zona radiata
interna; *m,* micropylar cell; *p,* egg protoplasm; *pp,* protoplasmic processes;
y, yolk in egg; *z,* zona radiata.

follicle. The chorion may be a thin flexible membrane, or a
tough resistant *shell,* as in Insects and Teleosts. Penetrating
the chorion there is nearly always a continuation of the micropyle. The formation of this results from the fact that one of
the cells of the follicle usually acquires a very intimate relation
with the ovum, through a fine pseudopodial process, so that
at this point no membranes are laid down (Fig. 50). When the

egg is fully formed and leaves the follicle, this process is withdrawn, leaving a funnel-shaped canal. In a few instances (some Insects) there are several micropylar perforations through the egg membranes. Such openings are to be regarded as specializations of the minute canals, mentioned above, which give the appearance of the zona radiata to the membranes.

Finally there is a great variety of *tertiary* membranes formed by the walls of the oviducts, or by special glands in connection with the reproductive system. These are applied outside the chorion, or if this is absent, directly upon the vitelline membrane. These envelopes may be of slime or jelly of an albuminous character, fibrous, or shelly coverings of chitin, lime, or other substance. In forms depositing the eggs in the water this is sometimes a thick jelly, holding the eggs together in strings or masses, or serving to attach them, either singly or in masses to plants, sticks, or other solid objects (*e.g.*, Amphibia). These tertiary membranes serve also, in special instances, as protection against drying, temperature changes, pressure, or mechanical injury, and against the attacks of food-seeking organisms, or infection by bacteria or other parasitic organisms. Frequently they are nutritive in character, as the albumin or "white" of the birds' eggs or the dense oily substance surrounding the eggs of the snails.

Eggs may possess none or all three of these classes of membranes; sometimes only primary and secondary, or primary and tertiary membranes are present. This usually depends upon the nature of the egg-laying habits, method and duration of development, and various other conditions.

The spermatozoa. when fully formed, bear little resemblance to ordinary cells, yet their individual history clearly shows them to be such. In a few forms, chiefly among the Crustacea, the sperms do resemble ordinary cells, and are often provided with long radiating processes, sometimes, though rarely, pseudopodial. But by far the most common form is that known as the *flagellate* spermatozoön, found in all groups of animals from Protozoa to man. These are minute thread-like cells in which three general regions can usually be made out (Fig. 51). One end is

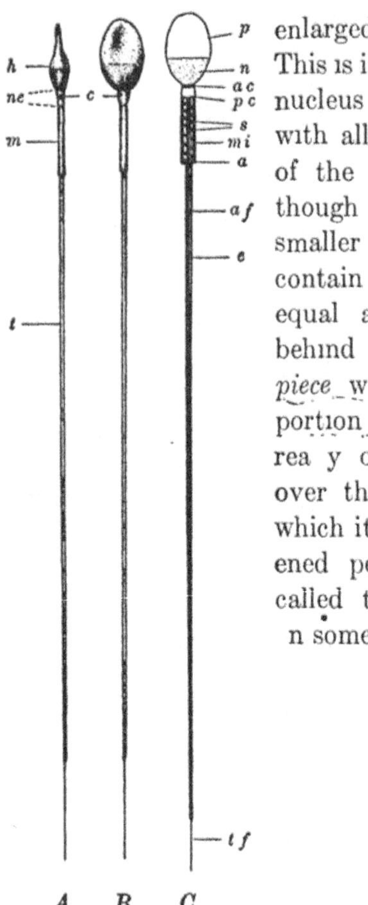

enlarged
This is i
nucleus
with all
of the
though
smaller
contain
equal a
behind
piece w
portion
rea y o
over th
which it
ened pe
called t
n some

Fig. 51.—Flagellate spermatozoön. *A, B.* Two views of human sperm cell. After Retzius. × 2000. *C.* Diagram of the structure of a generalized type of flagellate spermatozoön. After Meves. *a,* annulus; *ac,* anterior centrosome; *af,* axial filament; *c,* centrosomes (end knobs); *e,* protoplasmic envelope; *h,* head; *m,* middle piece; *mi,* mitochondria; *n,* nucleus; *ne,* neck; *p,* perforatorium (acrosome); *pc,* posterior centrosome; *s,* spiral filament; *t,* tail piece; *tf,* terminal filament.

frequently

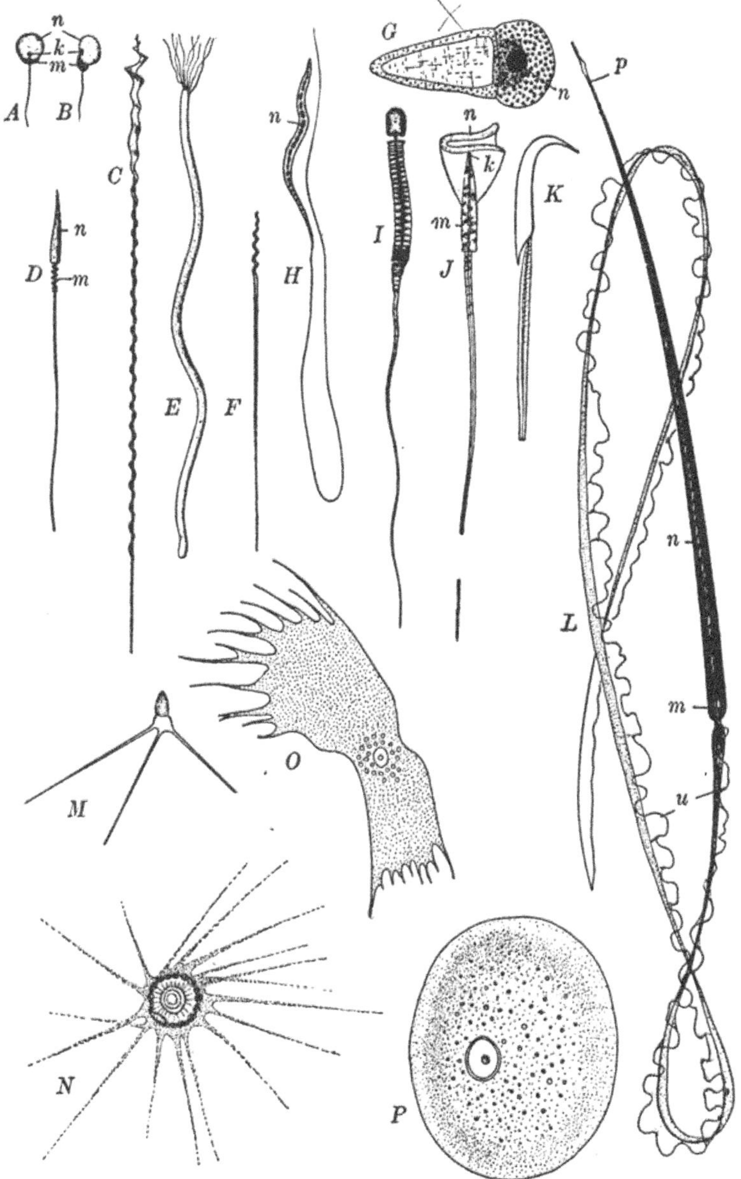

Fig. 52.—Various types of spermatozoa. *A, B.* The Teleost, *Leuciscus* (Ballowitz). *C, D.* The birds, *Phyllopneuste* and *Tadorna* (Ballowitz). *E, F.* Two forms of the sperm of the snail, *Paludina* (von Brunn). *G.* The Nematode, *Ascaris* (Van Beneden). *H.* The Annulate, *Myzostoma* (Wheeler). *I.* The bat, *Vesperugo* (Ballowitz). *J.* The opossum, *Didelphys* (Wilson). K. The rat (Wilson). *L.* The Urodele, *Amphiuma* (McGregor). *M.* The Crustacean,

Through the middle of the tail is an *axial filament* connecting with the centrosome of the middle piece, a relation which is very common in flagellate or ciliate structures. Proximally the axial filament passes through a ring-like structure, the *annulus,* in the end of the middle piece. Distally the filament may continue beyond the cytoplasmic envelope of the tail as a *terminal filament* or *end piece.* There is the greatest variety in details of form of the sperm, affecting chiefly the form of the head and acrosome, length of tail, and total size; a few of the more common or more striking forms are illustrated in Fig. 52. (In cases where a neck region is distinguished, the middle piece is often regarded as the proximal part of the tail.)

The smallest spermatozoa are found in Amphioxus and are only 0.016–0.020 mm. (16–20 *micra*) in length; the largest are found in some of the Amphibia where in *Salamandra* they are about 0.7 mm. (700 *micra*) in length, and in *Discoglossus* 2.0 mm. (2000 *micra*), the maximum length known. The sperm cells of the Crustacean *Cypris,* are also of this gigantic size (2 mm.). The spermatozoa of most of the Vertebrates are 25–75 *micra* in length. The human spermatozoön (Fig. 51) represents about the average size; the dimensions of this are (Krause): total length, 52–62 *micra* (1/400–1/500 inch); length of head, about 4.5 *micra* (between 1/5000 and 1/6000 inch); length of middle piece, about 6 *micra* (about 1/4200 inch); length of tail piece, 41–52 *micra* (1/500 to 1/600 inch); width of head 2–3 *micra* (1/12000 to 1/8000 inch); thickness of head, (this is one of the few instances where the head is flattened), about 1 *micron* (1/25000 inch).

The number of sperm formed by a single individual is very large in most organisms and can be only roughly estimated. It has been computed (Lode) that the total number formed in man may average about three hundred and forty billion, or

---

Ethusa (Grobben). *N.* The Crustacean, *Inachus* (Grobben). *O.* The Crustacean *Sida* (Weismann). *P.* The Crustacean, *Bythotrephes* (Weismann). *k,* end knob; *m,* middle piece; *n,* nucleus; *p,* perforatorium; *u,* undulatory membrane. Not drawn to same scale. A–F, I–K, from Wilson.

approximately eight hundred and fifty million sperm for each one of the four hundred ova matured during the reproductive period of the female. The volume of the human sperm is roughly only about 1/195000 that of the egg (egg = 0.25 mm. in diameter). The sperm of the sea-urchin contains only about 1/400000 to 1/500000 the material in the egg (Wilson).

In several forms, both Vertebrate and Invertebrate, atypical "giant" spermatozoa are occasionally found (Fig. 52, *E*). In most instances these are abnormalities resulting from some deviation from the usual course of events during sperm formation. But in a few instances (*Euschistus*) a dimorphism of the sperm seems entirely normal (Montgomery). In such cases the larger sperm heads contain the normal amount of chromatin, but an excessive amount of linin and karyolymph.

In marked contrast to the egg, the sperm cells are in very active movement on account of the rapid vibration of the tail. They are always contained within a fluid medium, the *seminal fluid*, which is either the fluid of the cavities of the body, or a special secretion of certain glands in connection with the reproductive system. In the latter case this fluid is equivalent to a tertiary egg envelope, and like this is sometimes of nutritive value to the germ cells.

The form differences between ova and sperm give a nice illustration of the modification of structure accompanying a physiological division of labor, which is so well marked here. Both cells contain equal amounts of nuclear substance, but the ovum possesses in addition a large amount of cytoplasm, and often a much larger amount of food substance, while the sperm contains an amount of cytoplasm which represents practically an irreducible minimum, and no deutoplasmic material whatever. The egg, therefore, provides practically the whole of the extranuclear substance of the developing organism. At the same time the ovum is a passive, non-motile structure. The spermatozoa, on the other hand, are not only extremely motile, but they are produced in very large numbers, conditions correlated with their function of finding the inactive ova and of ensuring the initial stimulus to activity (development) of the passive egg material. This differentiation affords the material basis for development while at the same time it ensures the fertilization of practically every egg produced, though the eggs and sperm may be shed freely into the water some distance apart, a distance often very great as compared with the size of the cells. The relatively very large amount of cytoplasm in the egg, and small amount in the sperm, constitute the most marked exceptions to the nucleo-cytoplasmic (*kern-plasma*) relation, mentioned in the preceding chapter; these conditions are entirely special and are to be regarded as adaptations to the very unusual functions of these cells.

The details concerning the form and number of the sperm and egg cells, the amount of yolk in the eggs, the character of their membranes, *etc.*, are significant only from the viewpoint of adaptedness to the conditions under which they must function. This adaptedness of the reproductive phenomena toward ensuring the final bringing to maturity of a number of organisms sufficient to maintain the specific group in undiminished numbers is a general biological topic of especial interest. In strictness this lies outside our province, but to omit entirely any reference to this subject, leaves without significance many of the details of structure and behavior mentioned in the preceding paragraphs. We may therefore suggest briefly a few of these relations, not only as regards the germ cells, but also the general processes of spawning, *etc.*, which are all concerned in finally bringing together an ovum and a spermatozoön.

All these varied, and often complex, phenomena of habit and morphological specialization of the reproductive cells are correlated with the special conditions of life which affect the chances that a single egg shall finally become a mature organism. They are conveniently grouped under three chief heads: (a) the ensurance of mating, (b) the ensurance of the actual meeting and fusion of the germ cells, (c) the chances of death before maturity, involving such factors as abundance of food, enemies, adverse conditions in the inorganic surroundings, necessity for reaching *special* conditions of development, food, *etc.*, duration of the period of development, and the like.

A few forms, especially in the warmer climates, appear to breed quite continuously throughout the year (many Cœlenterates, Mollusca, *etc.*), but commonly the germ cells are produced at regular periods, which may have a duration of only a few days or hours, or they may extend over several months. Breeding or spawning periods are nearly always seasonal and usually annual, but a few forms, particularly the Mammals, breed or spawn at shorter intervals.   In many Mammals it is true that there is only a single annual breeding season or period of *œstrus;* this condition, known as *monœstrous,* is characteristic of most Carnivora and is found also in the Chiroptera and Marsupials.   Others, however, are *polyœstrous,* and exhibit two or three annual breeding seasons (Insectivors), and in still others the period of œstrus may occur at intervals of a few weeks (man), or it may be quite continuous, as in most Rodents and some Carnivors.   (See Marshall, Physiology of Reproduction, 1910.)

Among the higher animals the breeding season is often preceded by a "nuptial season" during which, especially among the males, there may develop various special morphological and physiological peculiarities. The Fishes, Birds, and Mammals exhibit the frequent development of special external markings or colorings, special secretions, and unusual modes of behavior.   Both these and the breeding habits proper, are to be regarded as responses to stimuli, frequently climatic in origin, resulting from changes in temperature, light, moisture, food characters, *etc.*   These phenomena are commonly regarded as indications of an increased metabolism that affects not only the organs of reproduction, but secondarily the whole body.

In a few rare instances, chiefly among the segmented worms, the annual spawning season is very definitely fixed and varies within limits of only a few calendar days.   More usually the time of spawning is subject to wide variation and is dependent upon temperature and other seasonal conditions.   The species of the Palolo worm afford one of the best marked instances of a fixed spawning season.   It is not quite as regular and limited as tradition would have it, but in the Tortugas, most of the individuals of the Atlantic Palolo (*Eunice fucata*) swarm and spawn during one or two mornings which fall within three days of the moon's last quarter between the latter part of June and the end of July (Mayer).   The Pacific species (*E. viridis*) spawns similarly on and near the last quarter in October and November.   Somewhat similar relations have been determined for other Annulata, such as *Amphitrite* (Scott) and *Ceratocephale.*   The determining factor in these and similar cases seems to be the character of the tides, combined with factors of temperature and light.   Other organisms spawn at a definite time of day, individuals coming to maturity at any time during a longer breeding season.   Thus Amphioxus and some Hydroids spawn only about sun down or shortly thereafter.

During the intervals between the breeding periods the formation of the germ cells may almost or quite cease, to recommence shortly prior to the next period.   In some creatures, however, the eggs are formed continuously and are stored in secondary reproductive cavities pending the time of their production.   This is more likely to be the case in sperm production. But in all such cases the rate of formation of the germ cells is rhythmic, increasing just before the breeding period.

The sperm cells are always passed outside the body of the organism forming them (save in self-fertilizing hermaphrodites); the eggs may or may not be thus extruded. Animals used to be described (even classified) as "oviparous" or "viviparous," according to whether the female extruded undeveloped eggs or living "young," but these terms have now lost all precise meaning, for in any case eggs are formed, and in different species the developing organisms may leave the body or reproductive cavity of the parent at almost any stage.

The unfertilized eggs may be simply thrown outside the body of the female, as in most aquatic animals, the sperm being thrown out at the same time and in approximately the same place; in such cases fertilization is ensured chiefly by the production of immense numbers of spermatozoa. Such a process is very common among the Sponges, Cœlenterates, Echinoderms, Annulata, Mollusca, Fishes, and many Amphibia. Eggs thus thrown off into the water may float at or near the surface, as *pelagic* eggs, or they may sink to the bottom among the debris (*demersal*). Or the extruded eggs may be deposited with reference to definite and often very special conditions affording, to the new organisms, protection, food, *etc.* Among land animals which deposit the eggs outside the body, these are usually very definitely placed with reference to such conditions; the Insects afford a great variety of excellent illustrations of relations of this kind. In some cases, among both aquatic and terrestrial forms, definite "nests" are constructed in which the eggs are deposited, and where the newly hatched organisms may remain for some time. The eggs and young then may or may not be guarded or fed, by either or both of the parents. The nests may vary from simple depressions or pockets in the mud or sand, like those of many fresh water Fishes, to the structures of very complex architecture of many Birds.

Among the forms which do not liberate their eggs at an early stage in their development, there is a great variety of habit. In some Crustacea and Amphibia, for example, the eggs are first extruded, but are immediately placed upon the surface of the body, of either the male or female parent, and develop there. Or they may become embedded in the skin (many Amphibia) or may be deposited in some cavity not primarily a reproductive cavity, such as the pharyngeal cavity, in some of the Siluroid and Cichlid Fishes. In one of the Cyprinoid Fishes (*Rhodeus*) the eggs are placed in the mantle cavity of a clam, where they are fertilized and develop on the gills. In most cases where the eggs are retained in "brood cavities," these are modified portions of some part of the reproductive system proper; here the eggs may remain until a comparatively late period in their development. In such cases fertilization must be internal, and the sperm are then definitely introduced into some reproductive cavity of the female.

An interesting series of relations may be traced illustrating the gradual increase in the certainty with which fertilization shall be accom-

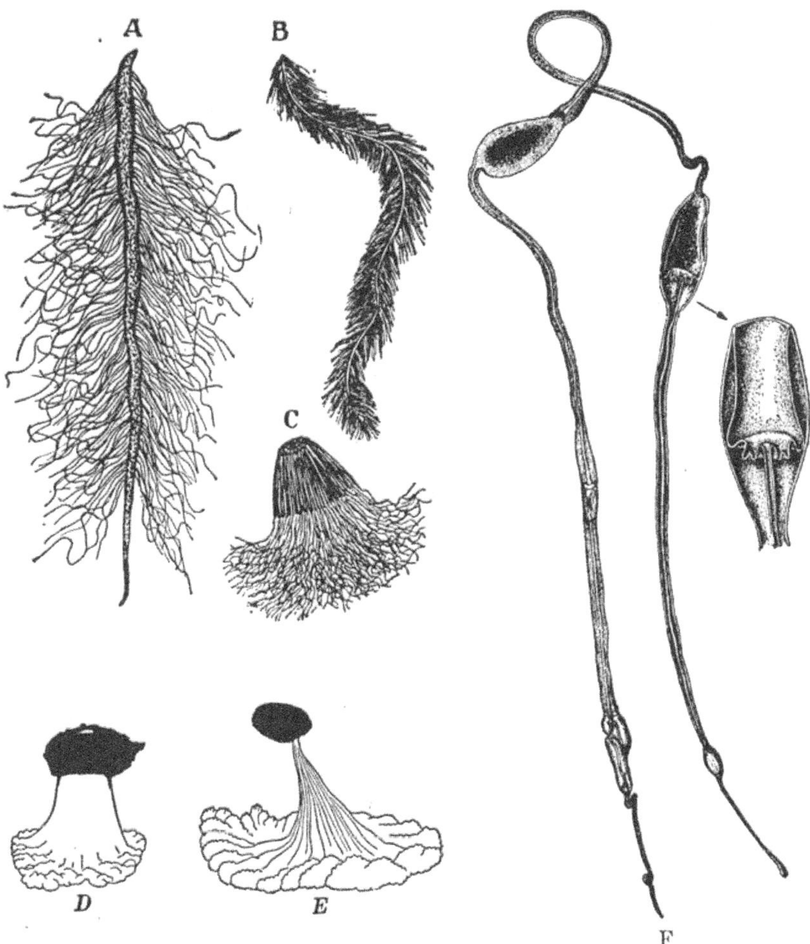

Fig. 53.—Forms of spermatophores. *A, B, C.* The Insects, *Loricera, Locusta,* and an Ichneumonid. From Korschelt and Heider, after Gilson. *D, E.* The Urodeles, *Amblystoma* and *Diëmyctylus.* From Bertram Smith. × 3. *F. Peripatus edwardsii* (incompletely developed). After von Kennel.(Korschelt and Heider).

plished, whether it be external or strictly internal. First, among many of the Crustacea, Annulata, Gasteropods, Cephalopods, and Amphibia, there is a process of *pseudo-copulation* or *amplexus*, where, sometimes after a complicated "courtship" (Urodeles, Jordan), the sperm are received by the.female in or near the reproductive cavities or openings. Usually in such instances the sperm are not scattered freely, but are contained within definite packets or cases called *spermatophores* (Fig. 53). In the Urodeles these are simple masses of spermatozoa enclosed in a thin envelope; they are discharged by the male and then picked up by the cloacal margins of the female and stored until the eggs are ready to be fertilized. In the amplexus of the earthworm and many Gasteropods, there is a mutual exchange of spermatozoa between two hermaphroditic individuals, the sperm being received into storage cavities and retained until the eggs are deposited. Such a receipt of sperm or sperm packets into storage cavities is quite common, and the sperm may in these cases remain alive for long periods, even for three or four years (honey bee, some snails). The spermatophores are sometimes very elaborate affairs containing a complex mechanism arranged so as to discharge the sperm just at the time the female is depositing the eggs (Fig. 53, *C*). Among the more complex are those formed by the male

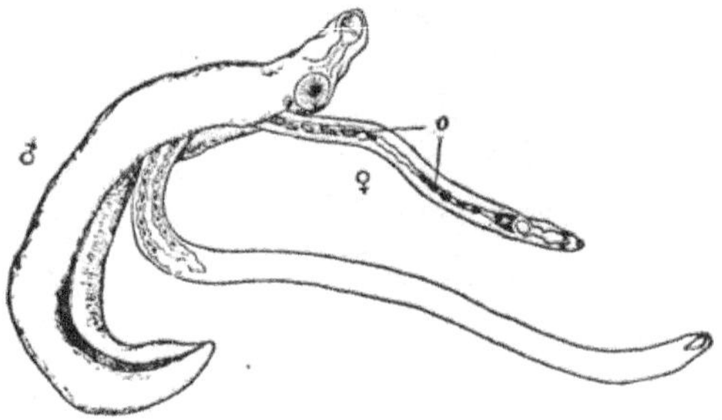

Fig. 54.—The Trematode, *Bilharzia hæmatobia.* Two individuals living *in copula.* After Fritsch. × 14. *o,* ova in oviduct. ♂, male; ♀, female.

squid (*Loligo*) and transferred to the buccal membrane of the female, where they remain attached pending the time of spawning (Drew).

Among forms where internal fertilization is the rule, this is more frequently the result of a definite act of *copulation,* by which the sperm are transferred directly to a reproductive cavity of the female through a male intromittent organ. This occurs in most Insects, many Turbellaria, Crustacea, Molluscs, and in most of the higher Vertebrates. This general relation, carried to an extreme may result in symbiosis, or even in the parasitic character of the male upon or within the body of the female. Thus in some of the Cirripedia, degenerate "complemental" males are found living semiparasitically within the body of the female. Several of the Trematoda live in pairs within a single cyst. In *Bilharzia* (Trematoda) the female lives permanently in a groove on the body of the male (Fritsch) (Fig. 54).   In *Diplozoön* (Trematoda) two hermaphrodite individuals, at first entirely separate, become permanently fused so that the openings of the reproductive ducts are in apposition.   In *Syngamus* (Nemathelminthes) also, the male lives permanently attached to the female.   The climax of this relation is afforded by *Trichosomum* (Nemathelminthes) where several (2–5) dwarfed males live within the uterus of the female (Leuckart).   Or it might be said that the climax is to be seen in the cases of *self-fertilizing* hermaphrodites; these are usually internal parasites, where the chances of the meeting of males and females would be practically *nil* (*e.g.*, many of the Trematodes and Cestodes).

The retention of the eggs during their development, within a brood cavity is primarily a protective arrangement, but it often leads to the establishment of an organic nutritive relation between the embryos and the wall of the cavity.   This is the case in some Tunicates, Elasmobranchs, *etc.*, and of course it reaches its climax in the intimate and extensive organic relation between embryo and oviducal (uterine) wall among the Eutherian Mammalia.

The number of eggs produced during each reproductive or spawning period varies enormously, and is related to a variety of conditions of development.   In general the number of eggs is larger when there are few or no other means of ensuring the complete development of a number of organisms sufficient to maintain the species numerically. Any structure of the egg or habit of deposition, adapted to ensure development, is likely to be associated with a reduction in the number of eggs formed.   The number is largest among forms which discharge the eggs at random or where they are subject to unfavorable external conditions, to liability to the attacks of parasites, or use as food by other organisms.   For example, the marine fishes produce very large numbers of pelagic ova, the codfish is said to form 8 to 10 millions in one season; and in a species of sea-urchin (*Echinus*) a single female is said to have discharged, in a single season, as many as 20 million ova.   Where very special conditions of development are essential, as in the complicated life histories of many internally parasitic worms, the number of eggs is very large and fertilization is ensured by hermaphroditism.

The number of eggs is smallest where they develop within a brood cavity, or where some degree of parental care is exercised.   In a few Mammals and Birds only a single egg is formed at each breeding period, and in these groups the number rarely exceeds eight or ten.   Further the number of eggs produced, in general varies inversely with the amount of food substance contained, or with the chances of the young finding food for themselves by the time they become free living.

The number of sperm cells formed is always larger than the number of eggs, and often reaches many millions.   The number is likely to be smaller among forms in which the sperm are directly or indirectly introduced into the reproductive cavities of the female. Some of the Crustacea afford interesting illustrations of this. In some Ostracods only a few hundred very active spermatozoa are formed; these are inserted directly into the seminal receptacles of the female. They are most remarkable for their size—2 mm. in length in a few, or more than twice as long as the body of the male. In *Daphnia* the number of sperm may be only twenty, or even less, six to eight in some species. These too are liberated directly into the brood cavity of the female, which forms only two eggs at a time; these sperm are very adherent, and are said to be somewhat amœboid. Indeed the spermatozoa of many of the Crustacea are unusually interesting on account of their atypical form and behavior (Fig. 52). Some (*Bythotrephes*) are quite like ova, large (0.1 mm.), rounded, and quiescent, depending upon a peculiar viscid or adherent quality for their likelihood of attachment to the egg. Others (some Decapods) have a number of stiff radiating processes which seem to function by catching in the hair-like bristles surrounding the openings of the oviducts, where they are placed in amplexus.

The amount of food yolk contained in the egg is related also to the duration of the embryonic period of development, or to the rate of development, a prolonged embryonic life requiring an abundant supply of food materials, and an unusually rapid rate of development depending upon a supply of easily assimilable nutritive substance. Such a relation is illustrated by the difference between summer and winter eggs, formed by Rotifers, and many Insects and Crustacea; the winter eggs, subject to unfavorable conditions and passing a longer period in development, contain considerably more yolk, and are covered with much tougher and more resistant membranes, than the summer eggs which develop rapidly and under favorable surroundings, indeed often within the brood cavity of the female. Thus in *Daphnia* the small summer eggs are formed by only three nurse cells (see below), while the large winter eggs are supplied with food by forty or more nurse cells. When the developing embryo acquires special nutritive relations with the parental tissues, the eggs are of course practically yolkless.

The provision of egg membranes is associated not only with a reduction in the number of eggs formed, but also with the duration of the embryonic period, liability to unfavorable external conditions, prevalence of food-seeking enemies, *etc.* The membranes which are functional under such conditions are of the secondary and tertiary classes described above. The nature of these varies from the common thin fibrous coverings, to tough and impervious membranes capable of resisting extreme dryness, or the leathery or calcareous "shells" of the Sauropsids. Among the most complex and perfectly adapted membranes or shells, are the egg cases of many Elasmobranchs, and particularly those of

the Holocephali (Dean) (Fig. 55), which often remain intact and functional, in their passive way, for more than a year. In some cases the
egg membranes may have a nutritive value and may augment or re-
place the food supply in the form of yolk; the eggs
of birds and snails are good illustrations of this
relation.

We must now trace briefly the steps leading
to the formation of the ova and spermatozoa
as the highly specialized cells we have described. In the lowest Metazoa, Sponges and
some Hydroids, the germ cells are scattered
through the tissues of the organism as separate, free cells, which may migrate from place
to place, feeding and growing, often at the
expense of the other cells (Fig. 56). But in
other Hydroids, and in all forms above these,
the germ cells are localized in a definite reproductive tissue and organ, or series of
organs, the *gonads—ovaries* and *testes*. The
simplest gonads are merely masses of rapidly
proliferating cells (Fig. 57), usually bordering
a cavity which is the cœlom, and which is
supposed to be primarily this reproductive
cavity simply. In most of the higher forms
the cœlom comes to have many secondary relations, and forms in addition to the reproductive and other smaller cavities, the very
extensive body cavity. In the embryos of
the Craniates there is a pair of longitudinal
ridges, either side of the attachment of the
dorsal mesentery, through a considerable extent of the body cavity; these are the *genital
ridges*, and the peritoneum covering these becomes thickened by the enlargement and pro-

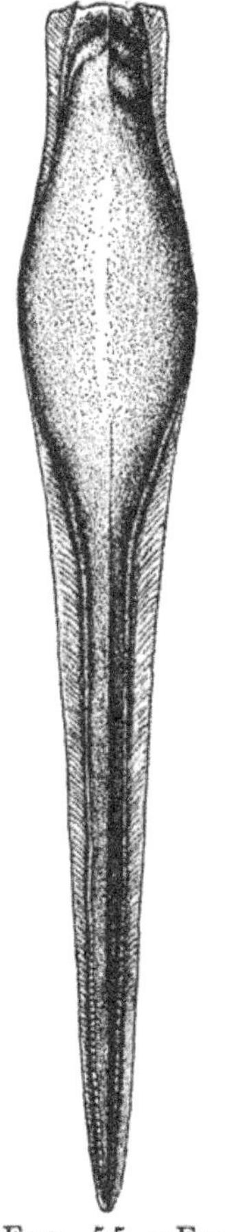

Fig. 55.—Egg
capsule of the Holo-
cephalan, *Chimæra
monstrosa.* Ventral
aspect. Two-thirds
natural size. After
Dean.

liferation of the cells (Fig. 58). These are the rudiments of
the gonads. The cells composing these rudiments are often of
two kinds. Some of them, indifferent cells composing in gen

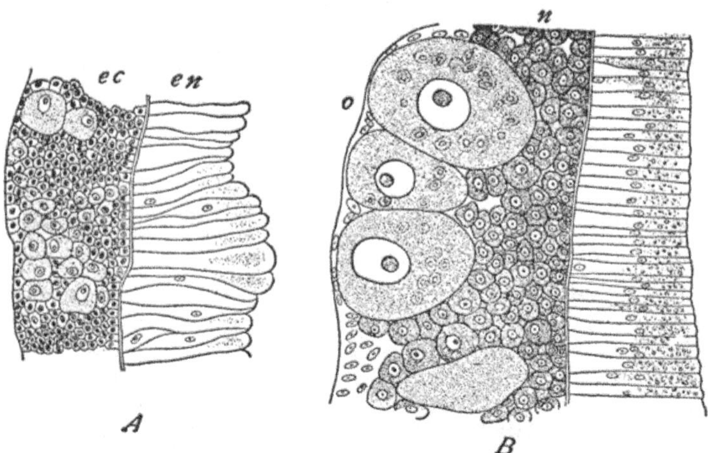

Fig. 56.—Origin of the germ cells in the Hydro-medusa, *Cladonema*. From Wilson, "Cell," after Weismann. *A*. Young stage: section through the wall of the manubrium. Ova developing in the ectoderm, *ec*; *en*, endoderm. *B*. Older stage, showing ova, *o*, and nutritive cells, *n*. The ova contain small nuclei probably derived from ingested nutritive cells.

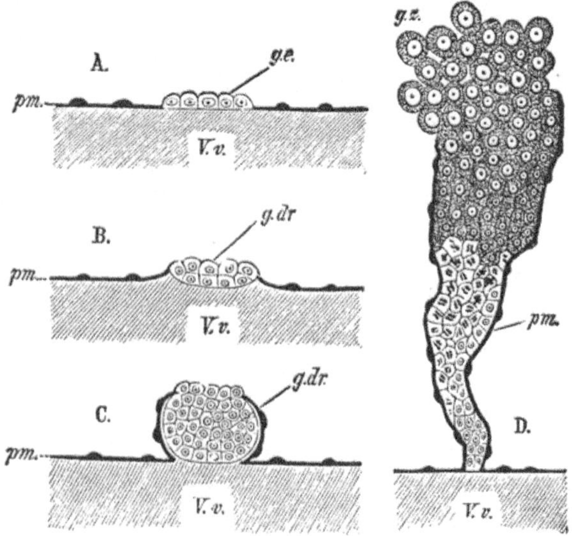

Fig. 57.—Diagram of the structure of the developing ovary of the Annulate, *Amphitrite rubra*. From Korschelt and Heider, after E. Meyer. *g.dr.*, rudiment of ovary; *g.e.*, germinal epithelium; *g.z.*, fully formed ova, scattering; *pm.*, peritoneum; *V.v.*, vas ventrale.

eral the frame-work of the gonad, have been formed *in situ* from the proliferating peritoneal cells. Others, the *primordial germ cells*, are often first distinguishable in some other region of the developing embryo (Fig. 59, *A*); they then make their way into this germinal epithelium as development proceeds.

The gonad may consist almost entirely of the reproductive cells proper, and may be then a more or less periodic structure,

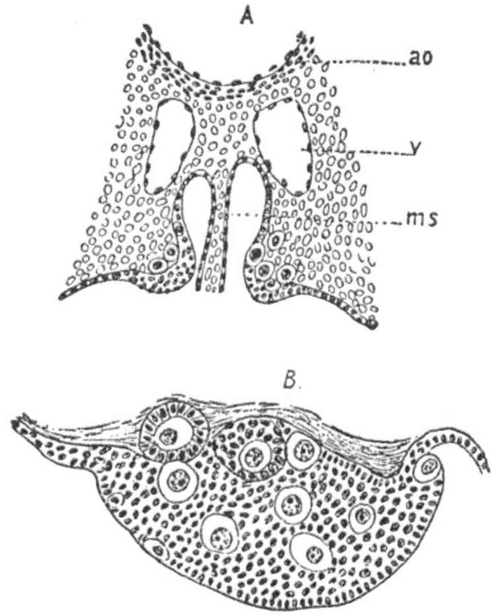

FIG. 58.—*A*. Part of a section through the body of a young lizard, *Lacerta agilis*, showing the genital ridges and associated structures. *B*. Genital ridge, enlarged, showing young follicles containing ova. From Korschelt and Heider, after Braun. *ao*, dorsal aorta; *v*, cardinal vein; *ms*, mesentery.

almost if not quite disappearing between the periods of reproductive activity. Or it may be a permanent organ of considerable though varying size, consisting of a complex *stroma* of a variety of cells—nutritive, vascular, nervous, connective tissue, and other cells, in addition to the true germ cells. When highly developed the gonad may also contain various cavities, of cœlomic character, into which the germ cells are passed when ripe. Thence they may pass directly into the body cavity from which exit is made to the outside through simple perforations in the body wall or through special tubes or ducts. In the testes these ducts may lead directly to the outside from the cavities of the gonads. The structure of the gonad can nearly always be reduced to that of a complexly folded epithelium the essential elements in which are the germ cells; the cœlomic surface is the free surface of the germinal epithelium. This relation becomes important in describing the fundamental morphology of the germ cell.

It is a question whether the germ cells are to be considered as originally undifferentiated cells, which become modified during the life of the organism for the reproductive function, or whether they are set apart from the beginning of the organism's multicellular existence as reproductive cells, and become visibly modified only in later stages. In those few forms where the

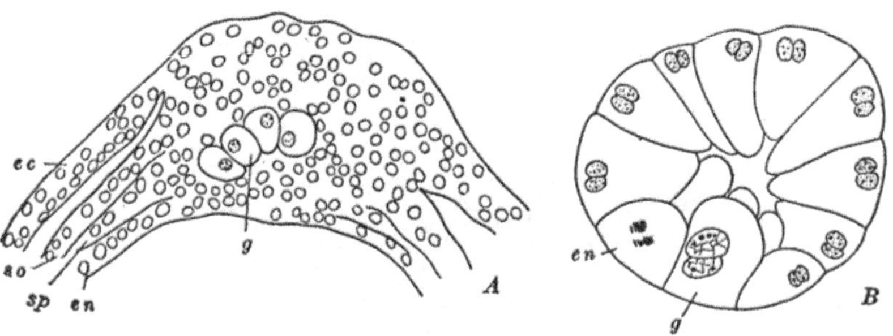

Fig. 59.—*A*. Section through an early embryo of the Teleost, *Micrometrus aggregatus*, showing the distinct germ cells. After Eigenmann. *ec*, ectoderm; *en*, endoderm; *0,* germ cells; *so*, somatic layer of mesoderm; *sp*, splanchnic layer of mesoderm. *B*. Section through forty-cell stage of the Crustacean, *Cyclops brevicornis*, showing, *0,* the cell that gives rise to the germ cells. *en*, the primitive endoderm cell in the process of its first division. After Häcker.

germ cells are diffused it seems that any tissue cell which is not completely specialized in some other direction may assume reproductive characters. In forms which develop special gonads, however, there are many reasons for believing that the germ cells are always to be distinguished as such very early in the history of the individual organism. In *Ascaris* the history of the primordial germ cells has been traced back to one of the two cells resulting from the very first division of the fertilized ovum; not all of the descendants of this cell are germinal however (Fig. 32).   In some of the bony fishes germinal cells are recognizable in the fifth cell generation, *i.e.*, in the thirty-two cell stage.   And in many other forms, including some of the Mammalia, the germinal cells can be distinguished from the somatic cells very early, even in the blastula (Fig. 59).   This may indicate that, although not visibly distinct, the germ tissue is, after all, in reality distinct from the somatic, in most if not in all forms.   It would seem more consistent with the present conception of development, however, to say that this distinction exists only potentially and comes about as a real differentiation in the developing organism, for however early this differentiation may occur, a stage is always found where germinal and somatic substances are contained undifferentiated within a single cell and are then indistinguishable.

The visible distinction between the gonads of different sexes may occur very early.   In some forms this distinction between sexes can be made out in the fertilized ovum.   And in many forms the two kinds of gonads can be distinguished soon after they are first marked out, though there is reason even here for supposing that the distinction is really, though not visibly, present in the fertilized egg.

The processes involved in the later differentiation or *histo genesis* of the eggs and spermatozoa are collectively termed *oögenesis* and *spermatogenesis* respectively.   They are conveniently divided for description into three periods or phases. These are (1) the period of *cell multiplication*, during which the simple epithelial cells, or primordial germ cells, divide more or less continuously, increasing the bulk of the gonad; (2) the period of *growth*, when cell division is less rapid or altogether inhibited, and the cells enlarge rapidly, the egg-forming cells much more considerably than those forming the spermatozoa; (3) the period of *maturation*, when the germ cell nuclei undergo profound modifications during their last two divisions as germ cells.   Sometimes the terms oögenesis and spermatogenesis are used to indicate only the events of this third period, which are of such importance that we shall make them the subject of the next entire chapter. In the history of the spermatozoa a fourth period is to be distinguished, namely, the period of *transformation* or *metamorphosis;* for the highly differentiated structure of the spermatozoön is rapidly assumed after the process of maturation is completed. This period is not marked in the history of the ovum, for this, with the exception of its unusual size, is not so markedly differentiated in structure.

During the first of these periods, that of multiplication, the cells of the reproductive tissue are termed *oögonia* and *sper*

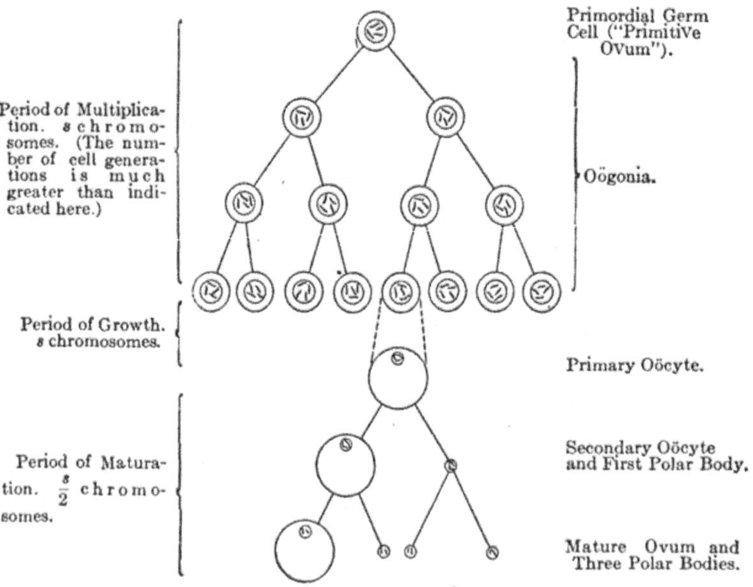

Fig. 60.—Diagram of the chief events of oögenesis. Adapted from Boveri. Compare with Fig. 61.

matogonia, and of these there may be a great many generations during this period, before growth commences. As the oögonia and spermatogonia become older, division becomes slower and ceases as the cells enter upon their growth period. At the close of the growth period, while still contained within the ovary or testis, the cells are known respectively as the *primary oöcytes,* or ovarian eggs, and the *primary spermatocytes.* From this point onward the histories of the eggs and sperm are not quite identical, although entirely equivalent (Platner, O. Hertwig).

As said above, the chief events concern the nuclear structure and we can only point out here that during the period of maturation two more cell divisions occur.

In the testis each primary spermatocyte divides once, forming two cells called the *secondary spermatocytes*, and then each of these divides again, forming altogether four cells called the

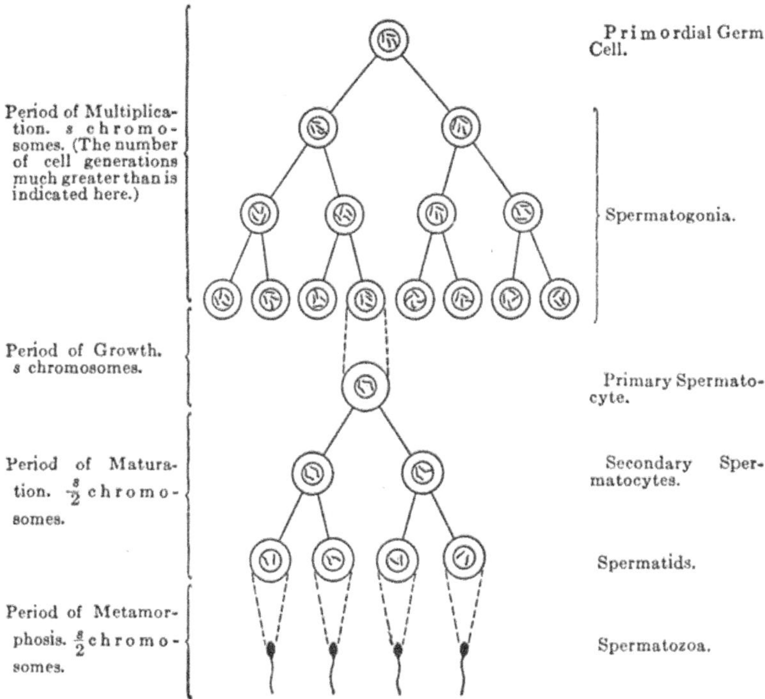

FIG. 61.—Diagram of the chief events of spermatogenesis. Adapted from Boveri.

spermatids. These are all alike and each becomes metamorphosed into a *spermatozoön* without further division. Thus are formed, from each primary spermatocyte, four similar spermatozoa. In the ovary, or sometimes after the primary oöcyte has left the ovary, this too divides, but here the division is very unequal, resulting in the formation of one large cell, the *secondary oöcyte*, and one very small cell, called the *first polar body.* Typically each of these then divides again. The secondary oöcyte divides unequally as before, forming a large cell the *mature ovum*, and another small cell, the *second polar body*. Meanwhile the first polar body divides equally forming two similar polar bodies. In some cases the division of the first polar body is suppressed. Thus each primary oöcyte typically gives rise, like the primary spermatocyte, to four cells, but these are not all alike in form and size, although they are fundamentally equivalent, *i.e.*, homologous, to each other, and to the four spermatids. Of these four cells, however, only one, the ovum, is functional; the polar bodies degenerate without functioning. The parallel events of **spermatogenesis** and oögenesis are shown diagrammatically in Figs. 60–61. (See also Figs. 76, 90, 94.)

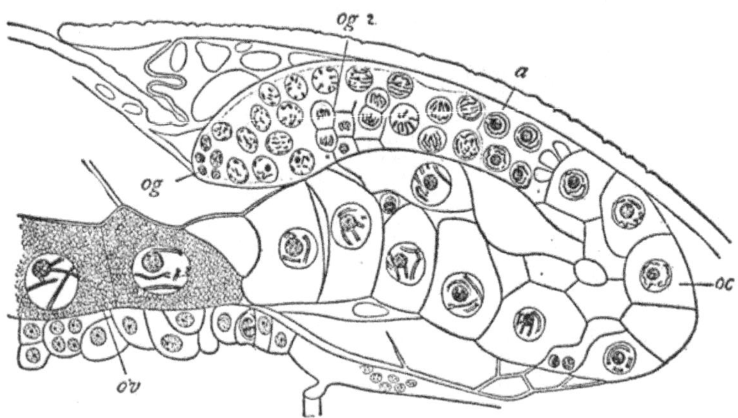

FIG. 62.—Longitudinal section through the ovary of the Copepod, *Cantho-camptus*. From Wilson, "Cell," after Häcker. *og*, the youngest germ-cells or oögonia (dividing at *og.*[2]); *a*, upper part of the growth-zone; *oc*, oöcyte, or growing ovarian egg; *ov*, fully formed egg, with double chromatin-rods.

All these processes may be going on in the ovary or testis at the same time, occurring progressively from the basement membrane of the germinal epithelium toward its free surface, so that a section through such an epithelium shows practically every step in the history of a single cell (Figs. 62, 68, 69). Before considering in detail the nuclear changes involved in the maturation processes we must consider the more important facts concerning the history of the cytoplasmic parts during this phase of the genesis of the germ cells.

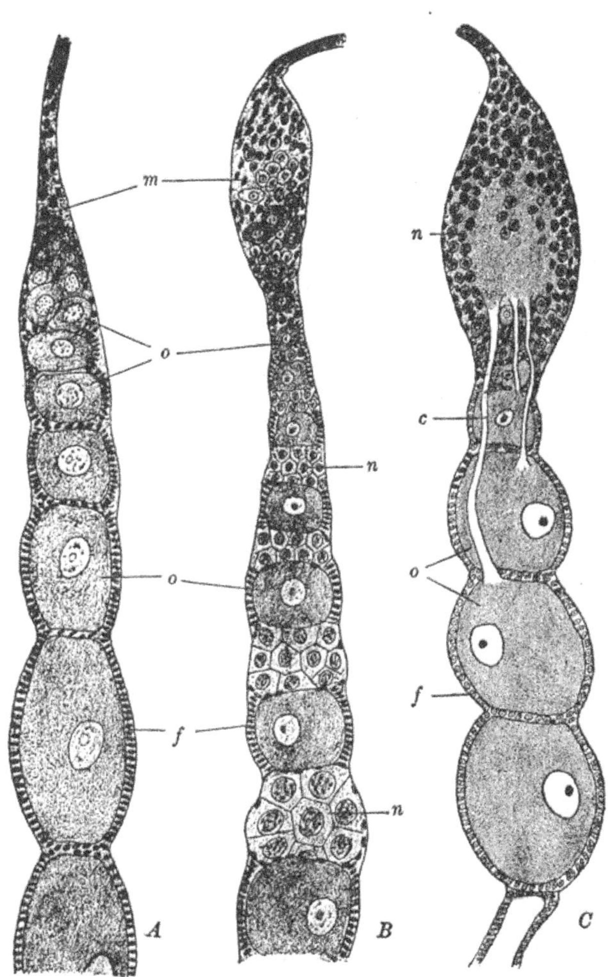

Fig. 63.—Diagrams of the egg-tubes (ovaries) of Insects. After Korschelt and Heider. *A*. Orthoptera, without groups of nutritive cells. *B*. Coleoptera, with many such groups. *C*. Hemiptera, with terminal nutritive group and nutritive channels extending to the ova. *c*, nutritive channel; *f*, egg follicle; *m*, zone of multiplication; *n*, nutritive cells; *o*, ova.

In the growth of the egg the chief aspects are those associated with nutritive relations of the developing ovum to the adjacent cells, especially in those forms whose eggs contain a considerable amount of yolk. In the non-localized ovary such as that of the Sponges and some Hydroids, the ovum grows at the expense of whatever cells happen to be adjacent to it (Fig. 56). In the common *Hydra*, as the ovum grows to be considerably larger than these tissue cells, it becomes amœboid and actually ingests these neighboring cells, digesting their substance and growing rapidly (Fig. 44). The nuclei of these ingested cells are relatively indigestible and remain for some time scattered through the egg cytoplasm. Among all those forms with definitely localized ovaries growth of the ova is accomplished very differently. When the eggs are small and contain relatively little food, no special nutritive mechanism is developed, the egg forming the food substances in its own cytoplasm from materials drawn from the circulating fluids in the cavities of the ovary. Such eggs develop independently of the neighboring cells intermingled with the ova.

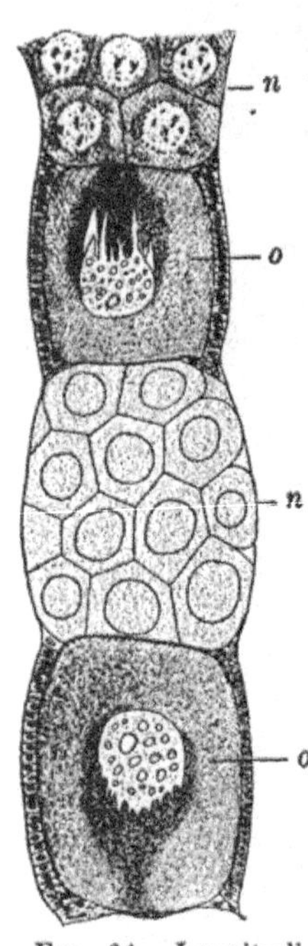

Fig. 64.—Longitudinal section through part of the egg-tube (ovary) of the beetle, *Dytiscus marginalis.* After Korschelt. *n,* groups of nutritive cells; *o,* ovum containing amœboid nucleus partly surrounded by n u t r i t i v e substance (deutoplasm).

When the fully formed eggs contain large amounts of food substance this is usually obtained by one of two chief methods. In the simplest cases certain of the ovarian cells adjoining the egg take on the characteristics of *nurse cells*. These may either contribute their own substance directly to the ovum or they may become intensely active, forming deutoplasm which is then drawn from them by the growing ovum (Figs. 63, 64). There may be a single nurse cell for each ovum (Fig. 65), or the nurse cells may be scattered irregularly through the ovary so that several may be related to each ovum. The nurse cells in many or even in most cases, are cells which were potentially germ cells, but which have lost their germinal potentiality and

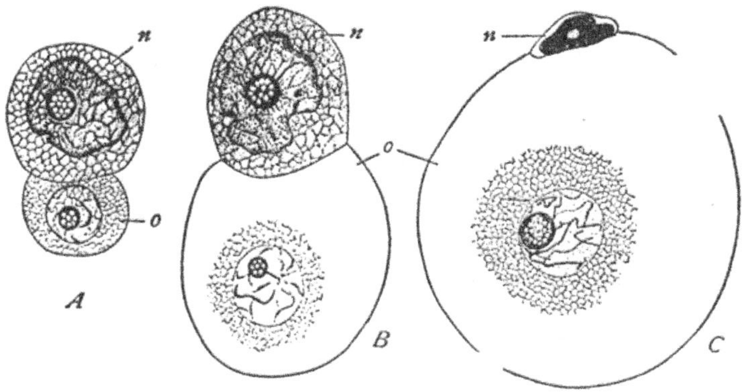

FIG. 65.—Sections through ovarian ova and nurse cells in the Annulate, *Ophryotrocha*. From Wilson, "Cell," after Korschelt. *A*. Young stage, the nurse cell, *n*, larger than the egg. *B*. Growth of the ovum, *o*. *C*. Late stage, the nurse cell degenerating.

become wholly nutritive, contributing to the formation of the true germ cell, degenerating and disappearing completely after the ovum is grown and has left the ovary. In the other

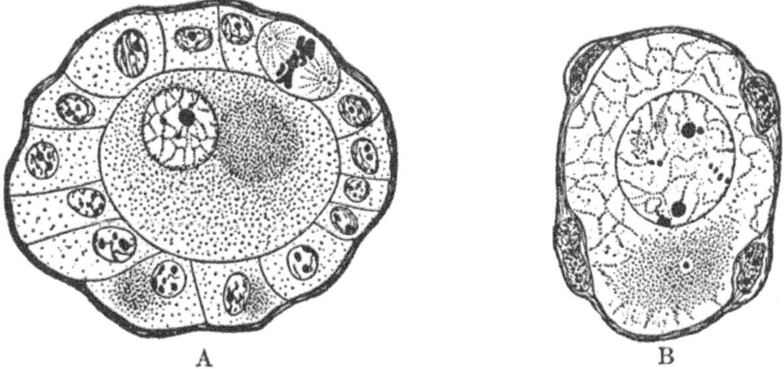

FIG. 66.—Sections through ovarian eggs and their follicles in, *A*, young magpie; *B*, newly born child. From Wilson, "Cell," after Mertens.

cases, which are commonest among the Vertebrates, almost universal among them in fact, the ovarian cells adjoining the ovum —themselves potentially germ cells originally, form around the ovum a definite layer termed the *follicle* (Fig. 66). The follicle cells have the arrangement of an epithelium; they may form either a single layered, simple epithelium or in other cases a many layered stratified epithelium (Fig. 43). They not only provide for the nutrition of the enlarging ovum, with which they are frequently connected by definite intercellular protoplasmic strands, but toward the close of the growth period of the egg they may become secretory and form certain egg envelopes of the secondary type, *i.e.*, chorionic. We have already mentioned how the micropyle is formed through the chorion and vitelline membrane by the insertion of one of these follicle cells with a long process, preventing the membranes from forming at that point. When the eggs are fully formed and ready to be laid the follicle ruptures allowing the eggs to escape freely. Often there develops in the follicle a definite region along which it bursts; this weakened region is called the *cicatrix* (*e.g.*, common fowl). In a few instances some of the follicle cells are actually taken into the egg and absorbed, much as in the case of those ova which ingest adjacent interstitial cells. This is the case in many of the Tunicata where some of the so-called "test-cells" lose their cell outlines and are directly taken into the cytoplasm of the egg; their nuclei remain distinct for some time (Fig. 67). The remaining test cells then form a distinct follicle outside the whole structure. These nuclei no longer function as nuclei of course; as the growth of the egg is completed they are extruded again, along with a portion of the superficial protoplasm forming then a thick vitelline membrane resembling a chorion and often so called.

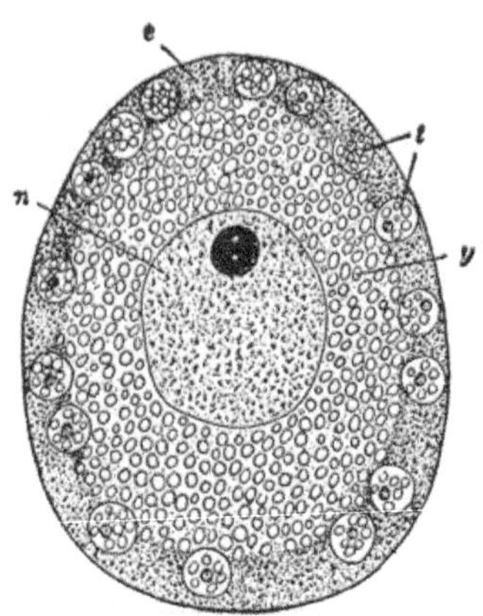

Fig. 67.—Section through the egg of the Tunicate, *Cynthia partita*. After Conklin. In the periphery of the egg are the nuclei of the test cells which have been ingested by the ovum (for their later history see Fig. 91). × 357. *e*, exoplasm or cortical layer; *n*, egg nucleus or germinal vesicle, with large nucleolus; *t*, nuclei of test cells or follicle cells; *y*, yolk.

Turning now to the formation of the sperm we find, as we should expect, processes on the whole entirely comparable with those of egg formation. The chief differences result from the fact that in the ovary conditions are associated with the formation of a few relatively large cells, while in the testis small cells are formed, but in very large numbers. In Sponges and Hydroids there is the same non-localized formation of the sperm as of the ova, the germ cells being distinguished not so much by position as by size. Apparently any ectoderm cell may enlarge and become reproductive. In all forms above these there are definite testes. Among many Cœlenterates and Echinoderms the testis is composed purely of germ cells, but usually the testis, like the ovary, contains other interstitial or accessory cells, and frequently these are directly nutritive in function. The general structure of the testis differs from that of the ovary in that its epithelium is thrown into folds, forming either simple *columns* or *acini*, each with an efferent duct or pathway which is to be considered cœlomic in origin. In the testicular epithelium, which is ordinarily reducible to the stratified type, we find sperm cells in all stages of formation, from spermatogonia to fully formed spermatozoa. As we have already said, the process of sperm formation is usually continuous, though frequently periodic (seasonal) in its rate or intensity. When the testis is composed of acini, each is usually surrounded by a follicle, equivalent in function to the egg follicle. But when arranged in lobules and columns, each of which may be derived from a single primordial cell or *prespermatogonium*, the nutritive cells do not show this follicular arrangement, but are fewer in number and scattered along the basement membrane of the epithelium, or along the connective tissue partitions bounding the spermatic columns of the lobule, and to some extent among the germ cells proper.

Among the Craniates the typical arrangement is that shown in Figs. 68, 69. Here, along the basement membrane, are several generations of spermatogonia with the scattered nutritive *basal cells*, sometimes called also the *Sertoli cells*, usually larger than the spermatogonia. As the spermatogonia increase in

number, through continued mitosis, they begin to increase in
size, though not nearly to the same extent that the oögonia do.
There is not always the same distinctness between the phases
of multiplication and growth here, and the two final divisions
of the full grown spermatogonial cell, then known as the

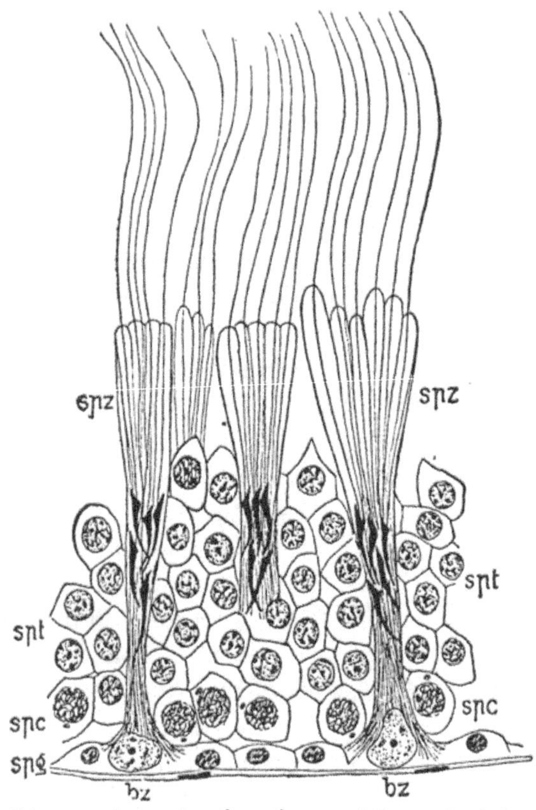

FIG. 68 —Diagram of a section through part of the testis of the rat, showing
some stages in spermatogenesis.　From Korschelt and Heider, after Lenhossék.
bz, basal cells; spc, spermatocytes; spg, spermatogonia; spt, spermatids; spz,
spermatozoa.

primary spermatocyte, are the reducing or maturation divisions.
These lead, as we have seen, to the formation, from each primary
spermatocyte, first of two secondary spermatocytes, both alike,
and then to four spermatids, all alike.　The cells of the column
then become arranged so that groups of spermatids become
related with each of the basal cells, which often leave their

original position and move out toward the free surface of the epithelium. Through their attachment to the basal cells the spermatids draw a supply of food and energy during their rapid and extensive metamorphosis into spermatozoa. In some cases a very close relation is established by the actual embedding of one end of the spermatid in the substance of the basal cell. It should be noticed that the function of the follicle or basal cells of the testis is to supply nutrition to the germ cells, not so much during their period of growth as after that is completed, during the period of metamorphosis; while in the ovary the corresponding cells function during the growth period; this is correlated with the smaller size of the spermatocyte and with

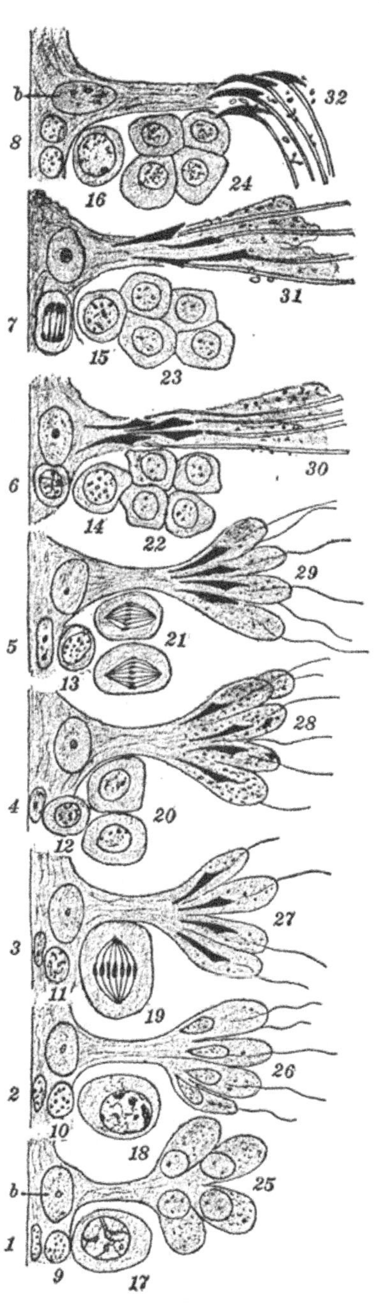

Fig. 69.—Diagrammatic outline of the spermatogenesis of the rat in thirty-two stages. After v. Ebner. Basement membrane toward the left. 1–8. Period of multiplication (the number of cell generations is actually very large). 9–18. Period of growth. 19–24. Period of maturation. 25–32. Period of metamorphosis. *b*, basal cells or Sertoli cells. 1–16. Spermatogonia. 17, 18. Primary spermatocytes preparing for division. 19. First spermatocyte division. 20. Secondary spermatocytes. 21. Secondary spermatocyte division. 22–25. Spermatids. 26–31. Transformation of spermatids. 32. Fully formed spermatozoa.

the need for an easily available food supply for the large number of sperm cells during their metamorphosis.

Probably the most interesting phase of the cytoplasmic aspects of spermatogenesis is this metamorphosis of the sper-

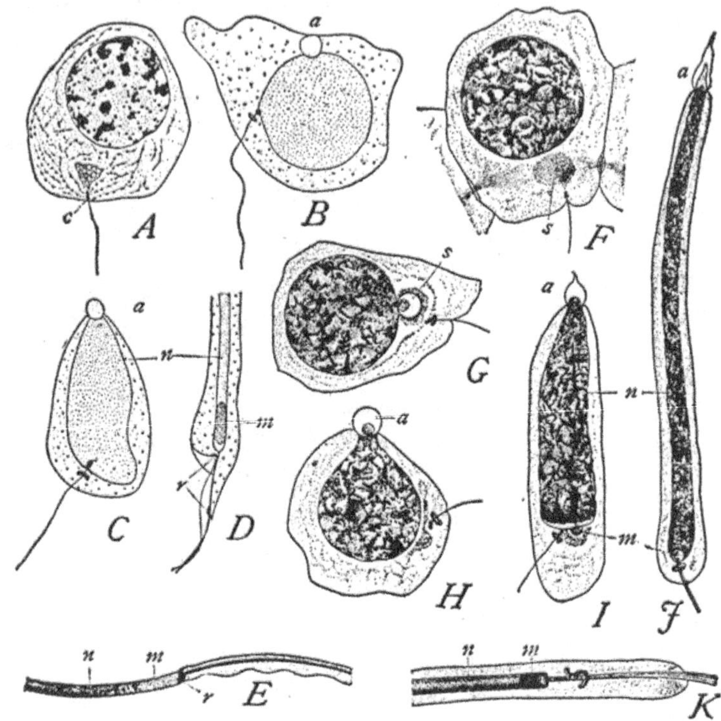

Fig. 70.—Formation of the spermatozoön in Urodeles. From Wilson, "Cell," *A–E*, *Salamandra*, after Meves; *F–K*, *Amphiuma*, after McGregor. *A*. Spermatid with peripheral pair of centrosomes (*c*) lying outside the sphere, and axial filament. *B*. Centrosomes near the nucleus, outer one ring-shaped; *a*, acrosome. *C*. Inner centrosome inside the nucleus, enlarging to form middle piece; *n*, nucleus. *D*. Portion of much older spermatid, showing divergence of two halves of the ring, *r*. *E*. Portion of mature spermatozoön, showing upper half of ring at *r*, and the axial filament proceeding from it. *F*. Spermatid of *Amphiuma*, showing sphere-bridges and ring-shaped mid-bodies. *G*. Later stage; outer centrosome ring-shaped, inner one double; sphere, *s*, converted into the acrosome, *a*. *H*. Migration of the centrosomes. *I*. Middle-piece at base of nucleus. *J*. Inner centrosome forms the end-knob within the middle-piece, which is now inside the nucleus. *K*. Enlargement of middle-piece, *m*, end-knob within it; elongation of the ring.

matids into spermatozoa. After growth and maturation the spermatids have much the same external appearance as any typical cell; they are more or less spherical cells with a pair of centrosomes or centrioles, and a large spherical nucleus with a dense chromatin network. Internally we know that the nuclei are unlike those of the somatic cells on account of the presence of only $\frac{s}{2}$ chromosomes. Without any further division each is converted into the special form of the spermatozoön typical of the species. While the details of this metamorphosis vary considerably in different groups, the essentials of the process are everywhere the same. The spermatid (Figs. 70, *A*, *F*; 71, *A*) contains, in addition to the typical cell organs just mentioned, a modified region of the cytoplasm which is sometimes a *centrosphere* or *idiosome*, sometimes of rather doubtful character and origin, which for convenience may be termed the *spermatosphere*. Close by lie the remains of the last division spindle. The spermatids are further characterized by the presence of a collection of chromidial structures termed the *mitochondria* (Fig. 71).

The details of the metamorphosis of these structures into the parts of the spermatozoön are subject to wide variation; the following account is based upon the history of the mammalian spermatid (Fig. 71). The first step in the process is the migration of the centrosomes to the surface of the cell, and at the same time the migration of the nucleus to the opposite side of the cell. In most cases it is difficult to determine the relation of the axis thus marked out, but in many instances this is perpendicular to the basement membrane of the germinal epithelium, thus expressing a polarity which coincides with that of the ovum; the nucleus lies toward the membrane, *i.e.*, the attached surface of the cell, the centrosomes toward the free surface. As the centrosomes and nucleus are taking these new positions the spermatosphere moves up to the nucleus and around it to the side opposite the centrosomes, quite to the surface of the spermatid. The two centrosomes now separate, one approaching the nucleus, the other remaining peripherally.

Following these changes in the relative positions of the parts come the real modifications of structure. The nucleus becomes elongated or ovoid, and the chromatin condenses, first into a

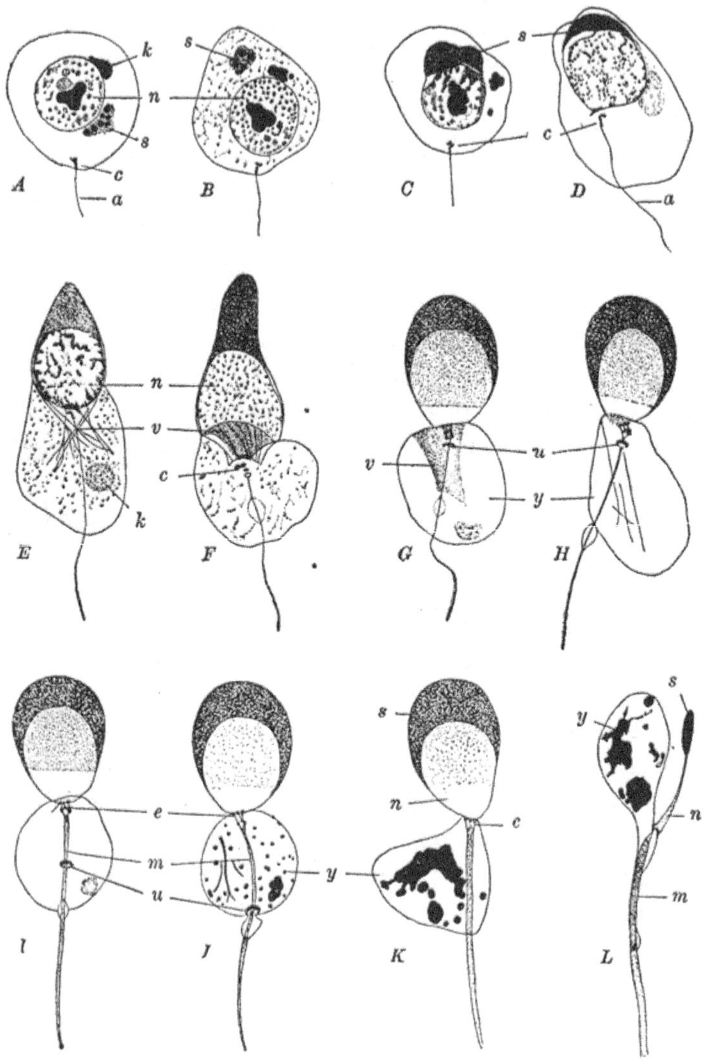

Fig. 71.—Metamorphosis of the spermatid of the guinea pig, *Cavia cobaya.*
After Meves. *L.* Side view, showing flattening of head. *a,* axial filament;
c, centrosomes (centrioles); *e,* neck, containing end knobs (proximal centriole);
k, chromidial "nebenkörper;" *m,* middle piece; *n,* nucleus; *s,* centrosphere (acrosome); *u,* annulus (posterior portion distal centrosome); *v,* mitochondria (partly
becoming a portion of the tail envelope); *y,* cytoplasmic portion of spermatid,
being thrown off in *K* and *L;* in *J, K, L* containing mitochondrial remains.

heavy reticulum, and then into a dense mass in which no visible structure remains; finally it acquires the form of the head of the mature spermatozoön. The spermatosphere meanwhile is largely converted into the acrosome or perforatorium, at the tip of the elongated nucleus; a smaller portion is transformed into a very delicate envelope covering a part of the head, in many cases scarcely distinguishable on account of its thinness. At the other end of the cell a fine flagellum begins to grow out in connection with the peripheral centrosome, either from the substance of the centrosome itself, or from the cytoplasm under its influence. Then the two centrosomes move farther in toward the nucleus. In the simpler cases the distal centrosome now divides into anterior (toward the head) and posterior portions. The posterior part grows out peripherally into a rapidly elongating fiber which becomes the axial filament of the flagellum, while at its base it becomes ring-like; then through this ring or annulus the axial filament grows in, finally connecting with the anterior portion of the centrosome. The anterior portion itself remains in the middle piece as a posterior centrosome of the spermatozoön. The proximal or anterior centrosome partly disappears, and partly is converted into that part of the middle piece which connects with the head (the neck).

The cytoplasmic part of the spermatid seems to be largely consumed in providing the energy for this transformation, in addition to that drawn from the nurse or Sertoli cells. But the cytoplasmic membranes of the middle piece and tail, including the undulatory membrane when this is present, are derived directly from the cytoplasm of the spermatid. The mitochondria of the spermatid seem to be transformed into the spiral layer of the middle piece. In many instances, particularly among the Mammals, the larger part of the cytoplasm remains for a time connected with the middle piece of the developing spermatozoön and then is cast off, and finally degenerates without taking any further part in the structural formation of the functional sperm cell. The chief structural correspondences between spermatid and spermatozoön are shown in the accompanying table.

Comparison of the Structures of the Spermatid and Spermatozoön

With particular reference to the mammalian condition. (Partly from Gegenbaur-Fürbringer, *Lehrbuch der Anatomie des Menschen*, Leipzig, 1909)

Spermatid	Spermatozoön
Nucleus.	Head.
Spermatosphere (centrosphere).	Acrosome (perforatorium) and sheath covering the anterior part of the head.
Proximal centrosome (centriole).	Forms an undifferentiated part of the middle piece; in Mammals, the neck. In part may disappear.
Distal centrosome (centriole).	
(*a*) Anterior portion.	Centrosome of middle piece. ·
(*b*) Posterior portion.	Annulus and axial filament of middle piece and tail.
Cell body.	Partly used as source of energy during metamorphosis. Partly thrown off. Remainder forms cytoplasmic envelopes of middle piece and tail (including undulatory membrane).
Mitochondria.	Spiral membrane of middle piece.

Whatever the details of the metamorphosis of the spermatid may be, the facts of essential importance are always identical. These are, that the nucleus of the spermatid is directly transformed into the head of the spermatozoön; the centrosomes of the spermatid become the centrosomes and kinoplasmic structures of the spermatozoön and are contained within the middle piece, or partly in the tail; the cytoplasm of the spermatid in part goes to form a thin cytoplasmic investment of the spermatozoön or is in part cast off.

The fully formed spermatozoa now lose connection with the

nurse cells and pass by way of the canals or ducts of the testis into the efferent reproductive ducts, *vasa efferentia* and *vasa deferentia*, to the outside, either directly, or after being stored for a time in special cavities such as the *seminal vesicles*. The sperm may remain alive in these storage cavities for a long time, awaiting the period of extrusion, upon the approach of which they become very active.

REFERENCES TO LITERATURE

AMMA, K., Ueber die Differenzierung der Keimbahnzellen bei den Copepoden. Arch. Zellf. **6.** 1911.

BALFOUR, F. M., Treatise on Comparative Embryology. London. 1880, 1881.

BALLOWITZ, E., Untersuchungen über die Structur der Spermatozoen, *etc.* Arch. mikr. Anat. **32.** 1888. **36.** 1890. Also Zeit. wiss. Zool. **50.** 1890. **52.** 1891.

BARDELEBEN, K. v., Dimorphismus der männlichen Geschlechtszellen bei Säugetieren. Anat. Anz. **13.** 1897. Weitere Beiträge zur Spermatogenese beim Mensch. Jena. Zeit. **31 (24).** 1898.

BEARD, J., The Germ Cells. I. *Raja batis.* Zool. Jahrb. **16.** 1902.

BOVERI, T., Zellenstudien. I. Die Bildung der Richtungskörper bei *Ascaris megalocephala* und *Ascaris lumbricoides.* Jena. Zeit. **21 (14).** 1887. Die Entwicklung von *Ascaris megalocephala* mit besonderer Rücksicht auf die Kernverhältnisse. Festschr. f. von Kupffer. Jena. 1899.

CONKLIN, E. G., The Organization and Cell-Lineage of the Ascidian Egg. Jour. Acad. Nat. Sci. Philadelphia. **13.** 1905.

DEAN, B., The Chimæroid Fishes and their Development. Publ. Carnegie Inst. No. 32. Washington. 1906.

EIGENMANN, C., On the Egg Membranes and Micropyle of some Osseous Fishes. Bull. Mus. Comp. Zool. Harvard Coll. **19.** 1890. On the Precocious Segregation of the Sex-Cells in *Micrometrus aggregatus.* Jour. Morph. **5.** 1891.

FRITSCH, G., Zur Anatomie der *Bilharzia hæmatobia*, Cobb. Arch. mikr. Anat. **31.** 1888.

HÄCKER, V., Die Eibildung bei *Cyclops* und *Canthocamptus.* Zool. Jahrb. **5.** 1892.

HERFORT, K., Die Reifung und Befruchtung des Eies von *Petromyzon fluviatilis.* Arch. mikr. Anat. **57.** 1901.

HERTWIG, O., Experimentelle Studien am tierischen Ei vor, während und nach Befruchtung. Jena. Zeit. **24 (17).** 1890. Vergleich der Ei- und Samenbildung bei Nematoden. Eine Grundlage für celluläre Streitfragen. Arch. mikr. Anat. **36.** 1890.

HOEFER, P. A., Beitrag zur Histologie der menschlichen Spermien und zur Lehre von der Entstehung menschlicher Doppel (miss) bildung. Arch. mikr. Anat. **74.** 1909.

JORDAN, E. O., The Spermatophores of Diemyctylus. Jour. Morph. **5.** 1891.

KORSCHELT und HEIDER, Lehrbuch der vergleich. Entwick. der wirbellosen Thiere. II. Abschnitt. Die Geschlechtszellen, ihre Entstehung, Reifung und Vereinigung. Jena. 1902–1903.

LENHOSSÉK, M. v., Untersuchungen über Spermatogenese. Arch. mikr. Anat. **51.** 1898.

LILLIE, F. R., Studies of Fertilization in *Nereis*. I. The Cortical Changes in the Egg. II. Partial Fertilization. Jour. Morph. **22.** 1911.

McGREGOR, J. H., The Spermatogenesis of *Amphiuma*. Jour. Morph. **15.** Suppl. 1899.

MARSHALL, F. H. A., The Physiology of Reproduction. London. 1910.

MAYER, A. G., The Atlantic Palolo (*Eunice fucata*). Sci. Bull. Mus. Brooklyn Inst. Arts and Sci. **1.** 1902.

MEISENHEIMER, J., Entwicklungsgeschichte von *Limax maximus*. I. Th. Furchung und Keimblätterbildung. Zeit. wiss. Zool. **62.** 1896.

MEVES, F., Ueber die Entwicklung der männlicher Geschlechtszellen von *Salamandra maculosa*. Arch. mikr. Anat. **48.** 1897. Ueber Struktur und Histogenese der Samenfäden des Meerschweinchens. Arch. mikr. Anat. **54.** 1899.

MONTGOMERY, T. H., JR., The Spermatogenesis of *Peripatus* (*Peripatopsis*) *balfouri* up to the Formation of the Spermatid. Zool. Jahrb. **14.** 1900. On the Dimegalous Sperm and Chromosomal Variation of *Euschistus* with Reference to Chromosomal Continuity. Arch. Zellf. **5.** 1910.

PLATNER, G., Ueber die Bedeutung der Richtungskörperchen. Biol. Centr. **8.** 1889.

RETZIUS, G., Weitere Beitrage zur Kenntnis der Spermien des Menschen und einiger Säugetiere. Biol. Untersuch. Neue Folge. **10.** 1902.

SCHWEIGGER-SEIDEL, F., Ueber die Samenkörperchen und ihre Entwickelung. Arch. mikr. Anat. **1.** 1865.

WALDEYER, W., Die Geschlechtszellen. Hertwig's Handbuch, *etc.*, I, 1, 1. 1901–1903. (1906.)

WEISMANN, A., The Evolution Theory. New York. 1904.

WILSON, E. B., The Cell, *etc.* New York. 1900.

ZELLER, E., Ueber den Geschlechtsapparat des *Diplozoön paradoxum*. Zeit. wiss. Zool. **46.** 1888.

ZIEGLER, H. E., Lehrbuch der vergl. Entw. der niederen Wirbelt. Jena. 1902.

CHAPTER IV

MATURATION

In this chapter we shall describe certain events which are in reality essential steps in the processes of oögenesis and spermatogenesis, namely, the maturing of the nuclei of the definitive germ cells. In animals these maturation processes are the final steps in the complete specialization of the germ cells, and must be accomplished before the two gametes can fuse completely and thus begin the life of the "new" organism as an individual. As a matter of fact, the processes of maturation may be inaugurated before the growth and differentiation of the germ cells are entirely completed, and these processes may then all go on together. They are considered here separately, and without complete regard for their normal time relations, partly as a matter of convenience, looking toward simplicity of description, and partly to emphasize their great importance as a period in the development of the organism. Such a separation is easy because the maturation processes are not visibly connected with the genesis of the germ cells as such, for, morphologically at any rate, they concern only or chiefly the nuclei alone; the accompanying cytoplasmic modifications of structure have already been described.

That morphological characteristic chiefly distinguishing the fully matured germ cells is the possession of but one-half the number of chromosomes, and of but a smaller fraction of the amount of chromatic material, possessed by the somatic cells (Van Beneden). We should include under the term maturation, the whole series of events leading to this reduction in number of chromosomes and amount of chromatin. It should be noted, however, that the terms " oögenesis" and "spermatogenesis" have sometimes been used in a restricted sense to mean what we here term "maturation," but we have understood the former terms to include the whole history of the germ cells up to the time of their formation as completely specialized structures, and maturation therefore becomes a phase in oö- and spermatogenesis.

It is a familiar fact that in fertilization the union of an egg nucleus and a sperm nucleus is an essential step. The repeated union of nuclei of the usual type, in this way, would result in rapid and limitless increase in chromatic elements and material, but for the operation of some mechanism preventing such an accumulation, and yet permitting the fusion of germ-cell nuclei. Maturation is such a process; but it is much more than this. Consideration of all the phenomena of maturation raises many questions, important, even fundamental, in their biological significance. The full meaning of the phenomena can be appreciated only in connection with the fertilization processes to which they are introductory; we may most profitably, therefore, postpone much of our discussion of the general significance of maturation until we have become acquainted with the process of fertilization.

At this time, then, we shall describe the essential facts of maturation as we find them in typical, in some respects perhaps schematized, form, together with a brief comparative account of some maturation processes in a few special instances. Then after a similar account of fertilization in the next chapter, we shall be in position to consider some of the general aspects of both these processes taken together. The events of maturation and fertilization are really closely related in time, as well as in significance. While the spermatozoa are always fully mature before they enter the egg cells, the entrance of the sperm may occur either before, during, or after the maturation of the ovum, although of course the essential step in fertilization, namely, the union of the nuclei, does not occur (excepting in some Protozoa and plants) until after the maturation of the egg nucleus is completed.

The maturation of the germ cells is accomplished, in the Metazoa, by a modification of a mechanism common to all cells and already familiar, namely, mitosis. But the cell divisions which occur here are of a very special form, not found in the history of other kinds of cells. These unusual mitoses are known as the *meiotic* (*maiotic*, Farmer and Moore), or *reducing* divisions, and their chief peculiarity consists in the fact that they result in the formation of daughter nuclei containing the *reduced* or *haploid* number of chromosomes.

We shall review first the maturation of the spermatozoön, as this is less modified than the ovum, from those conditions which we regard as typical. Throughout the multiplication divisions of the spermatogonia, the mitoses are all of the usual character, except that the mitotic figures are relatively larger than in the somatic divisions. The number of the chromosomes is the same as in the somatic cells of the same organism; this is spoken of as the *diploid* number. In many, perhaps most, organisms the chromosomes differ from those of the somatic cells in form and size characters, so that the germinal tissue can usually be identified; the germinal tissue nuclei are in general larger and richer in chromatin than those of somatic tissues, a difference which, as previously noted, in a few forms (*Ascaris*, *Cyclops*, some Teleosts) can be traced from very early cleavage stages. But the constitution of the nuclei which pass into the inter-kinesis after the last spermatogonial division, *i.e.*, into the primary spermatocyte nucleus, is essentially normal. Some-times certain peculiarities become noticeable during the growth period of these cells; the nucleus frequently does not remain in the typical "resting" condition, but forms a more or less dis-tinct spireme (leptonema, Winiwarter), and sometimes, even at the beginning of this stage, there may be a fission of the chro-matin granules, forming a sort of double spireme (Fig. 73). At the close of this growth period, when the primary spermato-cyte prepares to divide, the nucleus begins to show a very unusual condition. The nucleus itself remains large, but the chromatin, as it begins to form a spireme, in those cases where this has not formed previously, condenses at one side of the nucleus, in the vicinity of the nucleolus, usually in the region near the centrosomes and perhaps through their influence (Schönfeld), into a dense mass in which little structure can be

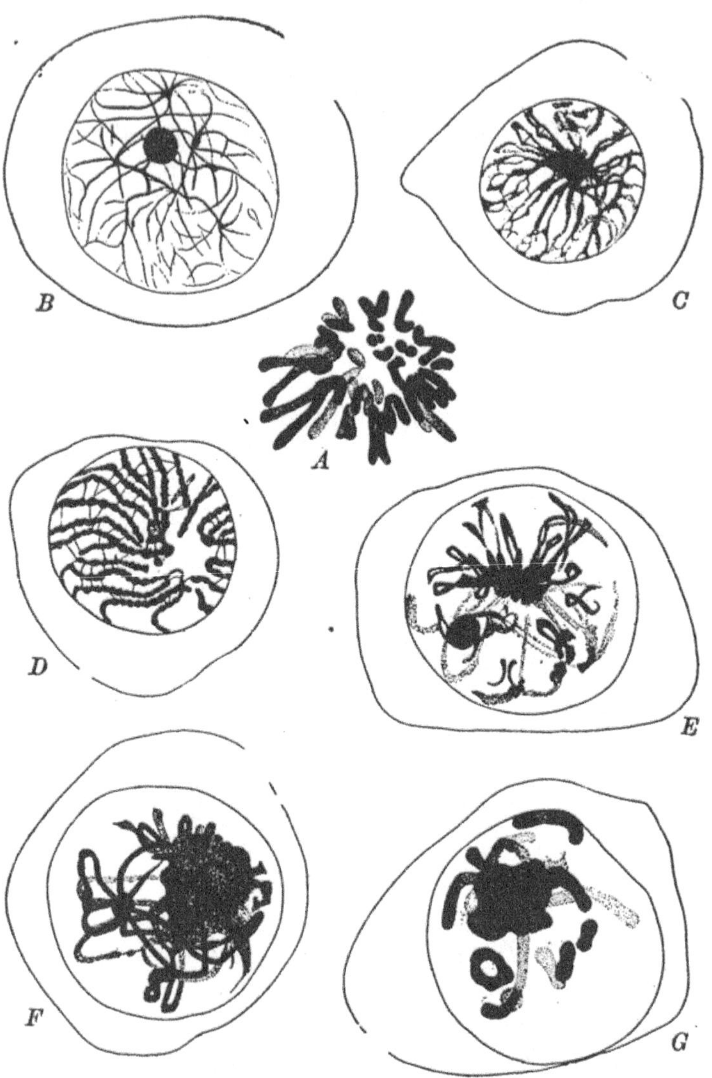

Fig. 72.—Early stages in the maturation of the Dipnoan, *Lepidosiren paradoxa*. After Agar. × 933. *A*. Polar view of equatorial plate of late spermatogonium, showing size and form differentiation and pairing of chromosomes. *B*. Spireme stage (leptonema) of primary spermatocyte. Only a few of the threads are shown. *C*. Nearly polar view, showing beginning of longitudinal fusion of chromatin threads; beginning of synapsis (zygonema). *D*. Nearly polar view of "bouquet stage" (pachynema). The threads are fused and condensed. *E*. Polar view showing beginning of contraction (synizesis) and splitting of the chromosomes (diplonema). Most of the thickened threads have split apart except terminally, where they remain fused, forming rings. In some, the

made out (Figs. 72, 73). This stage is called the *contraction phase*, or *synizesis* (McClung) (pachynema, Winiwarter). In some cases synizesis may occur near the beginning of the growth period, throughout which the nucleus then remains in this condition.

When growth of the spermatocyte is completed this knot of chromatin begins to disentangle and the spireme again becomes visible. This later spireme is not continuous, however, but is of the segmented type (Fig. 72), and the number of segments, *i.e.*, of chromosomes, *is but one-half the number of chromosomes that went into the nucleus at the close of the preceding division.* As regards the chromatic structures, this is the essential point in the whole maturation process; the number of chromosomes formed in the prophase of the first spermatocyte division is reduced to one-half the somatic number. This numerical reduction of the chromosomes is brought about in most, if not in all cases so far known, by an actual fusion, by twos, of the chromosomes contained in the last spermatogonial nucleus. This fusion of pairs of chromosomes is termed *synapsis* (Moore, McClung) or *syndesis* (Häcker) (zygonema, Winiwarter), and the resulting units are thus double or *bivalent* chromosomes. It seems entirely likely, if not definitely established, that the pairs of chromosomes which come together in synaptic fusion are each composed of one chromosome derived from the male parent, and the corresponding chromosome derived from the female parent, similar in size, and form, and also in function if we assume the fact of chromosomal specificity (Montgomery). These two groups of similar elements came into a single nucleus during the fertilization process which was the beginning of the new individual, whose cells are now preparing for reproduction; they have remained separate throughout the life of this organism until this event, through all the divisions of the ancestral germ-forming cells (Fig. 80). In a certain sense, therefore, this process of synapsis represents the real climax of the whole

splitting is less complete. One ring is cut through showing two free ends. *F.* Advanced synizesis. *G.* Chromosomes appearing after synizesis, shortened and thickened. The ex-conjugant chromosomes (univalent) have separated and show transverse constrictions, preparatory to the second maturation division.

series of developmental processes, and it is at the same time the starting point of the life cycle of a new organism of another generation.

In a few forms (*e.g.*, some Insects, *Lepidosiren*, Fig. 72), the double nature of these bivalent chromosomes is distinctly visible and is indicated by a split through the long axis of the chromosome, showing that the pair of univalent elements have fused side by side, a condition known as *parasynapsis*. In most cases observed (other Insects, Amphibia) the fusion is end to end, a condition known as *telosynapsis* (Wilson's terms). In many instances, however, the fusion seems to have occurred between the granules composing the chromosomes, so that in the bivalent body there is no visible indication of the duplex nature at this time; this is then only to be inferred from the fact of numerical reduction. It is very important to notice that the time relations between synizesis and synapsis may sometimes be just the reverse of that described above, and the synapsis stage may occur first, so that the numerical reduction of the chromosomes occurs at the close of the last spermatogonial division (some plants, Strasburger, Overton).

Following the period of synapsis the nucleus and cell may proceed at once to divide, or there may ensue another resting period, during which the chromosomes again become indistinct. In either case, when the new mitotic figure forms, always after an unusually long prophase which is characteristic of this division, the reduced (haploid) and bivalent chromosomes often show an unusual condition, in that they may prepare at once, not for a single ensuing division, but for two divisions which are to follow immediately, without an intervening resting period.

From this stage onward in the history of the sperm and egg nuclei, two general types of chromosome behavior are sometimes distinguished, although they are connected by transitional conditions and so are regarded as modifications of a single process. As one extreme condition we find a form of chromosome behavior called *tetrad formation*, which we may describe, not because it is a typical method of chromosome reduction, but because the facts of reduction come out more clearly in this form of maturation division (Boveri). In cases of tetrad formation, when the chromosomes appear in the primary spermatocyte, after the resting stage, each of the newly organized bivalent elements comes out in the form of four small bodies, the *tetrads*, arranged approximately in a square (Fig. 73, *E*). These bodies result from two successive splittings of each chromosome into four columns of granules, each of

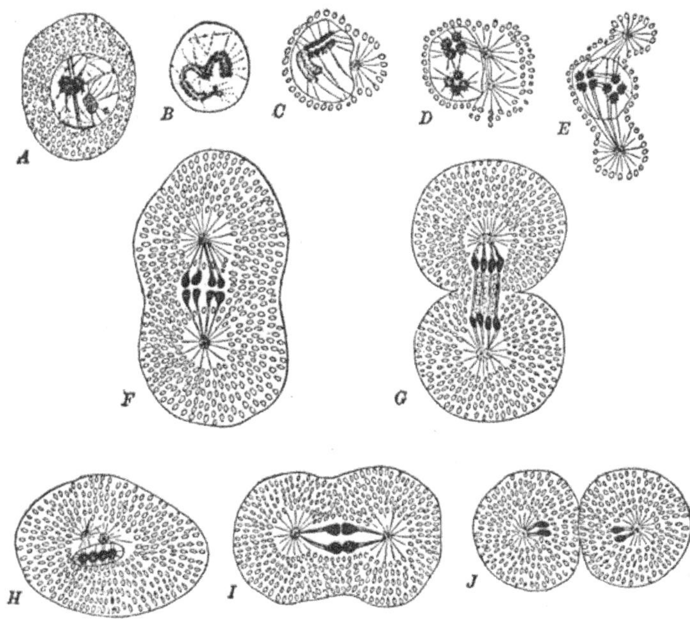

Fig. 73.—Tetrad formation in the spermatogenesis of *Ascaris megalocephala bivalens*. After Brauer. × 795. *A–G*. Stages in the division of the primary spermatocyte. *A, B*, splitting, and *C*, condensation of chromatin thread, seen in side view. *D* shows, in end view, that the splitting is double. Centrosome divided. *E*. Migration of centrosomes and formation of spindle. *F, G*. Separation of the two groups of dyads and division of the cell body. *H*. Secondary spermatocyte containing two dyads. *I*. Division of secondary spermatocyte. *J*. Two of the spermatids, each with two "monads" or single, univalent, chromosomes.

which is then condensed into a single element (Fig. 73, *A, B, C, D*). We may recognize here a precocious division of the chromosomes, which in these cases precedes considerably the division of the nucleus and cell as a whole. Not only this, but there are *two* chromosomal divisions, corresponding with two cell divisions, and these occur simultaneously, while the nuclear and cytoplasmic divisions occur consecutively at a later period, during which these chromosomes do not divide again. The number of tetrad groups is thus the same as the haploid number of chromosomes $\left(\dfrac{s}{2}\right)$, and the total number of elements composing the tetrads is four times the haploid or two times the diploid number $(2s)$.

The nucleus and cell now enter upon a mitosis in which each tetrad behaves as a typical chromosome. The division and migration of the centrosomes to opposite sides of the nucleus, the formation of the spindle and asters, and other details of this mitosis, have no unusual features and need not detain us. The tetrads, containing all told $2s$ elements, become arranged about the equator of the spindle and each separates into two pairs of elements called the *dyads* (Fig. 73, *F*, *G*). The groups of dyads then move to opposite poles of the spindle and the cell divides into the two secondary spermatocytes. Since the resting stage is now omitted, the dyads do not dissolve after this division, nor do they divide again in anticipation of the next mitosis—the division of the chromatic elements for this cell division has already occurred in the nucleus of the primary spermatocyte, as we have seen. The dyads, containing all told s elements, then move at once to the equator of the new spindle, and each separates into two *monads* (Fig. 73, *I*, *J*). The two groups of monads, each now containing $\dfrac{s}{2}$ elements, diverge to opposite poles of the spindle, and the division of the cell (secondary spermatocyte) results in the formation of two spermatids, each with $\dfrac{s}{2}$ chromosomes (Fig. 74). Each nucleus then reforms into a typical resting condition, and passes through the metamorphosis into the head of the spermatozoön, as described in the preceding chapter. The essential characteristic, therefore, of the nucleus of the spermatid and spermatozoön is that, while *each of the bivalent chromosomes of the spermatocyte is represented*, yet, as the result of the process of reduction

through synapsis, the chromosomes are now only half as numerous as in all the other cells of the organism. But while each
bivalent chromosome is thus represented, it may be a question

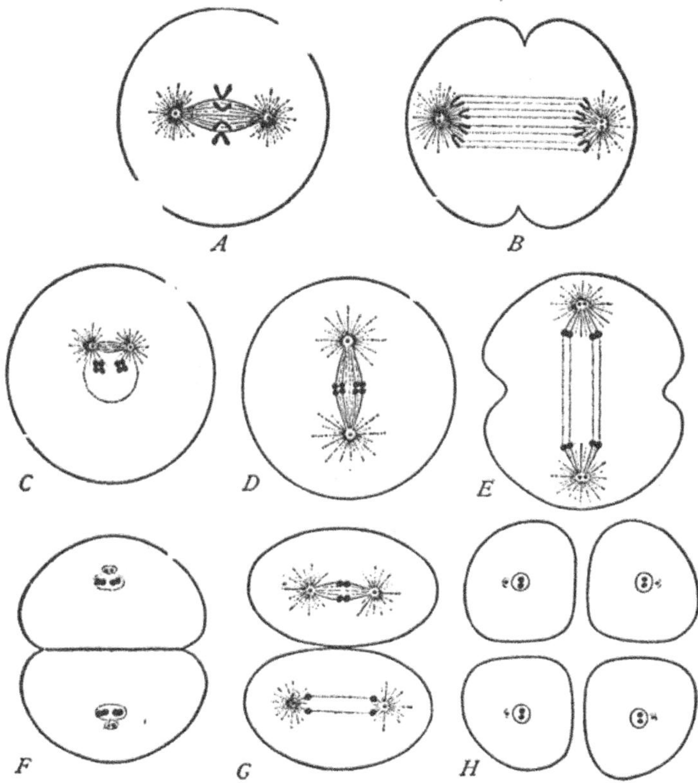

Fig. 74.—Diagram of reduction with tetrad formation in spermatogenesis.
From Wilson, "Cell." The somatic number of chromosomes is supposed to be
four. *A,B.* Division of one of the spermatogonia, showing the full number
(four) of chromosomes. *C.* Primary spermatocyte preparing for division; the
chromatin forms two tetrads. *D,E,F.* First division to form two secondary
spermatocytes each of which receives two dyads. *G,H.* Division of the two
secondary spermatocytes to form four spermatids. Each of the latter receives
two single chromosomes and a centrosome which persists in the middle-piece of
the spermatozoön.

whether each univalent chromosome is also somehow represented. To this we shall return later.

But the formation of tetrads is by no means of universal,
even of common, occurrence in reducing divisions. Tetrads are
commonly found only among the Nematodes, Annelids, and

Arthropods. In the great majority of animals the reduction divisions proceed without the formation of actual tetrads in their typical form, and when the bivalent chromosomes appear in the nuclei of the primary spermatocytes, as the result of synapsis, they have no unusual form. As they pass into the equatorial plate, however, it is seen that the two longitudinal halves, composing each bivalent chromosome, are united at their ends (Figs. 72, *E;* 75). When the anaphase begins each chromosome-half is drawn out first from its middle, so that the

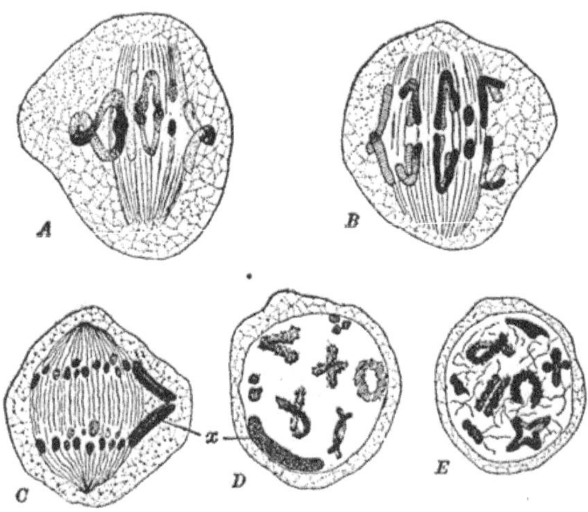

Fig. 75.—Maturation divisions in certain Insects, showing forms of chromosomes and their relation to tetrads. After de Sinéty. × 1125. *A,B.* Two stages in anaphase of primary spermatocyte division in *Stenobothrus parallelus.* Rings opening into *Vs,* which diverge. *C.* Anaphase of *spermatogonial* division in *Orphania denticauda,* showing differentiated chromosome, *x.* *D, E.* Preparation for first spermatocyte division in *Orphania,* showing "tetrads" in various stages of formation from rings and crosses.

whole original chromosome may appear as a ring or cross, or some related figure. As the halves separate each may assume a ⊃- or >-form, the limbs of which may come to lie parallel as the chromosome approaches the pole of the spindle, and thus may appear to be double (Figs. 75, 20). On account of this peculiar form assumed by the chromosomes in this division, it is known as the *heterotype division.* And it is to be noted that the separation of the halves of the bivalent chromosome here, is along the line of a split which is usually visible in the chromosomes when they appear out of the resting nucleus. As the result of this heterotype division each secondary spermatocyte receives the haploid number of chromosomes. The second maturation division commonly shows none of these rings or crosses, or other figures, and is known consequently as the _homotype division_ (Flemming's terms). The homotype division of the secondary spermatocyte follows the heterotype either immediately, or after a considerable pause, during which the chromosomes sometimes lose their definite outlines to some extent. This pause, which does not occur when tetrads are formed, is probably related to the fact that while in the tetrads both divisions of the chromosomes occur at the commencement of the process, in this form of reducing division the second splitting does not occur until after its first actual division. During the homotype division the chromosomes behave, then, essentially as in the divisions of the usual type, and the resulting spermatids receive, just as in tetrad formation, the haploid number of chromosomes, just as do the secondary spermatocytes. Numerically, the most important difference in reduction with and without tetrad formation is that in tetrad formation the secondary spermatocytes have the diploid number, and in the absence of tetrads, the haploid number of chromosomes. This is the result of the fact that when tetrads are formed, the division of the chromosomes actually belonging to the secondary spermatocytes (second maturation division) really occurs in the nucleus of the primary spermatocyte (first maturation division), while in the absence of tetrads the division of the chromosomes has the normal relation to cell division, and the haploid number persists from the primary spermatocyte, after synapsis, to the spermatid and spermatozoön.

Before mentioning any further details of chromosome behavior during maturation, we must compare the process of maturation as it occurs in the ovum with that in the sperm. We may say at the outset that in all essentials the two histories are identical, so that this comparison may be brief, but there are a few differences to be noted.

The divisions of the oögonia are normal; the diploid number of chromosomes appear, and the details of spindle, aster, and centrosome, call for no special mention. Aside from the chromosomal behavior, the divergences of the later maturation divisions from the normal are partly the result of the enormous growth of the egg cell, and partly in the nature of adaptation toward ensuring the practically undiminished size of the ovum at the end of the process; that is, an equal subdivision of the chromatic elements is accompanied by an unequal subdivision of the cytoplasm and deutoplasm. The general formation of the large primary oöcyte has been described. We should emphasize again the fact that the maturation of this cell frequently is not completed until after the sperm cell has actually entered its substance. If we were describing the events of the maturation of the egg in strict accordance with their usual, though not invariable, time relations we should next describe the ensemination of the egg—the first step in fertilization. For the sake of clearness, however, we shall describe maturation as if it occurred before the entrance of the sperm; as a matter of fact, there are a few forms in which this is really the normal course of events, as in the sea-urchin and most Echinoderms.

The nucleus of the oögonium is very large and lies toward one side of the cell—practically always toward the animal pole of the egg (Fig. 76, *A*). The first steps in the maturation of the ovum closely resemble those in the sperm. During the brief synizesis stage the chromatin condenses near the centrosome, closely around the large "nucleolus" which is commonly found in most oöcytes; the emerging spireme shows that synapsis has occurred for the spireme is segmented into the haploid number of elements, representing the bivalent chromosomes. That is, the actual reduction occurs in the primary oöcyte as in the primary spermatocyte. The oöcyte nucleus then passes through a condition not represented in the spermatocyte in that a large amount of chromatin leaves the chromosomes (spireme segments), either dissolving in the nuclear sap, or passing in the form of granules or small masses to some region of the nucleus quite apart from the chromosomes proper (Fig. 76). It is important to note that no chromosomes are lost in this way; the full haploid number of these bodies remains grouped at one, usually the distal, side of the nucleus. During the active preparation for the first maturation or heterotype division the nuclear membrane disappears, liberating the dissolved

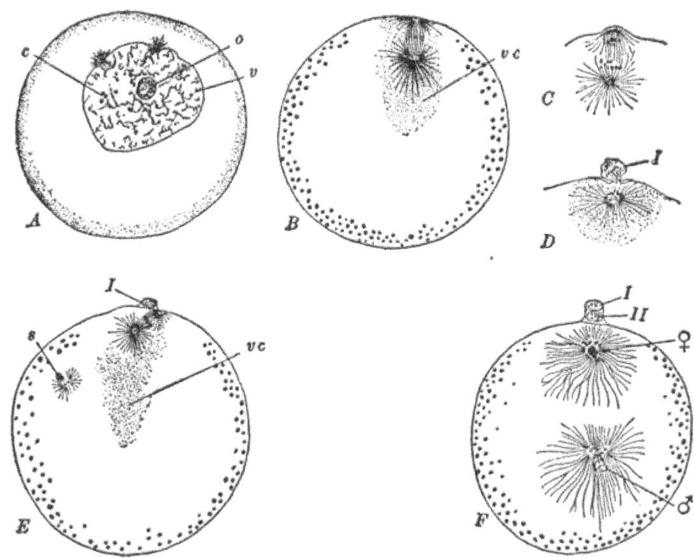

Fig. 76.—Maturation in the egg of the Nemertean, *Cerebratulus*. After Coe. C, D, × 375, others × 250. A. Primary oöcyte. Part of the chromatin has been condensed into chromosomes, only five of which are shown (the number present is sixteen). The remainder of the chromatin is thrown out into the cytoplasm. The centrosomes, each with a small aster, are diverging, and the nuclear membrane is commencing to disappear. B. First polar spindle fully formed and rotated into radial position. Chromosomes in equatorial plate. C. First oöcyte division; anaphase. D. First polar body nearly separated. E. First polar body completely cut off; second polar spindle formed and rotating into radial position. Spermatozoön within the egg. F. Second polar body completely separated. Egg pronucleus forming, surrounded by large aster. Sperm pronucleus, also with a large aster, enlarged and approaching the egg pronucleus. c, chromosomes; o, nucleolus, vacuolated and commencing to disappear; s, spermatozoön just within the egg; v, germinal vesicle; vc, contents (extra chromosomal) of germinal vesicle. I, II, first and second polar bodies; ♂, sperm pronucleus; ♀, egg pronucleus.

chromatin, or the extra-chromosomal masses, which then disappear gradually into the cytoplasm while the small chromosomes pass into the division figure. Sometimes this loss of chromatin is effected by a shrinkage of the chromosomes themselves, as in some Elasmobranchs, where the chromosomes at this time shrink to about one-fortieth of their previous length and one-tenth their previous diameter (Rückert, Fig. 34). Or the "nucleolus" of the oöcyte may be a karyosome or chromatin nucleolus, and in such cases (Echinoderms, for example) during synizesis the chromatin may be nearly all contained within this body. Then the chromosomes are formed singly or in groups out of this "chromatin reservoir." After they have all been given off, much the greater part of the chromatin still remains in the karyosome, which then may fragment before dissolution, or it may be dissolved directly (Fig. 35). The subsequent behavior of the chromosomes is closely similar to that of the spermatocyte chromosomes; tetrads may or may not be formed, according to the species, as the chromosomes pass into the division figure (Figs. 77, 78). The centrosome divides and the spindle and asters form typically in most respects save in size and position. The spindle is very small and in most eggs is close to the surface of the cell at its animal pole (Fig. 76). In alecithal and isolecithal eggs the nucleus and spindle are at first located centrally and then later move to the periphery. At first the spindle lies parallel to a tangential plane, but during the mesophase it rotates through ninety degrees, putting its axis in a radial direction (Figs. 76, 77). In many cases the division of the oöcyte is inhibited at this stage, until after the entrance of the spermatozoön, when it proceeds to completion; or this heterotype division may proceed without any interruption and the primary oöcyte cut at once into two cells. The extremely eccentric position of the nucleus in this stage leads to one of the most characteristic features of oögenesis, namely, the very unequal division of the cell body. One of the products of division, the *secondary oöcyte*, is of practically the same size as the primary oöcyte; the other cell—the *first polar body*—is very much smaller, indeed usually minute (Figs. 76, 77, 78). In essentials these two daughters of the primary oöcyte are equivalent; their nuclei are alike in size and composition, each contains a daughter centrosome, but with the polar body there is only the smallest amount of cytoplasm and practically none

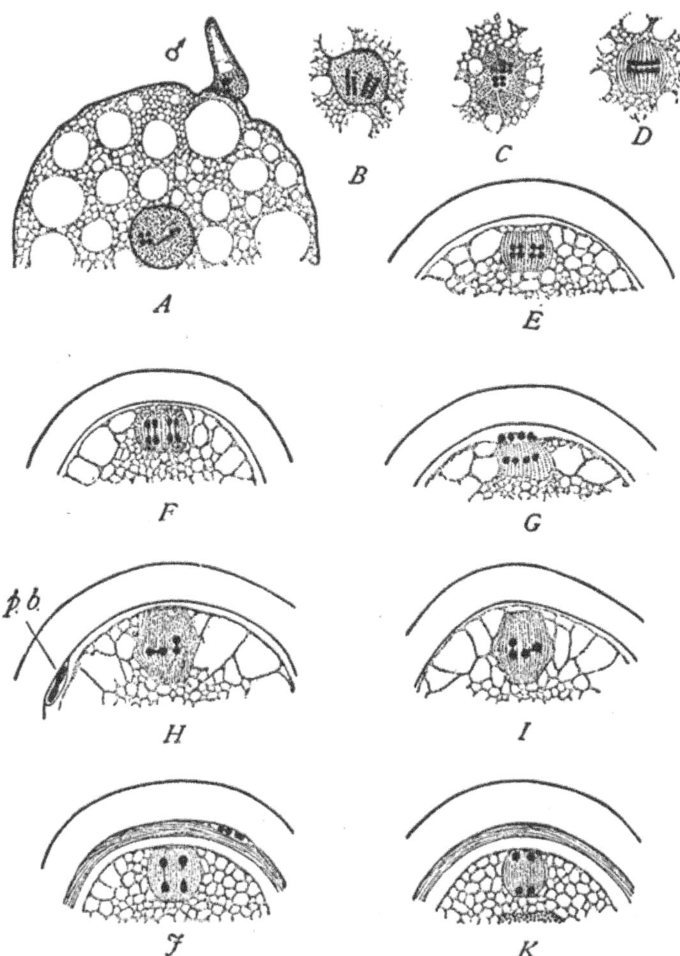

Fig. 77.—Maturation in the egg of *Ascaris megalocephala bivalens*. From Wilson, "Cell," after Boveri. *A.* Egg with spermatozoön just entering at ♂. The germinal vesicle contains two rod-shaped tetrads (only one clearly shown). *B, C.* Tetrads seen in profile and end views. *D.* First polar spindle forming (in this case within the germinal vesicle). *E.* First polar spindle in definitive position. *F.* Tetrads dividing. *G.* First polar body formed, containing, like the egg, two dyads. *H, I.* Dyads rotating into position for the second division. *J.* Dyads dividing. *K.* Each dyad has divided into two single chromosomes, as the second polar division approaches. (For final stages, see Fig. 94.)

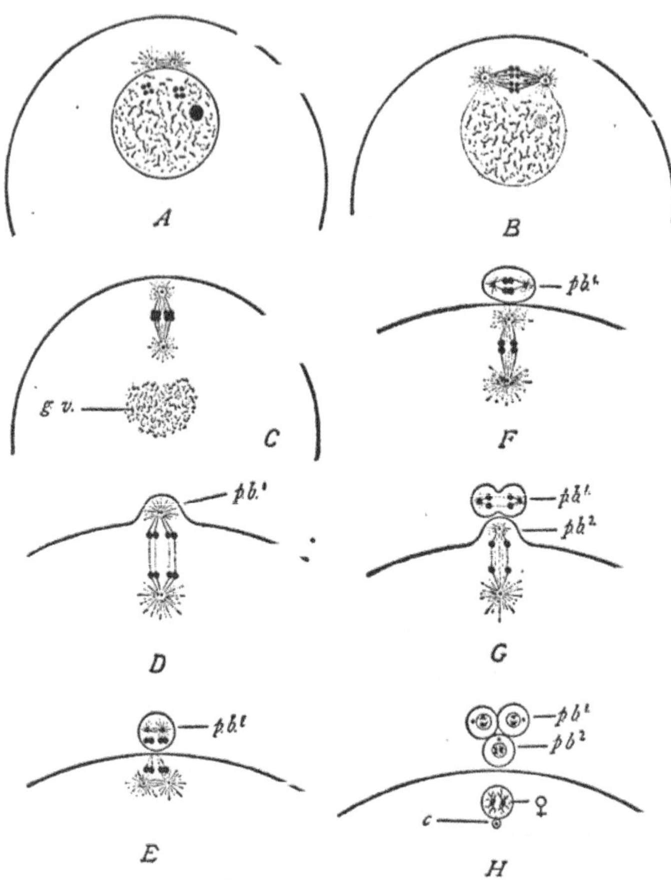

Fig. 78.—Diagram of reduction, with tetrad formation, in oögenesis. From Wilson, "Cell." The somatic number of chromosomes is supposed to be four. *A.* Initial phase; two tetrads have been formed in the germinal vesicle. *B.* The two tetrads have been drawn up about the spindle to form the equatorial plate of the first polar mitotic figure. *C.* The mitotic figure has rotated into position, leaving the remains of the germinal vesicle at *g.v.* *D.* Formation of the first polar body; each tetrad divides into two dyads. *E.* First polar body formed; two dyads in it and in the egg. *F.* Preparation for the second division. *G.* Second polar body forming and the first dividing; each dyad divides into two single chromosomes. *H.* Final result; three polar bodies and the egg-nucleus (♀), each containing two single chromosomes (half the somatic number); *c,* the egg-centrosome which now degenerates and is lost.

of the deutoplasmic substance. The equal division of the nucleus is thus accomplished without appreciable loss from the oöcyte of any of the cytoplasm and food reserve. ———

The second maturation or homotype division follows, either at once, or after a long pause in cases where the sperm normally enters at this time. The figure for the second maturation division is either a new figure or the reorganized remains of the preceding. In either case it appears in the region occupied by its predecessor, often in a radial position from its first appearance (Figs. 76, 77). Again the division is very unequal and the secondary oöcyte gives rise to the large *mature ovum* and a small *second polar body*, again alike as regards nuclear composition. The first polar body may or may not divide into two at the same time; we may assume that such a division is normal, but on account of the degeneration of the polar bodies such a division tends to disappear; in different forms various stages of this division are reached. In a few rare instances, as in some Rotifers and Insects, the second polar body may also divide.

The chromosomes remaining in the ovum then re-form a reticular nucleus, smaller than the original oöcyte nucleus, and with the haploid number of chromosomes; after forming a thin membrane the nucleus moves toward the cytoplasmic center of the ovum and there awaits union with the nucleus of the fertilizing spermatozoön (Figs. 76, 77). With few if any exceptions, the centrosome of the ovum is entirely lost as the nucleus is reconstructed; the absence of the centrosome is one of the peculiarities of the egg cell. It should be added that when the maturation is not completed until after the entrance of the sperm, the egg does not ordinarily re-form a typical nucleus but proceeds at once to unite with the sperm nucleus.

The final result of the two maturation divisions of the primary oöcyte is the formation of four cells whose nuclei are similar, and which are morphologically exactly equivalent to one another (Figs. 74, 78). Physiologically, however, there is the greatest difference among them, since of the four only one, the ovum, is able to function, or indeed even to remain alive,

and share in the development of a new organism. The other
three (or two), the polar bodies, after a brief time degenerate
and disappear.

The view that the ovum and the polar bodies are equivalent
cells morphologically, and that the latter are in reality to be
looked upon as degenerate egg cells (Mark) is now familiar.
Their identity in nuclear structure and history is of course a
decisive similarity.

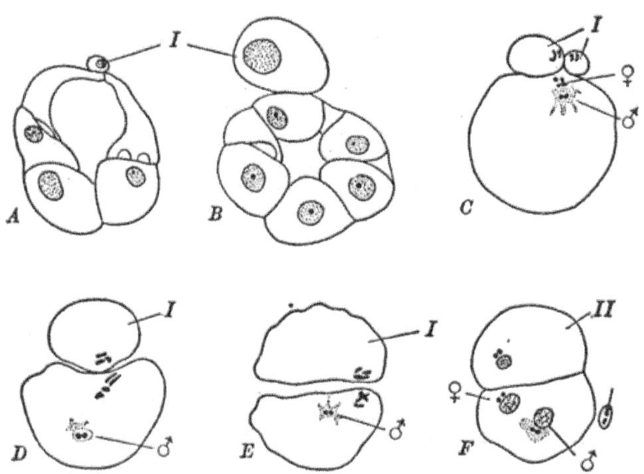

Fig. 79.—Variations in the size of polar bodies. *A, B.* Sections through
segmenting ovum of the Gasteropod, *Limax maximus,* showing polar bodies of
very different sizes. After Meisenheimer. *C–F. Ascaris megalocephala. C, D,
E,* after Sala, show influence of low temperature; *F,* after Boveri. *C.* Large
first polar body which has divided. *D.* Very large first polar body. *E.* Equal
division of egg during "first polar division." *F.* Equal division of egg during
"second polar division." *I, II.* First and second polar bodies; ♀, egg pronucleus; ♂, sperm pronucleus

The actual size of the polar bodies varies enormously. Those
of the Echinoderms are among the smallest, some being only
5–8 *micra* (1/5000–1/3000 inch) in diameter; in Amphioxus
they are about 7 *micra,* or about one-fourteenth the diameter
of the egg, while in the mouse they are relatively· much larger—
13 *micra,* or one-fifth the diameter of the egg. An interesting
series of forms can be arranged bridging over the size differences,
and to some extent also the physiological differences between
the ovum and the polar bodies. In a few Annelids, some

Turbellaria, and most Molluscs (Fig. 79) the polar bodies are very large, sometimes even one-fourth the diameter of the egg itself. In occasional instances, abnormal "giant" polar bodies are formed, which really approach the ovum in size (*e.g.*, Amphioxus, and the Turbellarian, *Prosthecerœus*). Some of these large polar bodies form a definite membrane, like the vitelline membrane of the ovum; and in rare instances they are actually fertilizable, although their development never proceeds beyond an incomplete cleavage. The last step in the transition from polar body to egg cell is represented by an interesting condition occasionally found in *Ascaris* and some other forms, (*e.g.*, mouse) where an abnormally placed polar spindle may result in the division of the oöcyte into two equal cells, one of which should be called a polar body (Fig. 79, *E*, *F*). The last step in the opposite direction—the dissimilarity between egg and polar body—reaches a climax in some of the Insects where polar bodies are not really formed as such; the oöcyte nucleus divides as in polar body formation and daughter nuclei are formed but these remain in the periphery of the egg cell, for no cytoplasmic division whatever is accomplished. The nuclear phase is the only part of the maturation division remaining. The "polar nuclei" formed in this way degenerate without sharing in development, just as if they had been cast out of the cell.

We have already suggested that this dissimilarity in size between the polar bodies and the ovum is in the nature of an adaptation such that maturation of the egg nucleus may take place without reducing the amount of cytoplasm and food materials which are to form the chief portion of the material basis of the new organism. In accomplishing this, three of every four potential egg cells are totally deprived of these substances and lose all possibility of developing. In some few instances, the polar bodies may for a time remain alive and during the early cleavage show some signs of activity, such as the performance of amœboid movement, "spinning activity," *etc.* (Andrews).

The location of the polar bodies within or without the vitelline membrane depends upon the time relation between membrane formation and maturation. The polar bodies may form before the membrane, in which case they usually are lost from the egg soon after their formation. More frequently they form after, and therefore within, the membrane and so can be seen for some time after development begins, when they form useful points for orientation, for in nearly all cases they mark the animal pole of the egg. The only exceptions to this location are among the Insects and Copepods, in which their position is variable.

Morphologically there is also complete correspondence between the ovum and the three polar bodies formed from the primary oöcyte, and the four spermatids and spermatozoa formed from the primary spermatocyte (Platner, O. Hertwig). But again there is physiological divergence, in that all of the derivatives of the spermatocyte are capable of functioning as germ cells, while only one of the oöcyte derivatives may do so. This physiological divergence is an expression, in another form, of the physiological division of labor between the egg and the sperm already referred to.

The chief points of similarity and difference between ovum and spermatozoön are summarized in the accompanying table.

Little is known regarding the nature of the stimuli which lead to the process of maturation, but it is clear that they are quite varied in different eggs. In some of those cases where the eggs are discharged freely into the water, contact with the water seems to initiate the process. But maturation may be begun previously to such a discharge. In some cases of this kind, as well as in others where the eggs are not thus freed, the rupture of the egg follicle seems to start the maturation process. In many cases the entrance of the sperm into the ovum is the effective stimulus to maturation, or to the completion of maturation in many of those instances where it has been begun previously and has been inhibited, either just before or just after the formation of the first polar body.

While we must postpone a part of our brief discussion of the theoretical significance of maturation, for reasons stated

Comparison of Typical Ovum and Spermatozoön

Similarities:

> Nuclei contain the haploid number of chromosomes, the result of two similar meiotic divisions.
>
> Chromosomes alike in form, size, and, with a few exceptions of special significance, in number.
>
> Can function only after syngamy.

Differences:

Spermatozoön	Ovum
Little cytoplasm.	
No deutoplasm.	
Actively motile.	
Centrosome present.	
One of four similar products of the division of the spermatocyte, all of which are functional.	
Usually completely formed and matured in the gonad.	

above, we should indicate at this time that the process has significance from at least two points of view. As a preliminary we should note that two different processes are involved in maturation; first, a reduction in the *number of chromosomes,* second, a reduction in the *amount of chromatin.* The earlier idea that maturation is merely a process by which the germ cells are rid of a part of their chromatin, and one-half of their chromosomes, as a preparation for the restoring union of chromatin and chromosomes during fertilization, is only one and probably the least important aspect. Many cells other than germ cells gain and lose large amounts of chromatin, and without going through any such complex process as that outlined above. Frequently much more than half, sometimes fully nine-tenths, of the chromatin is lost from the oöcyte nucleus during maturation, while during spermatogenesis comparatively little may be lost. And the mere numerical reduction of chromosomes is fully accomplished in synapsis, before the actual maturation divisions occur. For these and many other reasons it seems that the chief importance of maturation is from the standpoint of inheritance. This is true particularly of most of the details regarding chromosome reduction, which become significant only when correlated with the facts, first, that the germ cells are the simplest phases in the life cycle of the organism, alternating with the mature phases, the complex characteristics of which are related to the simpler characters of the germ, and second, that in some way, as yet unknown, the structural and physiological characteristics of the new organism are, at least in part, primarily determined by the chromosomal structure of *both* of the germ nuclei, *i.e.*, to the fact of biparental inheritance. From the standpoint of inheritance then the details of the behavior of the chromosomes during the maturation divisions take on the greatest importance. One of the more important matters is the precise plane of division of the chromosomes. It seems necessary to assume that each chromosome is not entirely homogeneous, but that its qualities differ in different parts. Consequently, in chromosomal composition the four nuclei derived from each primary oöcyte or spermatocyte nucleus may be all alike or may be of different kinds, according to whether the original chromosomes separated into similar or dissimilar parts in one or both of the maturation divisions (Fig. 80). In those cases where the chromosomes separate into qualitatively similar halves the division is said to be *equational*, and when into qualitatively unlike halves the division is *reducing*. And it is commonly believed that one of the maturation divisions is equational and one reducing. When the equational division precedes, *post-reduction* is said to occur; when the reducing division precedes it is described as *prereduction* (Korschelt and Heider). It is by no means a simple matter to determine whether a given chromosome division is equational or reducing, since there is externally visible no indication, in a chromosome itself, of its qualitative differentiation; and further because the processes of rearrangement and redistribution of the chromatin granules making up the chromosome are usually very obscure, particularly when synizesis is pronounced.

This whole subject is in a rather more hypothetical state than one might wish, considering the importance of the conclusions to be drawn. Upon the assumption that the qualities' of a chromosome differ from end to end, a longitudinal fission of the chromosome would divide it into exactly similar halves, while a transverse fission would of course divide it dissimilarly. In many forms it is clear that the first or heterotype division is longitudinal and the second or homotype division is transverse so that the four resulting nuclei are of two categories. In other cases it appears superficially that both divisions are longitudinal, while in still others it is really impossible to say definitely whether a given division is longitudinal or transverse. A transverse division would be reducing, however, only upon the assumption of an end to end differentiation of the chromosome, and upon the further assumption, which should be clearly apprehended, that *no rearrangement of the chromatin granules occurs* during the maturation divisions. And the additional fact must be taken into consideration that the chromosomes of the primary spermatocyte or oöcyte are bivalent, that they represent two chromosomes, which as wholes have fused in either parasynapsis or telosynapsis, and the actual *chromatin granules composing them may have an arrangement in fusion which is not indicated by the behavior of the whole chromosome.* The question whether a given longitudinal or transverse division of a chromosome is equational or reducing then can be determined only by taking all of these preliminary arrangements into account, and in most cases this is extremely difficult or even at present impossible. The small size of the chromosomes themselves and the minuteness, often invisibility, of the chromatin granules, often put the facts of their arrangement and behavior beyond the possibility of observation, and we can only infer their arrangement and history from subsequent events. In most cases we know only that reduction occurs. And from the few cases in which the course of events seems clear, we infer that in all maturation divisions, one is equational,

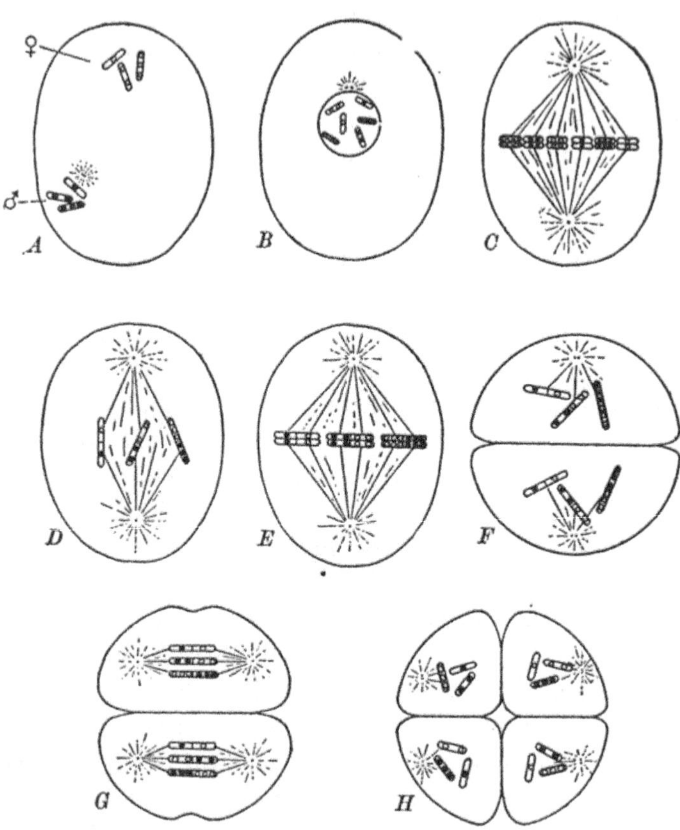

Fig. 80.—Diagrams representing the behavior of the chromosomes during fertilization and maturation. The differentiation of the three kinds of chromosomes is indicated by the number of small circles in each. ♂, chromosomes derived from the spermatozoön (sperm pronucleus) (black circles); ♀, those from the egg (egg pronucleus) (white circles). The somatic number of chromosomes is six. *A.* Entrance of spermatozoön. *B.* Fusion of egg and sperm pronuclei, forming the first cleavage nucleus. *C.* Splitting of chromosomes in equatorial plate, during the division of any *somatic, oögonial,* or *spermatogonial* cell. *D.* Primary oöcyte or spermatocyte in synapsis (telosynapsis). Fusion of similar chromosomes of maternal and paternal origin. *E.* Longitudinal splitting of bivalent chromosomes during first maturation division. *F.* First division completed forming the two secondary oöcytes or spermatocytes. The nuclei are alike in composition. *G.* Transverse division of chromosomes during second maturation division (reducing division—postreduction). *H.* The resulting four cells. With respect to each chromosome the cells are of two kinds, numerically equal.

one reducing, resulting in the formation of germ cells of two more or less unlike kinds, in equal numbers (Fig. 80).

Aside from its relation to the phenomena of heredity, the meaning of the maturation process is very problematic. In the Protozoa where definite chromosomes are formed only infrequently, maturation frequently involves a separation between reproductive and vegetative chromatin, as already suggested. Among the Metazoa there is usually a loss of chromatin from the nucleus, but it is very doubtful whether it has a similar meaning. In a great many cells, particularly those which are very active, *e.g.*, gland cells, oöcytes, *etc.*, the cytoplasm is constantly receiving substance from the nucleus. This material is frequently chromatic, and the granules of this kind have received a variety of special names, but collectively may be included under the term chromidia. (See Chapter II.) It may be that through some such process as this the nucleus exercises those forms of control and regulation of cell life that are its chief function. The loss and degeneration of the chromatin distributed to the polar bodies can have no significance here, for that process is involved in the degeneration of the entire polar bodies, which has an entirely different meaning. But during the growth period of most ova, just after synizesis, a relatively large amount of the chromatin is thrown out into the cytoplasm, and during the later stages of spermatogenesis a somewhat similar loss may be observed. And in the very early history of the germ cells of the organism, when this may consist of only a few cells, the primordial germ cells may often be distinguished by just this process of chromatin discharge from the nucleus. Such cells are often characterized by unusually large nuclei, and a large fraction of their chromatin content may be liberated into the cytoplasm at each mitosis. It may very well be, therefore, that this is a regular and highly significant process in the formation and maturation of the germ cells, having to do with the unusual activity of the sperm or with the development of various formed substances, both protoplasmic and deutoplasmic, present in the cytoplasm of the ovum. Indeed it may not be too much to suppose that the all-important "organization" of the ovum may in some way be related to this process of chromatin distribution.

The fact of maturation has been determined for all groups of many-celled animals and plants, and among the unicellular forms it is by no means uncommon. Among the Protozoa the phenomena of maturation are of considerable theoretical interest. In those forms in which the chromatin is not formed into definite chromosomes, but remains unor-

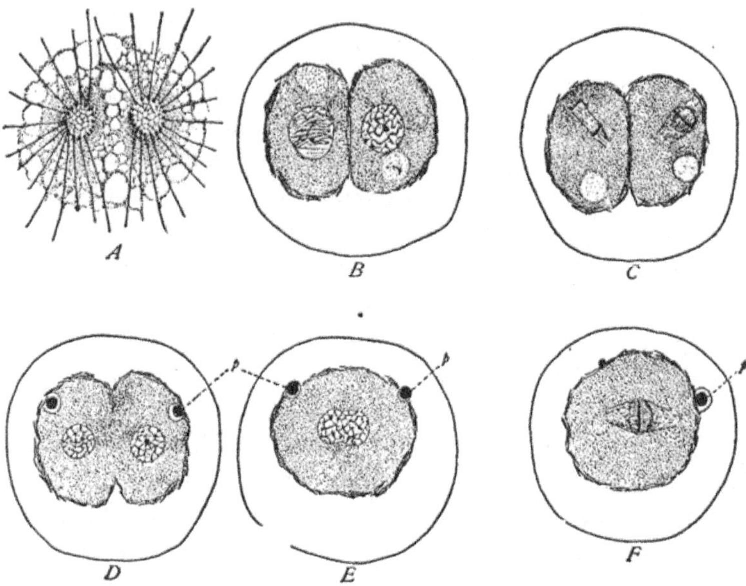

Fig. 81.—Maturation phenomena accompanying conjugation in the Rhizopod, *Actinophrys sol*. From Calkins, "Protozoa," after Schaudinn. *A.* Parts of two individuals, fused; the axial filaments terminate in granules on the surface of the nucleus. *B.* Nuclei in prophase. *C.* Formation of first polar spindle. *D.* Reconstruction of nuclei. *E.* Fusion of nuclei. *F.* First division spindle. *p*, polar body.

ganized, grossly, there seems to be a kind of division of labor between vegetative and kinetic (reproductive) forms of chromatin. The reproductive nuclei (idiochromidia) are frequently distinctly separate from the somatic nuclei (chromidia), and just before fertilization the former may divide twice in rapid succession. This process bears the greater resemblance to the maturation of the ovum since after each division one of the two products degenerates, often without actually being thrown out of the cell, leaving functional only one of the four products of the original idiochromidium. *Actinosphærium* is a typical example (R.

Hertwig), and there are several other forms where much the same thing occurs, *e.g.*, *Actinophrys* (Fig. 81), *Entamœba*. In all cases such as these, it is impossible to say whether *reduction*, in a strict sense, is accomplished, or whether this is merely an *elimination* of chromatin, for the chromatin is not organized into chromosomes whose precise behavior may be traced; it is quite likely that there is here no true reduction in the Metazoan sense. In some of the Infusoria, however, definite chromosomes are formed in the nucleus during these divisions and a definite chromosomal history may be made out. In *Paramœcium*, for instance, as described by Calkins and Cull, where the idiochromidia are known as the *micronuclei*, these alone are concerned in the "maturation" divisions. The micronucleus forms a fairly typical division figure consisting

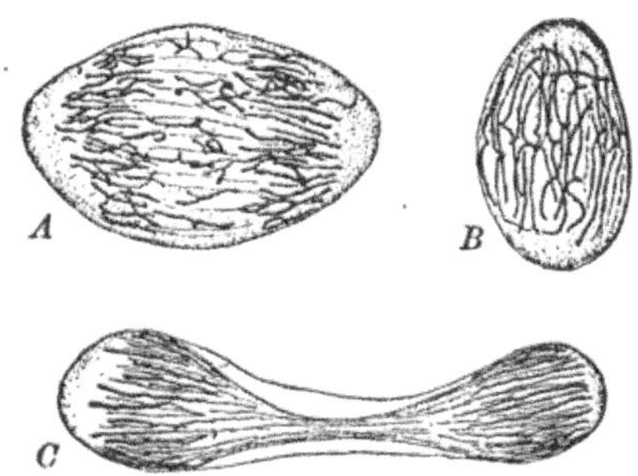

Fig. 82.—Maturation divisions in *Paramœcium aurelia* (*caudatum*). From Calkins and Cull. Only a few of the chromosomes are represented in each case. *A*. Late anaphase of first maturation division of micronucleus; some chromosomes incompletely divided. × 1000. *B*. Early anaphase of second maturation division. × 633. *C*. Telophase of second maturation division. × 900.

of a spindle and more than 200 separate chromosomes. During the first maturation division each of these divides longitudinally, the resultants passing in each case to the separate daughter nuclei, without any corresponding division of the cell body (Fig. 82). A second maturation division follows immediately and is precisely like the first giving four daughter nuclei (micronuclei, idiochromidia), three of which then degenerate as in polar body formation. The one remaining nucleus divides again, this time the chromosomes dividing transversely (reducing division?). This third division is not strictly comparable with anything to be found in the Metazoa and is apparently correlated with the character of the fertilization process in this form, for both parts share in reproduction. One-half remains *in situ* as the equivalent (analog) of the egg nucleus, and the other half migrates, as the equivalent (analog) of the sperm nucleus, to the body of another organism, fusing with (fertilizing) the stationary nucleus of that individual. In the majority of the Protozoa the so-called "maturation" or "reduction" divisions are not equivalent to these processes in the Metazoa, but are merely divisions by which a separation is effected between the reproductive and nutritive chromatin, *i.e.*, *idiochromatin* and *trophochromatin;* in nearly all known forms only the former takes any active part in the subsequent reproductive processes, while the trophochromatin usually dissolves and disappears.

Two very special modifications of the maturation process deserve just a word. The first is in connection with those few eggs which normally develop without fertilization (parthenogenesis), *i.e.*, without the union of equivalent egg and sperm nuclei. In such cases, which are

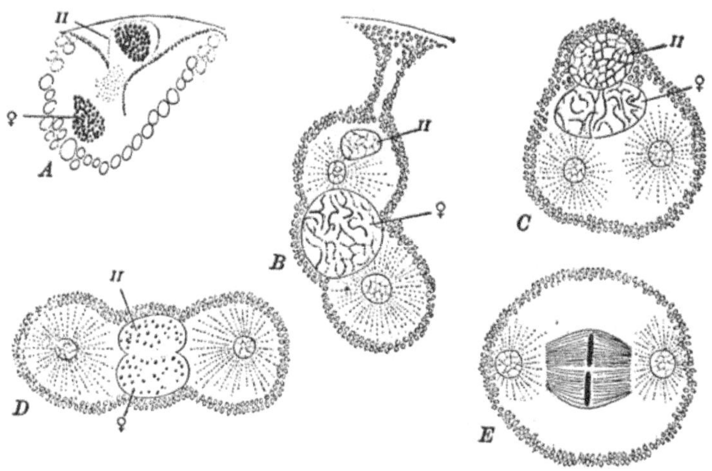

Fig. 83.—Maturation in the parthenogenetic egg of the brine-shrimp, *Artemia*. After Brauer. *A*, × 795, others, × 368. *A*. Second polar body incompletely cut off. *B*. Second polar nucleus reëntering the egg and approaching the egg pronucleus. *C, D.* Fusion of second polar body nucleus with egg pronucleus. *E*. First cleavage spindle with two groups of chromosomes derived from the two nuclei. *II.* Second polar body or nucleus; ♀, egg pronucleus.

known in the Aphids, many Crustacea, and Rotifers, for example, the normal course of maturation would lead to the formation of an organism with the haploid number of chromosomes $\left(\dfrac{s}{2}\right)$ in all of its cells. In most, if not in all, such cases which have been studied, it is now known that as a matter of fact the egg is not left with the reduced number of chromosomes. Thus in the brine-shrimp, *Artemia* (Brauer), which illustrates the usual course of events in parthenogenesis, the first maturation division proceeds as usual and is equational (reducing), leaving $\dfrac{s}{2}$ bivalent chromosomes in the secondary oöcyte nucleus. Then one of two courses may be followed (Fig. 83). In most normally parthenogenetic eggs a second polar body is formed typically, leaving the reduced number of univalent chromosomes in the egg nucleus, but then the second polar body immediately reënters the egg, apparently taking the place of an equivalent sperm nucleus and restoring the chromosomal characters to the normal somatic condition, after which development proceeds. The polar body need not be actually extruded from the egg cell in order to give the same history, as long as the nuclear events are equivalent (Fig. 84). In some eggs, even of species in which the history is at times similar to that just described, a different method gives the same result. Thus, while the second polar spindle may form typically, the chromosomes upon it do not divide, and the equivalent of the second

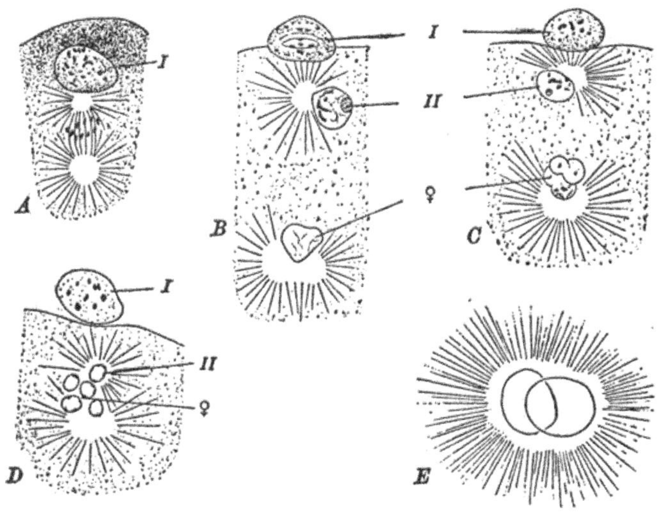

Fig. 84.—Maturation in the parthenogenetic egg of the Echinoderm, *Astropecten*. After O. Hertwig. *A*. First polar body formed but not extruded; second polar division in early anaphase. *B*. First polar body extruded; second polar division completed, the polar nucleus near the periphery. *C, D, E*. Stages in the gradual approach and fusion of the second polar nucleus and egg pronucleus, to form the cleavage nucleus. *I*. First polar body; *II*, second polar nucleus; ♀, egg pronucleus.

polar nucleus is never formed. The egg nucleus then re-forms with its $\frac{s}{2}$ chromosomes, but these are *bivalent* as shown by the character of the first maturation division, so that in effect the egg nucleus contains the somatic number of chromosomes, which actually appears in subsequent divisions. It is therefore clear that while such parthenogenetic eggs fail to receive a sperm nucleus they retain or receive back the equivalent of such a nucleus in the form of the second polar body nucleus, which is not lost as it is in eggs requiring to be fertilized.

The second unusual modification of maturation is to be seen in the spermatogenesis of many Arthropods, chiefly Insects. These species have already been mentioned as showing a numerical difference between the chromosome groups of the male and female individuals, the female having, in different species, one or several chromosomes more than the male. In these forms the first maturation division is typical and the two secondary spermatocytes are similar. But the second maturation division is asymmetrical in that one or more chromosomes known as the *accessory* or *idiochromosomes* fail to divide and are therefore distributed to only one-half of the spermatids and spermatozoa. Half of the sperm cells then have $\frac{s}{2}$ chromosomes, the other half $\frac{s}{2}$ plus one, or more as the case may be in different species. These striking phenomena and their relation to the question of sex determination are described more fully in Chapter VII.

In conclusion we should mention briefly the place of the maturation divisions in the life histories of different organisms. In any many-celled organism the life cycle as a whole may be said to consist of two phases, one characterized by the possession of the diploid chromosome group, the other by the haploid group. Among all of the Metazoa, and many of the Protozoa, there are invariably only two cell generations with the haploid number, and further, these always are the two final generations in the process of gametogenesis. Here they seem bound up with the process of fertilization and are to be understood only from the point of view of what is involved in this process. Considering these forms alone it is difficult to understand how it should have come about that numerical reduction of the chromosomes should occur in advance of the condition out of which arose the necessity for reduction, namely, the fusion of the germ nuclei. But this arrangement is by no means invariable. In *Amœba diploidea* (Hartmann and Nagler) reduction does actually occur *after* conjugation. And in some of the lower plants, such as many of the green Algæ (Chlorophyceæ), the relation between fertilization and numerical reduction is that which apparently must have been the more primitive. In these forms the gametic nuclei contain the same number of chromosomes (s) as do the somatic or vegetative cells; these fuse forming a zygote with double this number ($2s$). This fusion is then followed immediately by two maturation divisions, the first of which is usually heterotypic, which result in the formation of four cells, each again with the original vegetative number (s). Certain or all of these four cells then produce the body of the new organism, all the cells of which, including the germ cells when these form, have this same chromosome number (s). That is, numerical reduction of the chromosomes *follows* syngamy, a relation which seems more understandable than the more common precedence of reduction. In describing cases like these

we might say that the somatic or vegetative cells and the germ cells all
have the haploid chromosome group.   In fertilization the diploid group
is formed, but is then retained through only two generations, after which
the haploid condition is restored.   In other words the predominating
stage in the life cycle is that with the haploid chromosome group, the
diploid group occurring only in the divisions following fertilization;
"haploid" is here synonymous with "somatic."

In many other plants, such as the ferns and mosses among others, the
life cycle is more equally divided into two distinct periods, one carried
on with the haploid (in the usual sense of the word), one with the diploid
chromosome group.   The cells of the fern while it is in the typical
"fern-plant" stage have the diploid group, but during the formation
of spores by this plant, reduction occurs, the reduced number appearing
in the spore mother cell.   And in all of the cells of the prothallus,
derived from the spore, the haploid number remains; no further reduction occurs when the prothallus forms gametes.   The diploid number is
only restored by the union of two gametes in the formation of the new
fern plant, throughout the existence of which it is retained.   It is a
matter of considerable theoretical interest that the familiar alternation
between the sporophyte and gametophyte generations, between fern
plant and prothallus, for example, should be accompanied by a corresponding alternation between the diploid and haploid chromosome
groups.   We may relate this to the condition in the green Algæ by saying
that the number of cell generations following fertilization, in which the
diploid chromosomes are retained, is greatly increased and the number
with the haploid group correspondingly diminished, indeed in most
cases here, the diploid stage is of greater duration than the haploid.   In
the higher plants (Gymnosperms and Angiosperms) it is agreed that the
prothallus, *i.e.*, the stage with the haploid chromosome group, is represented only by certain vestiges—the pollen tube and embryo sac, and it
is significant that here, after the two maturation divisions leading to the
formation of the germ cells, two or more (but never more than a few)
additional divisions occur giving rise to these vestiges; the haploid
chromosome number is found in all of these divisions.   Here then the
phase with the reduced number of chromosomes is still more limited—
practically to the extent found in animals.   And whereas in the lower
plants the diploid stage is restricted to two cell generations, in the higher
plants it is the haploid stage which comes to be so limited.

Many consider the gametophyte generation, *i.e.*, the prothallus, or
its equivalent in other forms, as the primary form or phase; consequently
they regard the number of chromosomes in the cells of this phase, the
haploid number, as primitive or normal, and not as a reduced number.
Correspondingly the diploid number would result from a *doubling*, not
from a restoring of the normal.   The development of this point of view

in connection with the conditions in the higher plants (Strasburger) has led to the suggestion (Whitman) that even in animals the number of chromosomes in the secondary oö- and spermatocytes and mature germ cells, *i.e.*, the haploid number, is again in reality the normal, that this is doubled in fertilization, and remains doubled throughout the somatic divisions, only to be again reduced to normal by the subsequent maturation divisions. Upon this hypothesis, which also explains the present precedent relation of maturation to fertilization, the two cell generations immediately preceding fertilization are all that remain of the primary phase of the animal life cycle.

The alternation between the sporophyte and gametophyte in ferns and mosses is truly an alternation of generations and we may thus see an alternation of generations even in the higher plants where there may be a total of only four divisions with the haploid number—the normal according to some. If this is allowed, it is possible that in animals where there are but two of these corresponding cell divisions, we might still speak of an alternation of generations as well; the equivalent of the gametophyte would then be represented, vestigially, only by the cells with the haploid chromosome group, *i.e.*, primary and secondary oö- and spermatocytes which form the gametes proper—and the equivalent of the sporophyte generation would be represented by all the remaining generations of cells which we commonly think of as the true organism, and which forms "asexually" the oö- and spermatogonia—the equivalents then of the spore mother cells. Of course in animals the matter is complicated greatly by the separation of the two sexes as two separate individuals. Such a comparison as this must remain, at least for the present, as an interesting speculation merely, for none of the Metazoa offers any variations in the maturation process which shed any light upon the comparison.

REFERENCES TO LITERATURE

AGAR, W. E., The Spermatogenesis of *Lepidosiren paradoxa*. Q. J. M. S. **57**. 1911.

VAN BENEDEN, E. (See ref. Ch. II.)

BOVERI, T., Zellenstudien I. 1887. (See ref. Ch. III.)

BRAUER, A., Zur Kenntniss der Spermatogenese von *Ascaris megalocephala*. Arch. mikr. Anat. **42**. 1893. Zur Kenntniss der Reifung des parthenogenetisch sich entwickelnden Eies von *Artemia salina*. Arch. mikr. Anat. **43**. 1894.

CALKINS, G. N., and CULL, S. W. (See ref. Ch. II.)

CARDIFF, I. D., A Study of Synapsis and Reduction. Bull. Torrey Botan. Club. **33**. 1906.

COE, W. R., The Maturation and Fertilization of the Egg of *Cerebratulus*. Zool. Jahrb. **12**. 1899.

Davis, B. M., Nuclear Phenomena of Sexual Reproduction in Algæ. Amer. Nat. **44.** 1910.

Farmer, J. B., and Moore, J. E. S., On the Maiotic Phase (Reduction Divisions) in Animals and Plants. Q. J. M. S. **48.** 1904.

Fick, R., Ueber die Vererbungssubstance. Arch. Anat. u. Physiol. (Anat. Abth.). 1907.

Gates, R. R., The Mode of Chromosome Reduction. Botan. Gaz. **51.** 1911.

Grègoire, V., Les Cinèses de maturation dans les deux règnes. L'unité essentielle du processus méiotique. Cellule. **26.** 1910.

Hartmann, M., und Nagler, K., Copulation der *Amœba diploidea,* n. sp., *etc.* Sitz.-Ber. Ges. Nat. Freunde. Berlin. **4.** 1908.

Hertwig, O., Beiträge zur Kenntniss der Bildung, Befruchtung und Teilung des Tierischen Eies. I. Morph. Jahrb. **1.** 1875.

Jordan, H. E., The Spermatogenesis of the Opossum (*Didelphys virginiana*) with special Reference to the Accessory Chromosome and the Chondriosomes. Arch. Zellf. **7.** 1911.

Korschelt und Heider, Lehrbuch, *etc.* (See ref. Ch. III.)

McClung, C. E., The Chromosome Complex of Orthopteran Spermatocytes. Biol. Bull. **9.** 1905.

Montgomery, T. H., Jr., The Heterotypic Maturation Mitosis in Amphibia and its General Significance. Biol. Bull. **4.** 1903. (See also ref. Ch. II, III.)

Morse, M. W., The Nuclear Components of the Sex Cells of four Species of Cockroaches. Arch. Zellf. **3.** 1909.

Platner, G., (See ref. Ch. III.)

Schaffner, J. H., Synapsis and Synizesis. Ohio Naturalist. **7.** 1907.

Schönfeld, H., La Spermatogénèse chez le taureau et chez le mammifères en général. Arch. Biol. **18.** 1901.

de Sinéty, R., Recherches sur la Biologie et l'anatomie des Phasmes. Cellule. **19.** 1901.

Strasburger, E., Ueber periodische Reduktion der Chromosomenzahle im Entwicklungsgang der Organismen. Biol. Cent. **14.** 1894. English Translation in Annals Botany. **8.** 1895. Ueber Reduktionsteilung. Sitz.-Ber. Akad. Wiss. Berlin. 1904.

Wilson, E. B., (See ref. Ch. II.)

Winiwarter, H. v., Recherches sur l'ovogenèse et organogenèse de l'ovaire des Mammifères. (Lapin et Homme.) Arch. Biol. **17.** 1900. (See also Anat. Anz. **21.** 1902.)

Winiwarter, H. v., and Sainmont, G., Nouvelles recherches sur l'ovogenèse et l'organogenèse de l'ovaire des Mammifères. Arch. Biol. **24.** 1909.

CHAPTER V

FERTILIZATION

The complex processes of the formation, differentiation, and maturation of the germ cells, described in the two preceding chapters, are to be regarded as preliminaries to the process of fertilization. They can be understood only as preparatory steps leading to the final meeting of an ovum and a spermatozoön, and their fusion into a single cell. The cell thus formed is a "new" organism, which immediately commences a long series of reactions, collectively termed development, leading finally to the establishment of a form resembling that of the individuals from which the fusing germ cells were themselves derived.

Among animals, with the probable exception of a few of the simplest, an almost invariable condition of the continued existence of any specific form of protoplasm is such a periodic mingling of the living substance of two individuals of the same species. The few exceptions among the Metazoa are found in the rare self-fertilizing hermaphroditic creatures, and even here the mingled plasms may be said to have had somewhat separate histories, although formed within the body of a single organism. Animals which reproduce parthenogenetically, or by such methods as budding or fission, sooner or later in their life history exhibit these processes of germ cell formation and fusion.

Among the plants, on the other hand, while the union of germ cells may be a frequent, and in some cases a necessary preliminary to the formation of a new organism, yet other tissues and living masses composed of several kinds of tissue, taken from almost any part of the organism, may give rise to a new individual. Many species of plants are thus normally propagated by cuttings of leaf, stem, or root, or by runners, buds, *etc.* This is particularly true of the Begonias, where even a few cells, from almost any part of the growing plant, may be removed and, under proper conditions, be made to form a new complete organism capable of producing typical germ cells.

The questions why fertilization should be necessary, and how fertilization actually accomplishes the results which obviously follow it, are not easy to answer. But the essential facts regarding the process of syngamic fusion, and the visible results of it, are clear. We shall, therefore, confine our atten tion first to these; then, having described the phenomena of fertilization we may consider briefly some of the more theoret ical aspects of the process.

The word "fertilization" is a general, inclusive term, used to denote all of the various phenomena concerned in the meeting and fusion of the germ cells (gametes) or germ plasms, and even some of the results of such a fusion. The simpler fact of the mere fusion of gametes is more precisely termed *syngamy.* In all of the Metazoa syngamy may be defined as the meeting of two completely specialized, unicellular gametes, an ovum and a spermatozoön, derived in most cases from two individuals, and their subsequent fusion, nucleus with nucleus, and cytoplasm with cytoplasm, into a single, uninucleate cell, the zygote. This definition is not completely applicable to the unicellular organisms, for in these the gametes are usually not completely specialized, sometimes indeed not especially differentiated at all. Such forms may offer some suggestions as to the history and significance of the fertilization process, and we shall return to consider this subject later in this chapter.

We have seen in Chapter III some of the methods by which it is ensured that eggs and sperm shall be brought into the same general region or into fairly close proximity, but it remains to be seen how the ovum is actually encountered by the sperm cell. Taken altogether, the processes leading to this result often become very complicated and special, and in most species the probability is very high that practically every normal egg produced will be fertilized. The gametes are completely specialized cells, and if they cannot conjugate and develop, they soon perish, not able even to remain alive long, except in a few special instances. Spermatozoa discharged freely into the water, as in external fertilization, are usually able to remain active only a few minutes or hours. But when fertilization is internal and the spermatozoa are received into some reproductive cavity of the female, or into some storage cavity, they may remain alive and able to function under appropriate conditions for a much longer period: various observers give the following specific instances: dog and rabbit, eight days; man, seven to twenty days; fowl, twenty days; the bats and some snakes, from autumn to the following spring; *Salamandra maculosa*, from summer to the following spring; snails, three years; bees, four to five years.

Whatever the particular circumstances connected with and ensuring the meeting of sperm and ovum, the medium in which it occurs is a fluid. In this the sperm cells are in active, though apparently random, movement, due to rapid vibration of the flagellum or tail. In many instances their movement follows a spiral path (Buller, Adolphi), either close or open, such as is common among flagellated unicellular organisms. In a few rare instances (some Crustacea) the spermatozoa are amœboid.

Fertilization becomes more likely when direction is given to the active random movements of the sperm cells. In some instances where fertilization is internal, the movements of the sperm seem to be directed by ciliary currents of the oviduct or other passage. Spermatozoa tend to swim against such a current, and thus to ascend toward the eggs which are being carried down the passage toward the exterior (Lott). According to the interesting observations of Lott, human spermatozoa swim at the rate of 27 mm. (*i.e.*, about 550 times their own length) in 7.5 minutes. At the corresponding rate of progression a man 5.8 feet in stature would walk a mile in 12.4 minutes.

There appear to be two chief methods by which spermatozoa are finally brought to the surface of the ovum. In some few forms the egg is said to give off a chemical substance to which the active sperm are attracted; when the random movements of the sperm bring them within the sphere of chemical influence of the egg, their movements immediately become directed toward the unfertilized egg. Among some of the lower plants it is known that weak solutions of malic acid and its compounds attract spermatozoids; in others, solutions of cane sugar act similarly (Pfeffer). It is at present doubtful, however, whether in many animal eggs the control is also of a chemical nature (Buller). In some forms the stimulus is certainly not of a chemical sort, but is a contact stimulus. The sperm of many fishes, for example, swim at random until they touch some solid object, egg or other body, and from this they are apparently unable to escape. According to the observations of Drago the collection of the spermatozoa about an ovum is unusual, and when it occurs, it is the result of agglutination. It should be said that in most cases it is doubtful whether the movements of the sperm really are given direction toward the ovum, and what the nature of the stimulus may be, when such is the case. As we have seen, the contact between sperm and egg results chiefly from the large numbers of sperm produced, and from their general proximity to the egg resulting from special habits of spawning, copulation, *etc.*

When the ovum is naked, or surrounded by only a thin vitelline membrane, the sperm apparently may enter at almost any point on the surface of the egg. This is true of many forms among the Medusæ, Turbellaria, Nemertea, Annelida, Echinodermata, Gasteropoda, Cephalochorda, Amphibia, and Mammalia. Entrance is usually said to be effected by the active swimming movements of the spermatozoön, which force the sharp acrosome, adapted to this purpose, through the limiting surface of the ovum, into its superficial cytoplasm. But here again extended evidence is lacking, and in many forms the egg is known not to be a wholly passive recipient of the sperm, but to take a considerable share in accomplishing its entrance. Thus in the sea-urchin, when a sperm head approaches the egg closely, the superficial cytoplasm, at the point nearest the sperm, is elevated into a small cone or papilla called the *attraction cone* (Wilson). This not only rises to meet the spermatozoön, but seems to aid in drawing it into the egg (Fig. 85). In some instances (*e.g.*, *Julus*) this attraction cone may be quite high and may contain a part of the chromatic substance of the egg nucleus. According to the observations of Lillie, the spermatozoön of *Nereis* is clearly drawn into the egg through the activity of the latter, the sperm itself taking no active part in the process (Fig. 86).

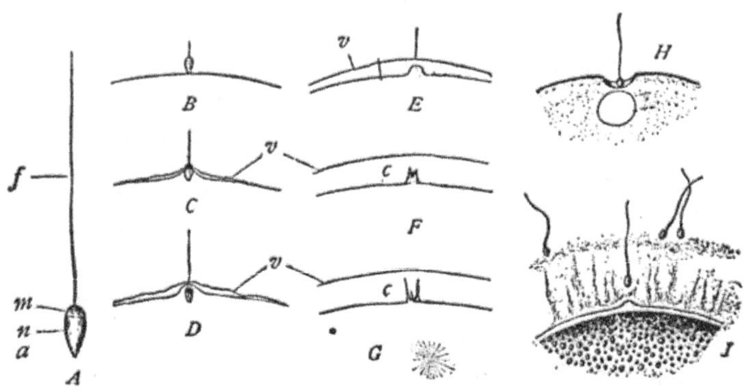

Fig. 85.—Entrance of the spermatozoön into the egg. From Wilson, "Cell," *H*, after Metschnikoff; *I*, after Fol. *A*. Spermatozoön of *Toxopneustes*, × 2000; *a*, the apical body; *n*, nucleus; *m*, middle-piece; *f*, flagellum. *B*. Contact with the egg-periphery. *C,D*. Entrance of the head, formation of the entrance-cone and of the vitelline membrane (*v*), leaving the tail outside. In some other Echinoderms, the tail may enter the ovum. *E,F*. Later stages. *G*. Appearance of the sperm-aster (*s*) about three to five minutes after first contact; entrance-cone breaking up. *H*. Entrance of the spermatozoon into a preformed depression. *I*. Approach of the spermatozoön, showing the attraction-cone.

When the egg is surrounded by membranes of some thickness or density, the spermatozoa are usually unable to penetrate them and the only path of entrance is then through the micropyle, the existence of which is an adaptation for this event. There is apparently no agent directing the sperm toward this perforation in the membranes; the finding of it is a matter of chance. The chance is not small, however, that some spermatozoön will enter the micropyle, for ordinarily the fluids around the egg are filled with a swarm of sperm cells. As already noted the micropyle is commonly at the animal pole of the egg, though at the vegetal pole in a few instances (some

Molluscs). Thus the region which is to receive the spermatozoön is already determined, and frequently the cytoplasm is considerably modified, in the region just beneath the micropyle, into a special substance concerned in the receipt of the sperm; this is known as the *entrance disc* (Fig. 86).

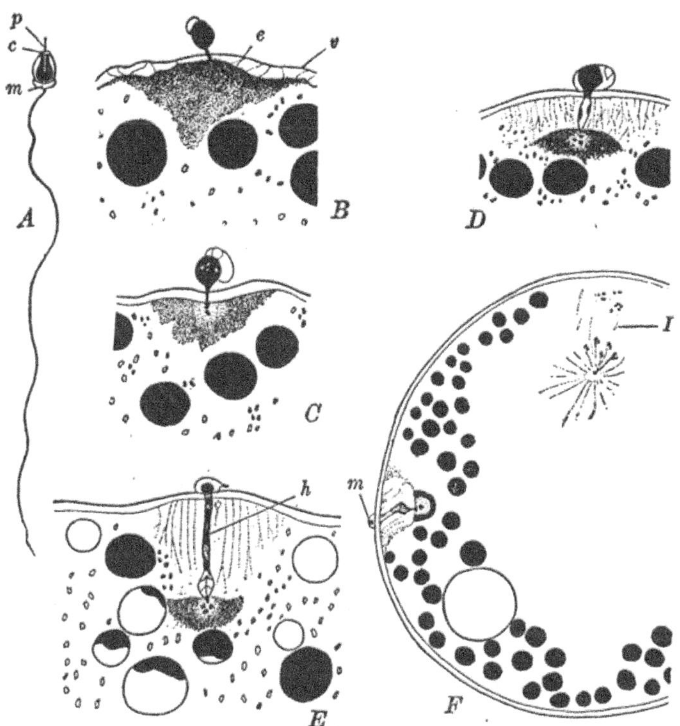

Fig. 86.—Entrance of the spermatozoön in the fertilization of the Annulate, *Nereis limbata*. After Lillie. *A.* Spermatozoön. *B.* Perforatorium has penetrated egg membrane; entrance cone well developed. Fifteen minutes after insemination. *C.* Thirty-seven minutes after insemination. *D.* Entrance cone sinking in and drawing the head of the spermatozoön after it. Forty-eight and one-half minutes after insemination. *E.* Head drawn in still further. Forty-eight and one-half minutes after insemination. *F.* Entrance completed. First maturation division in anaphase. Fifty-four minutes after insemination. The middle piece, as well as the tail, remains outside. *c*, head cap; *e*, entrance cone; *h*, head of spermatozoön (nucleus); *m*, middle piece; *p*, perforatorium; *v*, vitelline membrane; *I*, first polar division figure.

With but comparatively few exceptions, only one sperm cell normally enters a single egg (*monospermy*). This sperm is the first one to reach the egg or micropyle, and there are various methods of excluding additional sperm, and thus of preventing abnormal multiple fertilization. In some of the lower plants, after one sperm cell has entered, the egg gives off immediately a chemical substance which actually repels the other sperm congregated about the egg. A frequent method among animals is the secretion of an impenetrable membrane, or a layer of jelly, immediately upon the entrance of one spermatozoön. Or, if a membrane was previously present, its density may be suddenly increased, or an additional membrane formed

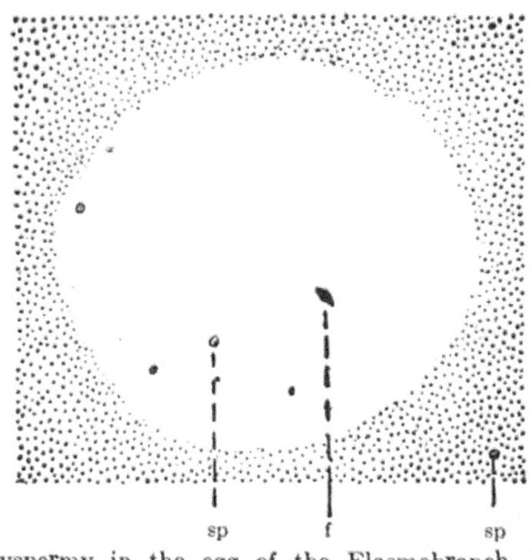

Fig. 87.—Polyspermy in the egg of the Elasmobranch, *Torpedo ocellata.* From Ziegler, after Rückert. Germ disc with first cleavage spindle, *f*, and accessory sperm nuclei, *sp*.

(Amphioxus), or the micropyle may be closed by the rapid swelling of the egg membranes. Although this process of membrane formation may really have this effect of excluding supernumerary spermatozoa, the general significance of the process renders it doubtful whether this is to be regarded primarily as an event adapted toward this end.

Occasionally two or more spermatozoa succeed in gaining entrance into the ovum (*polyspermy*). This ordinarily results in an abnormal course of development, which does not proceed very far before the egg ceases to develop and dies. A few forms, however, are adapted to the receipt of more than one sperm and polyspermy occurs normally (*physiological polyspermy*). Such eggs (Fig. 87) are usually yolk-filled, for example those of some Insects, *Petromyzon*, Selachians, Urodeles, Reptiles, Birds, and perhaps the toad and some Teleosts. Sometimes a few, sometimes many, sperm thus enter the ovum, but in any case only one of them ever takes any real part in the actual processes of fertilization. The others, known as *accessory spermatozoa*, may either remain quite inactive and soon degenerate, or they may give rise to "vegetative" nuclei, and perish after a brief period of activity. While active they seem chiefly to be concerned in the preparation of the yolk for ready absorption; they are then called *merocytes*. Rarely, if ever, do the nuclei derived from accessory spermatozoa contribute directly to the formation of any part of the embryo proper.

Apparently there is little specific adaptedness in the behavior of the germ cells such that an egg and a sperm of the same species tend to unite much more readily than do those of different species. With some eggs any spermotozoön that is *morphologically* capable of gaining entrance, can do so, apparently about as readily as the specific sperm. The limitations here are frequently due to the size of the sperm head as compared with the micropyle, or to the necessity for special perforating mechanisms or powerful swimming movements in order to penetrate the egg membranes, or the performance of appropriate reactions upon the part of the egg itself. As a rule the eggs and sperm of a single species unite, because, as the result of the breeding or spawning habits, only the sperm and ova of a single species are associated in time and space in any considerable numbers. When eggs are placed in a mixture of equal quantities of two or more kinds of sperm, there seems to be no appreciable *selective* fertilization, provided, as said above, that both or all kinds of the sperm are able to enter the egg at all.

The ease with which "foreign" sperm may enter an egg is affected in many instances by chemical treatment of the eggs and sperm; treatment with alkalies or with specific salts often renders penetration of the sperm readily possible in cases where normally it is difficult or impossible (Loeb, Godlewski).

And once within the ovum, a "foreign" sperm seems to act almost as efficiently as the proper sperm in *inaugurating* development. After the entrance of a foreign sperm the two germ nuclei may not, usually do not, fuse, and other internal developmental processes may not be entirely normal, but the external processes of cleavage and differentiation may proceed normally for some time, even to the formation of a free-swimming larva, as in many species of Echinoderm eggs fertilized by the sperm of other species, genera, or even of other classes of Echinoderms (Baltzer), or of other phyla (Mollusca, Kupelwieser).

In many species the entire spermatozoön enters the cytoplasm of the ovum (some of the Turbellaria, Annelids, Insects, Molluscs, and many Vertebrates) while in others the tail piece separates from the remainder of the sperm cell and is left outside of, or embedded within, the vitelline membrane, so that only the head and middle piece actually share in the formation of the zygote. In some instances the middle piece, too, fails to enter the egg (*e.g., Nereis*, Fig. 86). Once within the egg, the sperm continues its inward course for a short distance only. If the entire sperm cell has entered, one of the first events is a sharp bending or flexure between the tail and middle piece, often followed by a separation of the two, after which the tail piece is left behind as the remainder continues its migration (Fig. 88).

The entrance of the spermatozoön within the egg cytoplasm is the event which inaugurates a whole series of fertilization processes culminating in the formation of a typical mitotic division figure within the zygote. The precise character of the stimuli which start this chain of actions is still in doubt, but it seems likely that in many instances it acts, first, by bringing about the formation of a permeable membrane over the surface of the egg, through which may occur rapid and extensive osmotic interchanges leading to marked oxidations; and second, by reducing the amount of fluid in the egg cytoplasm, either actually, by its loss through the permeable membrane, or relatively by the addition of the much denser substance of the spermatozoön itself (Loeb). At any rate, whether it be a

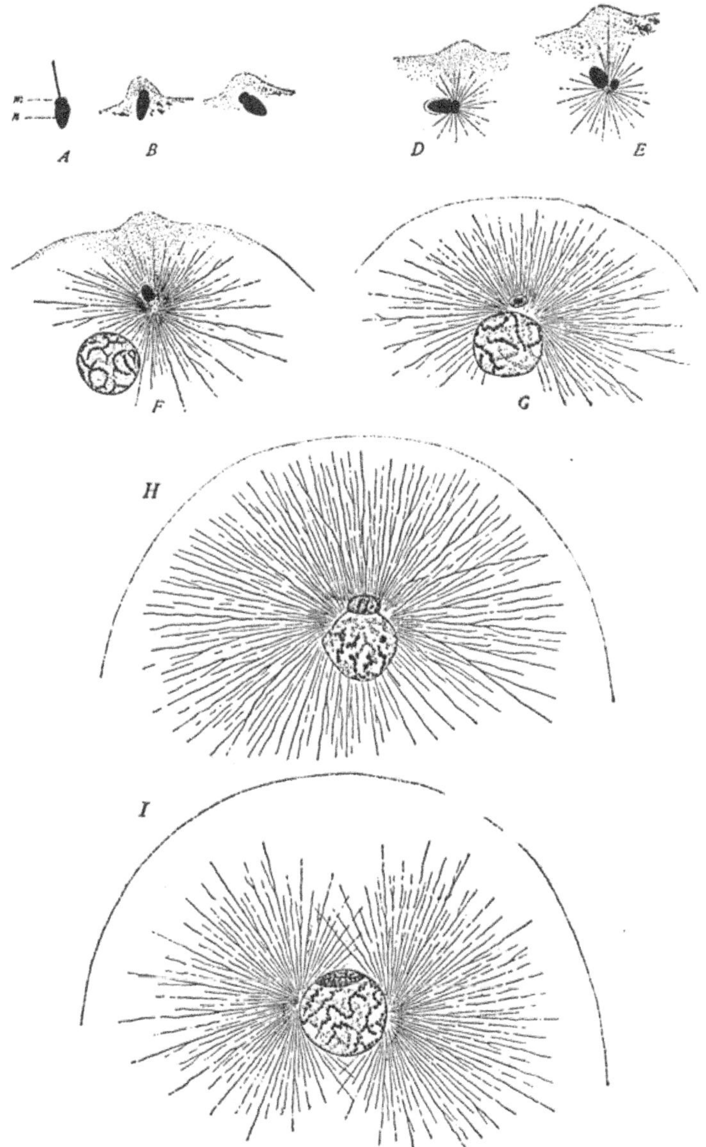

Fig. 88. Fertilization in the sea-urchin, *Toxopneustes.* From Wilson, "Cell."
A–F, × 1067; *G*, × 533; *H, I*, × 667. *A.* Sperm-head before entrance; *n*, nucleus;
m, middle-piece and part of the flagellum. *B,C.* Immediately after entrance,
showing entrance-cone. *D.* Rotation of the sperm-head, formation of the
sperm-aster about the middle-piece. *E.* Casting off of middle-piece; centrosome at focus of rays. *F,G.* Approach of the pronuclei; growth of the aster.
H. Union of pronuclei. *I.* Flattening of the sperm pronucleus against the egg
pronucleus; division of the aster.

primary or a secondary process, this loss of water following the entrance of the spermatozoön, appears as one of the important aspects of fertilization.

In the eggs of many species there is a peripheral layer of

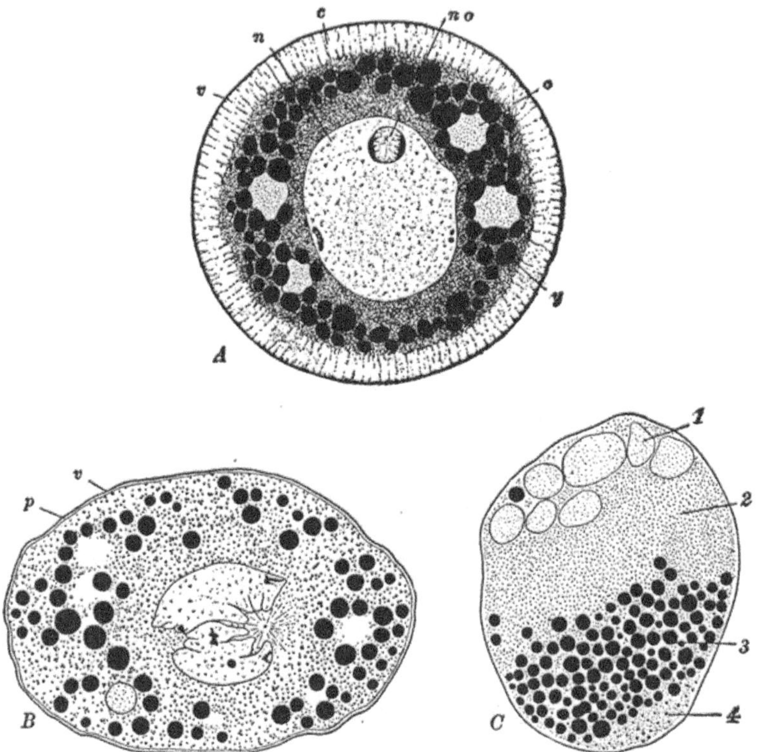

Fig. 89.—Changes in the structure of the ovum in *Nereis*, upon fertilization. After Lillie. *A*. Unfertilized oöcyte. Large germinal vesicle; cytoplasm contains oil drops and yolk spheres, and shows well marked cortical layer (exoplasm). *B*. Fifteen minutes after ensemination (the spermatozoön is not shown). The cortical layer has chiefly gone to form a jelly-like layer outside the ovum, and is not shown. *C*. Egg after centrifuging to show component substances. *c*, cortical layer (exoplasm); *n*, germinal vesicle or nucleus; *no*, nucleolus; *o*, oil drops; *p*, perivitelline space; *v*, vitelline membrane; *y*, yolk spheres; 1, layer of oil drops; 2, hyaline cytoplasm with small basophile granules; 3, yolk spheres; 4, hyaline cytoplasm with large basophile granules.

cytoplasm (exoplasm) which is comparatively clear, free from granules, and characterized by the presence of fluid vacuoles (Echinoderms, *Nereis* (Fig. 89), Amphioxus (Fig. 90), Teleosts; see Chapter III). Entrance of the spermatozoön leads to the breaking down of these vacuoles and the discharge of their substance from the surface of the ovum (Figs. 89, 90). This substance may be in part transformed into, or may carry before it a modified surface layer of material which appears then

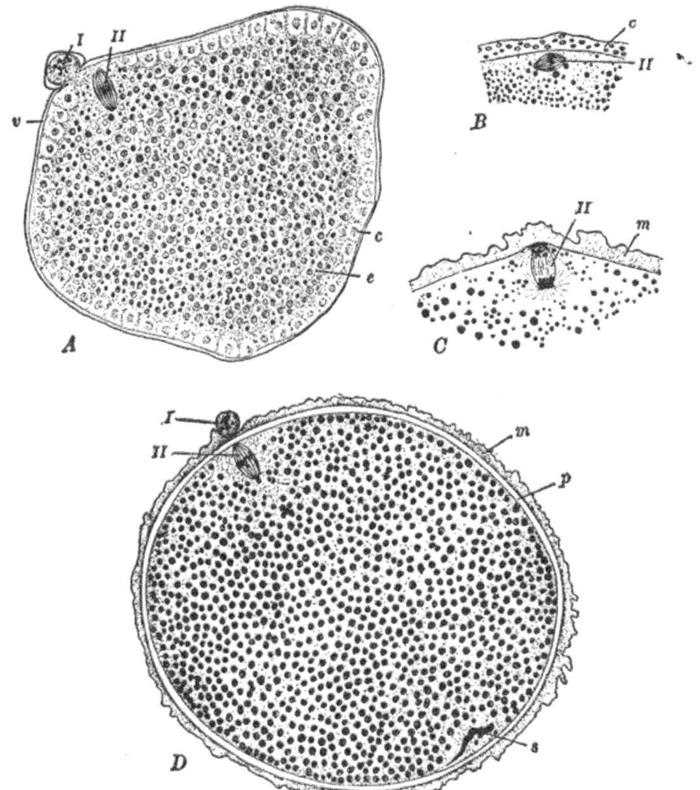

FIG. 90.—Fertilization in the egg of Amphioxus. *C*, after Cerfontaine, others after Sobotta. *A*. Ovarian egg showing cortical plasm. *B*. Cortical layer forming a membrane on the surface of the egg, within the vitelline membrane. *C*. Egg membrane fully formed but still attached to surface of egg. *D*. Extruded, fertilized egg. Membrane fully formed and beginning to leave the surface of the egg. *c*. Cortical layer; *e*, endoplasm; *m*, egg membrane; externally vitelline, internally a product of the exoplasm; *p*, perivitelline space; *s*, spermatozoön; *v*, vitelline membrane; *I*, first polar body; *II*, second polar spindle.

as a *fertilization membrane;* this may be the vitelline membrane or it may be an addition to a previously present vitelline membrane (Fig. 90). Then either by the shrinkage of the egg, or the expansion of the membrane, or both, or by the rapid absorption of water by the substance between the egg and the membrane, a space of widely varying dimensions in different species, is left between the egg and its membranes; this is the *perivitelline space* (Fig. 90).   Within this space the egg is free to

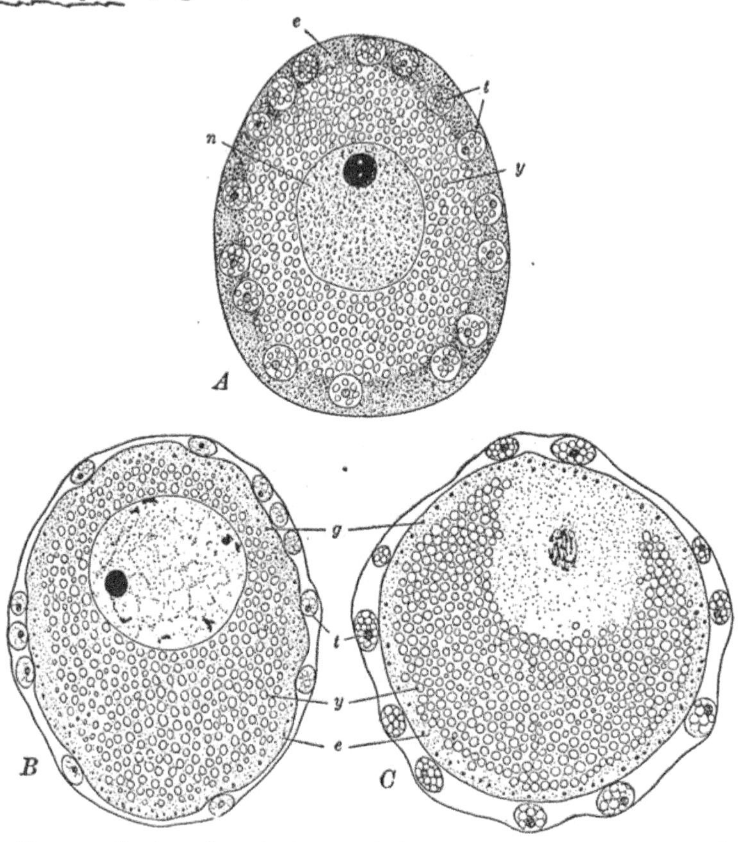

Fig. 91.—Sections through the egg of the Tunicate, *Cynthia partita*. After Conklin. × about 350. *A*. Ovarian egg fully formed. Germinal vesicle surrounded by yolk bodies; peripheral layer of protoplasm containing test cells and yellow granules (small circles). *B*. After extrusion of the test cells. Nuclear membrane still intact with chromosomes at periphery of nucleus (germinal vesicle). *C*. After laying (before fertilization, the egg remains in this condition until fertilized). Chromosomes and granular substance, from which the spindle is formed, lie in the center of the karyoplasm, now free in the cell. *e*, exoplasm or cortical layer; *g*, granules of yellow pigment; *n*, egg nucleus or germinal vesicle; *t*, nuclei of ingested test cells or follicle cells; *y*, yolk.

move or rotate, although the superficial membrane may be fixed to some foreign body.

Frequently this phenomenon of membrane formation is but

one phase of a general physical and chemical reorganization of the whole substance of the egg, following the entrance of the spermatozoön. The egg may exhibit more or less amœboid movement, or waves of contraction may pass over it. A frequent result is seen in the rapid streaming of differentiated cytoplasmic substances into certain regions, where these specific substances collect. Thus in many yolk-filled eggs like those of the Teleosts, the protoplasm, which before the entrance of the sperm is quite uniformly distributed over the surface of the egg as a very thin layer, now collects at the animal pole into a thick and fairly circumscribed disc called the _germ disc_ (Fig. 48). The Ascidian egg, as described by Conklin, offers one of the most marked examples of this rapid transformation and redistribution of the substances of the egg cytoplasm (Figs. 91, 92). In the secondary oöcyte of _Cynthia_ (_Styela_) the greater part of the cell is composed of a gray "endoplasm"; superficially there is a thin but complete layer of yellowish "mesoplasm"; while the large nucleus or germinal vesicle contains a clear "ectoplasm." During maturation, which here precedes sperm entrance, the ectoplasm collects at the upper pole of the oöcyte. Immediately upon entrance of the sperm the yellow mesoplasm streams from all directions toward the lower pole; this is followed by the clear ectoplasm which forms a stratum just above the mesoplasm, and leaves the upper half or more of the egg cytoplasm composed entirely of the gray endoplasm. Then this radial or rotatory symmetry gives place to a bilateral symmetry, for the mesoplasm and ectoplasm move up on one side (the posterior) of the egg, appearing on the surface in the form of a crescent just below the equator. Meanwhile the yellow mesoplasm and gray endoplasm have each become differentiated into two distinct substances, so that altogether five forms of protoplasm are distinguishable in the cytoplasm of the zygote (Fig. 92).

Very few eggs exhibit such marked differentiation as this, but the corresponding phenomena are of widespread occurrence, and it is quite likely that they have frequently been overlooked because the various substances are not often marked by

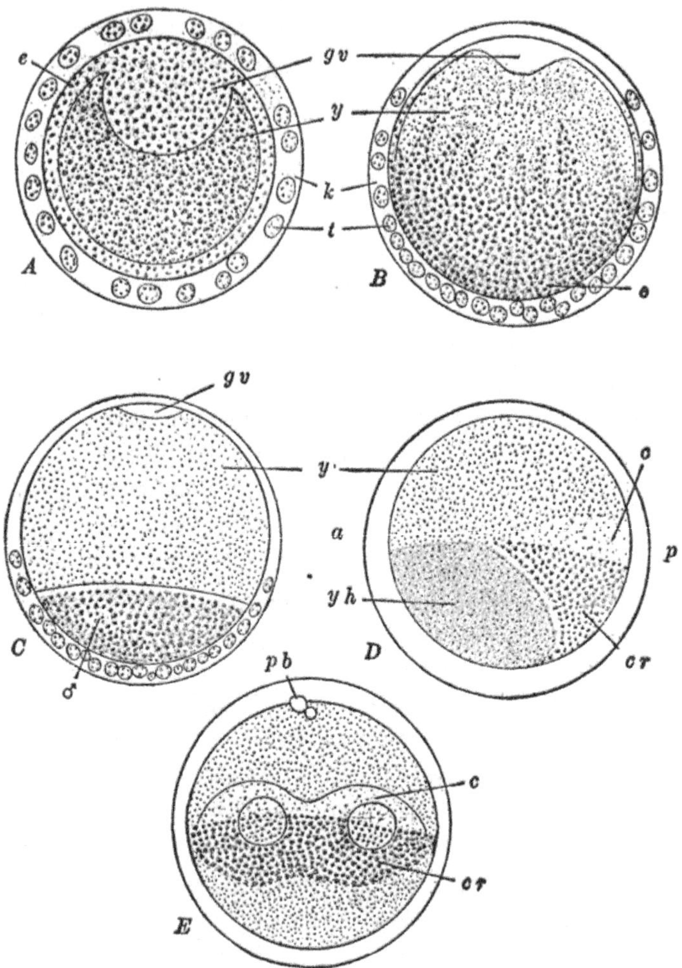

Fig. 92.—Total views of the egg of the Tunicate *Cynthia partita*, showing the changes in the arrangement of the materials of the egg subsequent to fertilization. After Conklin. × 200. *A.* Unfertilized egg, before the fading out of the germinal vesicle. Centrally is the mass of *gray* yolk; peripherally is the protoplasmic layer with yellow pigment, and surrounding the egg, the test cells and chorion. *B.* About five minutes after fertilization, showing the streaming of the superficial layer of protoplasm toward the lower pole, where the spermatozoön enters, and the consequent exposure of the gray yolk of the upper hemisphere. The test cells are also carried toward the lower pole. *C.* Side view of egg showing the yellow protoplasm at the lower pole; at the upper pole a small clear region where the polar bodies are forming. The location of the sperm pronucleus is also indicated. *D.* Side view of egg shortly before the first cleavage, showing the posterior collection of the pigmented protoplasm (yellow crescent) and the clearer area above it. *E.* Posterior view of egg during the first cleavage, showing its

characters so easily observed in the living egg. It is also note-worthy that frequently a radial or rotatory symmetry of the egg is changed to a bilateral symmetry by the entrance of the spermatozoön, and that usually the position of the plane of bilateral symmetry is determined by the point at which the sperm enters or by the path which the sperm takes through the cytoplasm. And further this new plane of symmetry of the zygote coincides closely with the plane of the first division of the zygote and with the median plane of the embryo and adult.

These aspects of organization and reorganization of the egg are among the highly important aspects of development, and largely determine many of the phenomena of subsequent differentiation. They also illustrate the statement made in the introductory chapter, that some of the most important aspects of development are *intra*-cellular processes. We shall return to this subject in a later chapter (Chapter VII).

After the entrance of the spermatozoön and the consequent redisposition of the substances of the cytoplasm, the course of the immediately subsequent events is determined largely by the state of the egg nucleus as regards its maturation process. For in most cases maturation proceeds only to a certain point, varying greatly in different forms, when it pauses, and is completed only after receiving the stimulus caused by the entering sperm. Ordinarily the state of the egg is such that sperm can gain admission only when this pause has been reached.

In some cases, such as *Nereis* (Fig. 86), *Ascaris* (Fig. 94), *Cerebratulus* (Fig. 95), and many Molluscs, it is true that the sperm does enter the ovum before the maturation process has even begun—while the egg nucleus (germinal vesicle) is still in the final resting stage preceding the first oöcyte division. In other cases—most Molluscs, *Thalassema*, *Sagitta*, Teleosts, the first polar spindle has formed and the first maturation division may have reached the metaphase or anaphase, when

---

relation to the symmetry of the egg. *a*, anterior; *c*, clear protoplasm; *cr*, yellow crescent; *e*, exoplasm or cortical layer, with yellow pigment; *g.v*, germinal vesicle *k*, chorion; *p*, posterior; *p.b*, polar bodies; *t*, test cells; *y*, yolk (central gray material); *y.h*, yellow hemisphere; ♂, sperm pronucleus.

it pauses to await the sperm entrance. In *Sycandra*, *Lepas*, Amphioxus (Fig. 90), and many Amphibia and Mammalia the first polar division is completed and the second polar spindle formed before the pause. And finally in some Cœlenterata and most Echinodermata and Ascidians, the maturation of the egg is entirely completed before the entrance of the spermatozoön. In such cases as these, contact with sea water seems

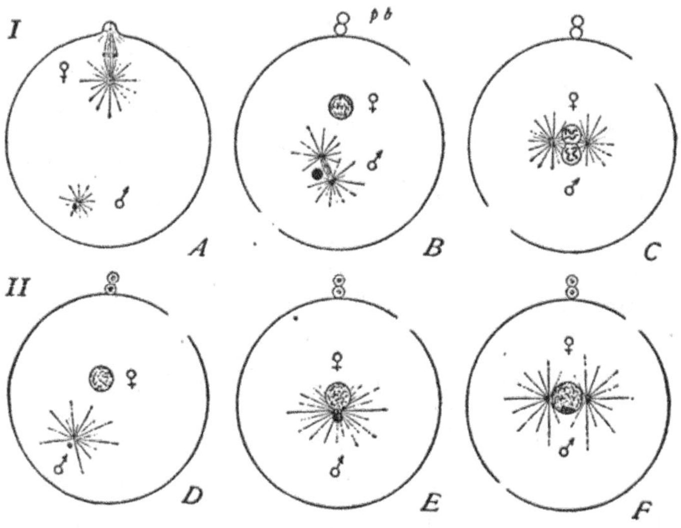

Fig. 93.—Diagrams of the two most frequent relations between the events of maturation and fertilization. From Wilson, "Cell." *I.* Polar bodies formed after the entrance of the spermatozoön (Annulates, Molluscs, Turbellaria). *II.* Polar bodies formed before the entrance of the spermatozoön (Echinoderms). *A.* Sperm pronucleus and centrosome at ♂, first polar body forming at ♀. *B.* Polar bodies formed; approach of the pronuclei. *C.* Union of the pronuclei. *D.* Approach of the pronuclei. *E.* Union of pronuclei. *F.* Cleavage nucleus.

to furnish the stimulus to complete the maturation process, which may be begun before the eggs are produced.

We may distinguish two general types of behavior on the part of the germ nuclei according to whether maturation has or has not been completed at the entrance of the sperm (Fig. 93). We shall consider first, and at greater length, those cases in which maturation is not yet completed, for this would seem the more usual course of events. Otherwise maturation occurs precociously, apparently before the necessity for it has arisen.

We left the sperm head and middle piece lying a short distance below the surface of the egg. We may disregard the tail piece now, for even in those cases in which it enters the egg it is left behind the head and takes no active part in subsequent processes. The head and middle piece now move more slowly, along a path which is, for a short distance at any rate, a radius of the ovum. Then they separate slightly, and the two rotate through approximately 180°, so that the middle piece is placed in advance of the head (Figs. 94, 95). There ensues a considerable metamorphosis of these elements. The sperm head loses its sharp outline and gradually enlarges; its outline soon becomes very irregular and indistinct, and vacuoles appear. Soon it has expanded into an organ of considerable size and has acquired a typical nuclear structure with linin network, chromatin granules, and nuclear membrane. In the meantime the middle piece has undergone an even more extensive transformation (Fig. 94). Before the sperm halts in its inward progress, even before the rotation in some cases, the middle piece has begun to dissolve and in connection with it appears a centrosome, surrounding which a small aster appears. During the pause of the sperm nucleus, the centrosome and aster each divide into two, and the daughters diverge slightly while the asters grow somewhat larger.

It will be remembered that during the metamorphosis of the spermatid into the spermatozoön, one or both of the centrosomes of the former either passed into the middle piece, or actually formed the larger part of it, although frequently no centrosome is actually visible in the middle piece of the fully formed spermatozoön. When the centrosome appears in the egg, in connection with the middle piece of the entering spermatozoön, it is possible, though not likely, that this is really the same centrosome that was present in the spermatid, and that there is consequently a genetic continuity of centrosomes, from generation to generation, as well as of nuclear components. Such a continuity has, however, not been definitely observed. On the other hand, it may be that the middle piece forms, either from its own substance, or from that of the egg, a new centro-

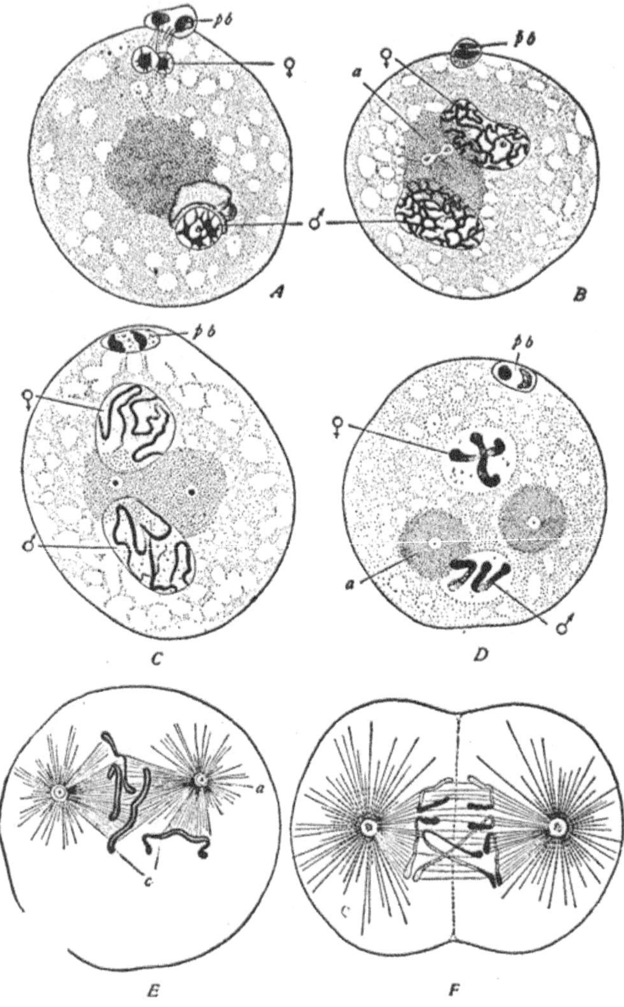

Fig. 94.—Fertilization in *Ascaris megalocephala bivalens*. From Wilson, "Cell," after Boveri. (Later stages are shown in Fig. 36.) *A*. The spermatozoön has entered the egg, its nucleus is shown at ♂; beside it lies the granular mass of "archoplasm" (attraction-sphere); above are the closing phases in the formation of the second polar body (two chromosomes in each nucleus). *B*. Germ-nuclei (♀, ♂) in the reticular stage; the attraction-sphere (*a*) contains the dividing centrosome. *C*. Chromosomes forming in the germ-nuclei; the centrosome divided. *D*. Each germ-nucleus resolved into two chromosomes; attraction-sphere (*a*) double. *E*. Mitotic figure forming for the first cleavage; the chromosomes (*c*) already split. *F*. First cleavage in progress, showing divergence of the daughter-chromosomes toward the spindle-poles (only three chromosomes shown).

some. The aster doubtless is formed out of the egg cytoplasm, by the influence of the centrosome or centrosome-forming substance of the spermatozoön; and it is quite possible that the centrosome itself may be similarly formed from the cytoplasm through the action of some chemical substance introduced by the sperm. For it is known that the cytoplasm of the egg does possess the property of forming asters with typical centrosomes, under the influence of appropriate "artificial" stimulus (Yatsu). And recently Lillie has shown, in *Nereis*, where the middle piece does not enter the egg at all, that the centrosome forms in association with the sperm nucleus, even when only a small portion of this is allowed to enter the egg. This indicates that the centrosome, as well as the aster, results from the redisposition of substances of the egg cytoplasm following the entrance of the spermatozoön. A conclusion as to whether or not the law of genetic continuity applies to the centrosome in fertilization is less important than recognition of the uniformity of its chemical and physical actions, in either case. And although the centrosome as an organized body may disappear in the spermatozoön, this still contains kinoplasmic substance of an equivalent function. Here, as elsewhere, the essential continuity may be chemical rather than morphological, but for that reason it is not to be regarded as any less actual or important.

While the spermatic structures have been thus active, the egg nucleus has been completing its maturation, at the conclusion of which the egg centrosomes and asters have disappeared (Figs. 94, 95). The egg nucleus is left near the surface of the animal pole, either near the sperm nucleus or at some distance from it. There are now present in the ovum all of the chief elements which are to take part in the essentials of fertilization and development. These are (1) the egg nucleus with its $\frac{s}{2}$ chromosomes, either distinct or formed into a characteristic nuclear reticulum, and with or without a nuclear membrane; (2) the sperm nucleus, also known to contain $\frac{s}{2}$

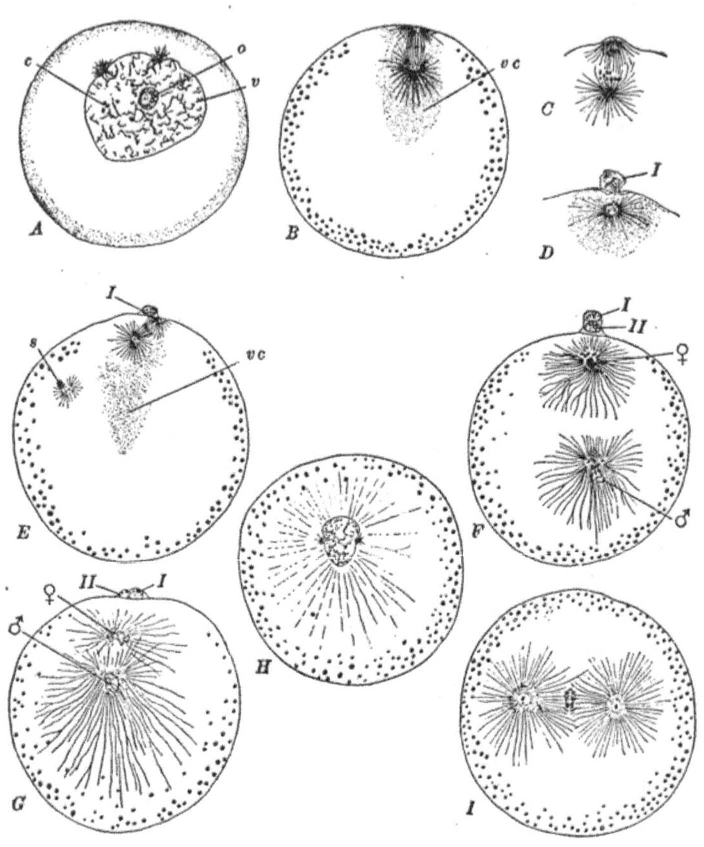

Fig. 95.—Fertilization in the Nemertean, *Cerebratulus*. After Coe. *C*, *D*, × 375, others × 250. *A*. Primary oöcyte. Part of the chromatin has been condensed into chromosomes, only five of which are shown (the number present is sixteen). The remainder of the chromatin is thrown out into the cytoplasm. The centrosomes, each with a small aster, are diverging, and the nuclear membrane is commencing to disappear. *B*. First polar spindle fully formed and rotated into radial position. Chromosomes in equatorial plate. *C*. First oöcyte division; anaphase. *D*. First polar body nearly separated. *E*. First polar body completely cut off; second polar spindle formed and rotating into radial position. Spermatozoön within the egg. *F*. Second polar body completely separated. Egg pronucleus forming, surrounded by large aster. Sperm pronucleus, also with a large aster, enlarged and approaching the egg pronucleus. *G*. Approach of the two pronuclei. Egg aster reduced, sperm aster greatly enlarged and centrosome divided. *H*. Fusion of pronuclei; divergence of the sperm centrosomes. *I*. First cleavage figure in early anaphase. Chromosomes divided and beginning to diverge; centrospheres enlarged. *c*, chromosomes; *o*, nucleolus, vacuolated and commencing to disappear; *s*, spermatozoön just within the egg; *v*, germinal vesicle; *vc*, contents (extra-chromosomal) of germinal vesicle; *I*, *II*, first and second polar bodies; ♂, sperm pronucleus; ♀, egg pronucleus.

chromosomes; (3) the centrosomes and asters derived in some way, either directly or indirectly, from the spermatozoön. In syngamy, therefore, the ovum supplies the great bulk of the cytoplasmic basis of the zygote, together with one-half the nuclear material, while the spermatozoön furnishes the other half of the nuclear substance, and produces the centrosomes, which here as elsewhere are to be regarded as the dynamic centers for division. It should not be overlooked that a small amount of cytoplasm, including certain mitochondrial structures, does accompany the sperm, particularly when the tail piece enters the egg; it is by no means impossible, though not at all demonstrated, that this cytoplasm from the sperm may contain substances of great importance in later development and differentiation. In brief, however, it is true that in this union of gametes the ovum is the material factor, the spermatozoön the dynamic, and each contributes equally to the nuclear or controlling mechanism.

But these structures are as yet distributed in different parts of the cell. The association of the scattered elements into a typical mitotic figure now follows and constitutes the final step in fertilization and the formation of the new organism. Maturation completed and the sperm nucleus dissolved, the two germ nuclei commence to approach one another, the sperm nucleus following the centrosomes and asters. The paths of their approach are seldom directly toward one another, as they are in some of the Nematodes, but are more or less curved (Fig. 96), and seem in a way determined by some factors other than mere mutual attraction, though this is doubtless the essential factor in their movement (Wilson).

The entrance path of the spermatozoön is frequently marked by cytoplasmic modifications, often of a very pronounced character, giving evidence of intense metabolic (katabolic) activity; thus we might note the frequency of the accompanying formation of pigment, which is usually regarded as a by-product of protoplasmic decomposition.

We have already seen that the path of the sperm nucleus may be an important factor, either in determining or in making

evident, the position of the plane of symmetry of the zygote,
and hence of the embryo. In a few cases the nuclei are somewhat amœboid, in others they seem to be carried by protoplasmic currents, and in still others they seem directed by the

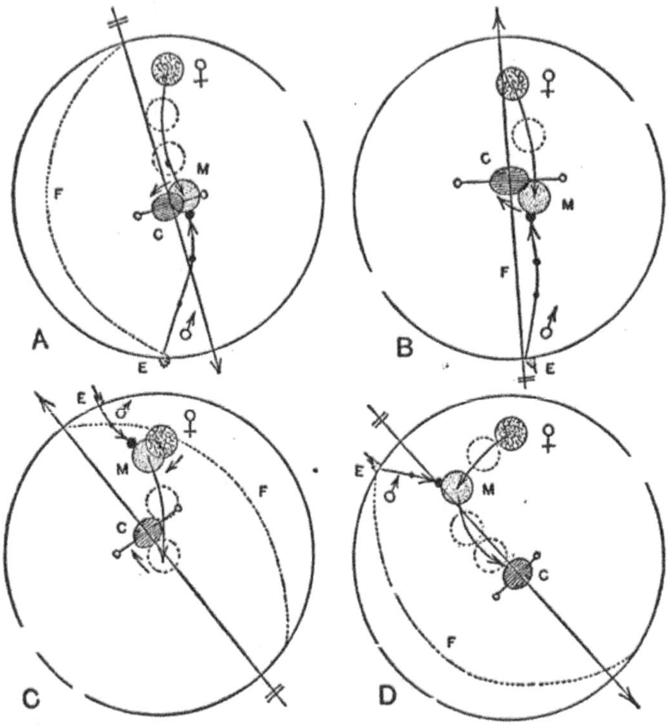

FIG. 96.—Diagrams showing the paths of the germ-nuclei in four different eggs
of the sea-urchin *Toxopneustes*. From camera drawings of the transparent
living eggs. From Wilson, "Cell." In all the figures the original position of
the egg-nucleus (reticulated) is shown at ♀ ; the point at which the spermatozoön
enters at *E* (entrance-cone). Arrows indicate the paths traversed by the nuclei.
At the meeting-point (*M*) the egg-nucleus is dotted. The cleavage-nucleus in
its final position is ruled in parallel lines, and through it is drawn the axis of the
resulting cleavage-figure. The axis of the egg is indicated by an arrow, the
point of which is turned away from the micromere-pole. Plane of first cleavage,
passing near the entrance-point, shown by the curved dotted line.

location of the asters. These have grown to a considerable
size now, and seem mechanically forced into that position where
the tensions between the cytoplasm and the more or less rigid
asters reach an equilibrium. Most frequently this is the center
of the cytoplasmic mass, so that ultimately the two nuclei

approach and meet in this region. As the nuclei come into contact the two asters diverge in such a way as to lie at opposite ends of a tangent drawn through the point of nuclear contact (Figs. 94, 95). The nuclear walls then dissolve; a spireme forms in each nucleus and segments, in each, into $\frac{s}{2}$ chromosomes, and these, as a typical spindle forms from the egg cytoplasm, become arranged at its equator. The result is the formation of a typical mitotic figure with s chromosomes. This is the first cleavage figure, and here, for the first time in the existence of the new organism, substances of paternal and maternal origin are associated, on equal terms, in a common structure.

Before we mention any of the further details in the history of this cleavage figure, we must return to consider briefly the course of events in the fertilization of those eggs which are already fully mature when the sperm cell enters (Echinoderms, Ascidians). In such cases (Figs. 88, 93) the chief divergences from the account just given result from the absence of any pause of the sperm nucleus and middle piece after their entrance. The egg nucleus is ready for fusion, and immediately upon the entrance of the sperm the two nuclei proceed toward one another as described above. The sperm nucleus thus does not have time to be dissolved to any considerable extent, so that when the two nuclei meet they are by no means of equal size (Fig. 88), for the egg nucleus nearly always returns to its "resting" state after its maturation is completed. Nevertheless it is known from their history that the two nuclei are equivalent in chromosomal composition. Frequently, too, the centrosome does not divide until just as the two nuclei meet, or even after they have begun to fuse. The two centrosomes accompanied by asters then move to opposite poles of the combined nuclei and there establish the mitotic figure. The sperm nucleus in these cases does not become resolved into a typical nuclear condition until after its fusion with the egg nucleus. It often results from this that the two nuclear substances seem to mingle quite completely before the spindle is formed and it is not so easy, as in the cases previously described, to distinguish at this time between the elements derived from the egg and sperm nuclei. When this duplex nucleus forms its spireme, this segments into the somatic number of chromosomes immediately, and the mitotic figure for the first cleavage forms typically.

In the formation of the first cleavage figure we see the net result of all the complex processes of the formation and maturation of the germ cells, and the union of the two gametes. In a word, what has been accomplished is the reëstablishment of a single typical cell with specific organismal characteristics. But this cell now has a nucleus derived in equal parts from two separate individuals of the species. Into this nucleus the events of maturation have made it possible that there should have been brought a complete and equivalent series of chromosomes from each parent, for the haploid group is composed of one of each pair of chromosomes of the diploid or somatic series. And from this nucleus are derived all of the nuclei of the developing organism; hence every cell of the adult body may, probably does, contain substance derived from both its parents.

We may regard the organization of these two haploid series into a single nucleus as the culmination of the whole process of fertilization. Or, on the other hand as previously suggested, we may consider the final step in fertilization as not occurring until these pairs of chromosomes actually fuse in synapsis during the maturation of the succeeding generation of germ cells. From this point of view the actual union of maternal and paternal structures and substance never occurs in the somatic cells, for in these synapsis is not known. There is involved in this process of fertilization much more than these simple morphological facts express, and to this subject we shall return presently.

We shall find it profitable to consider now, as briefly as may be, the phenomena of fertilization and accompanying gamete formation in a series of unicellular organisms of increasing complexity and resemblance to the Metazoa. This subject has been in part postponed from Chapters I and III, and it should be stated again that the series to be described is not supposed to represent a phyletic relation. It is now too late to state with any considerable degree of probability the course of evolution of the germ cells and the process of syngamy. Apart from this, however, the consideration of such a series as this brings out many important and interesting facts regarding the general process of fertilization, and emphasizes the idea that this complicated process as we see it in the higher organisms to-day, is all the product of an evolutionary history.

As a preliminary distinction of a general and underlying character we should note that among the unicellular forms the cells which meet or fuse may be of the same race or family, that is, closely related by descent from a comparatively recent common ancestor; or they may be only distantly related, so distantly as to be regarded as unrelated, coming from races or families that have long been distinct, and that have had different histories. The former condition is termed *endogamy*, the latter *exogamy*. These two relations may be distinguished in all forms of syngamy or conjugation; exogamy is much the more usual, and involves the more complicated reproductive processes among the Protozoa, but no such relation seems really necessary, for conjugation may occur with equal facility between any two different individual protoplasms, whether closely related or not.

In a general way we may arrange the varied phenomena of conjugation and syngamy in three classes with reference to the nature and extent of the fusion which occurs. In its simplest form this fusion is not morphological, but is expressed by the congregation of cells in groups; this is to be regarded as a form of *cytotropy*. Occasionally large collections of cells result and the elements come into close and extensive contact without really fusing or losing cell limits. Whatever exchange of substance there may be occurs through an osmotic process. After a temporary association of this kind the cells scatter and resume vegetative and reproductive processes. Such a process of cytotropy has been observed in *Amœba* (Rhumbler).

The simplest form in which a real fusion of plasmas occurs is that known as *plastogamy*. Here two or sometimes more (2–30 in *Actinophrys*) vegetative cells meet and flow together so that the cytoplasms mingle completely; the cell nuclei remain separate, though osmotically they may affect one another and the fused cytoplasms. The result of this is the formation of a physiologically bi- or multinucleate cell. Plastogamy may be only temporary; in such a case the cells come into relation only through comparatively limited contact surfaces and the original cell outlines are not lost. Then after a brief period, during which chemical interchanges may occur, the cells separate again. More frequently, however, plastogamy is permanent, and the fusion of the cells is so complete that the original cell outlines are completely lost. Then, following plastogamy, the nuclei of the combined cells usually divide several times forming a considerable number of smaller nuclei; finally the cytoplasm divides correspondingly, producing thus a group of zoöspores (brood formation). Such processes as these occur most typically in the Mycetozoa (Myxomycetes) and also in such forms as *Arcella, Actinophrys,* and some Foraminifera (Fig. 97).

Finally we come to a third general form of conjugation known as *karyogamy,* which as the word indicates involves primarily a process of nuclear fusion of the conjugating cells, although accompanied, of course, by cytoplasmic fusion which may be of hardly secondary importance. For instance, in some of the Infusoria the achromatic spindles fuse, as well as the nuclei and undifferentiated parts of the cytoplasm; in some

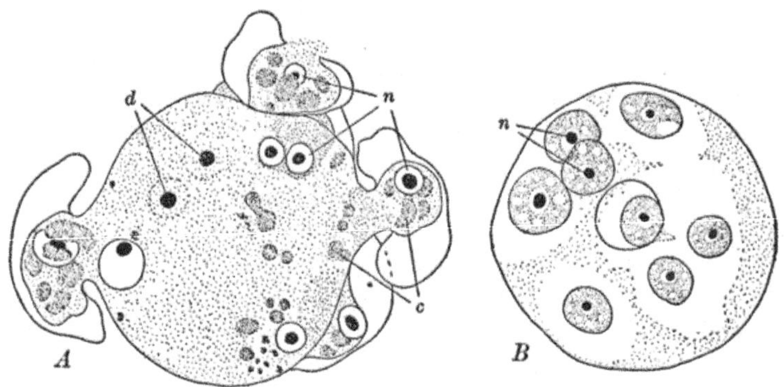

Fig. 97.—Plastogamy in the Rhizopod, *Arcella vulgaris.* After Elpatiewsky. *A.* Plastogamic union of about five individuals, apparently preparatory to the formation of zoospores ("pseudopodiospores"). *B.* Reproduction (formation of "macroamœbæ") following plastogamy. *c,* chromidia; *n,* nuclei; *d,* degenerating nuclei.

of the lower plants, even the plastids of the gametes, perhaps also their centrosomes, fuse together during conjugation.

Most of the more familiar fertilization processes of the Protozoa are essentially karyogamic, but, as a rule (as in the Metazoa), not all of the nuclear substance of the cell is involved in the process. For usually, as a preliminary to conjugation, the vegetative nucleus gives off, into the cytoplasm, portions of its substance (chromidia). These may be formed as a result of general nuclear disintegration, or the nucleus may remain quite intact and extrude chromidia, either directly through its membrane, or by a process of nuclear budding. Some of these chromidia are concerned in reproduction; such are termed *idiochromidia.* Karyogamy, consequently, involves only a portion of the nuclear substance ordinarily, and the remaining chromidia and vegetative nuclear structures may even break down and disappear during the process of fertilization. Altogether these processes of chromidia formation are diverse and often very complicated and the details cannot be given here. (Many of these details, and references to the literature of the subject, are given by Calkins, "Protozoology," New York, 1909).

As a preliminary to the description of typical karyogamic union we may refer to the very special form known as *autogamy*, which occurs in many of the simplest Protista (*e.g.*, *Entamœba*, *Amœba*, some Myxosporidia). In autogamy there is really no fusion of cells at all; the characteristic event is the separation of the nuclear chromatic

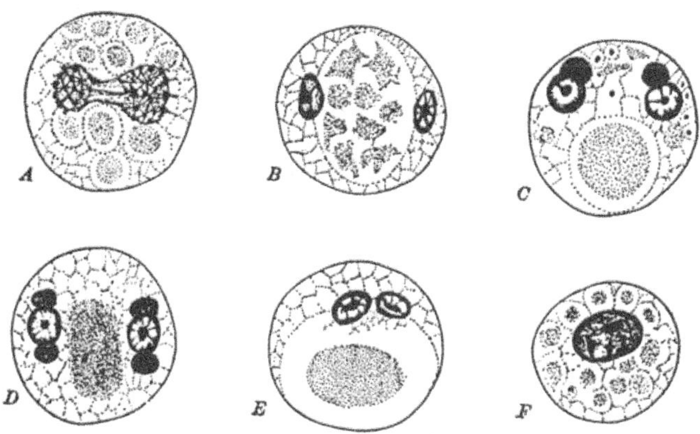

Fig. 98.—Autogamy in the Flagellate, *Trichomastix lacertœ*. After Prowazek. *A.* First nuclear division in the encysted form. *B.* The two nuclei completely separated. *C.* First "reducing" division. *D.* Second "reducing" division. *E.* Approach of the "reduced" nuclei. *F.* Fusion of the nuclei to form a single nucleus (synkaryon).

substance of a single cell into a number of separate bodies (Figs. 98, 99), which become scattered through the cytoplasm as chromidia, or rather as idiochromidia, for they are concerned in reproduction. After their formation is completed these idiochromidia fuse, by twos, or sometimes in larger groups, forming in effect "new" nuclei, or at any rate new combinations of chromatic substance. These fused chromatin masses then commonly move to the surface of the cell and are budded off with small bits of cytoplasm, as small cells or spores (Fig. 99) (Schaudinn, Calkins). Here then the nuclei which fuse together are the direct derivatives of a single nucleus, and they remain within the same cytoplasmic mass throughout their formation and fusion. This might readily be regarded as an extreme form of endogamy; it suggests the roughly analogous process of the reëntrance of the nucleus of one of the polar bodies which occurs in a few of the parthenogenetic Arthropods. Fertilization by autogamy is considered by some as a primitive method of fertilization preceding all processes of gamete formation or cell fusions; others regard it as a derived condition in which the nuclei act precociously. However this may be it seems more instructive to classify the process as karyogamic.

Coming now to the consideration of typical karyogamic fusion we find that all of the fertilization processes common to the Metazoa, as well as those of most of the Protozoa, belong here. And here again fertilization may be either endogamous or exogamous.

Two general forms of karyogamy proper are usually distinguished, although they are so clearly connected by transitional conditions that they must be regarded as merely convenient groupings. These are *isogamy*, where the pairing cells are similar in size, form and behavior; and *anisogamy*, where the pairing cells are markedly dissimilar in size,

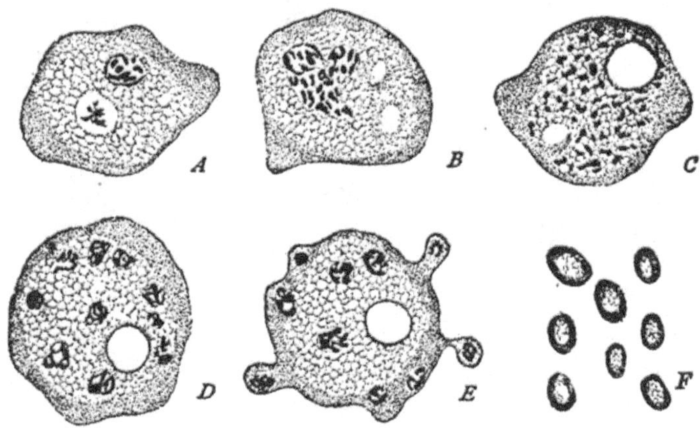

Fig. 99.—Autogamy in the Rhizopod, *Entamœba histolytica*. From Calkins, "Protozoology," after Craig. *A*. Organism showing rods and granules of chromatin in the nucleus, vacuole with some stained substance, and dense ectoplasm. *B*. Chromatin of the nucleus passing into the cytoplasm, as chromidia, shown in *C*. *D*. Aggregation of chromidia to form secondary nuclei. *E*. "Spore formation" by budding. *F*. Spores formed from buds.

form, and behavior. That is, this distinction is not based upon differences in nuclear structure, or behavior during union, indeed, these are essentially the same throughout karyogamy, but upon external characters of the conjugating cells.

Considering first isogamy, as the simpler and less modified process, we find it restricted to the unicellular forms. Isogamic union frequently occurs between two individuals of the usual vegetative type which do not show, externally at least, any structural modifications usually associated with gametic behavior (Fig. 100). This is most common among the Flagellates, such as the familiar *Copromonas*, and *Noctiluca*, but it occurs also in *Actinophrys* and in some species of *Amœba*. In other cases the conjugating individuals show some modification in form as compared with vegetative individuals (they are modified similarly of course), but there is no reduction in size. Thus in *Cercomonas* and *Tetramitus*, the flagellate bodies of the conjugants become more or less amœboid and distinctly plastic and viscous. Other genera exhibit various stages of reduction in size (Fig. 101), although in form they still resemble vegetative cells. This is the case in most of the Foraminifera

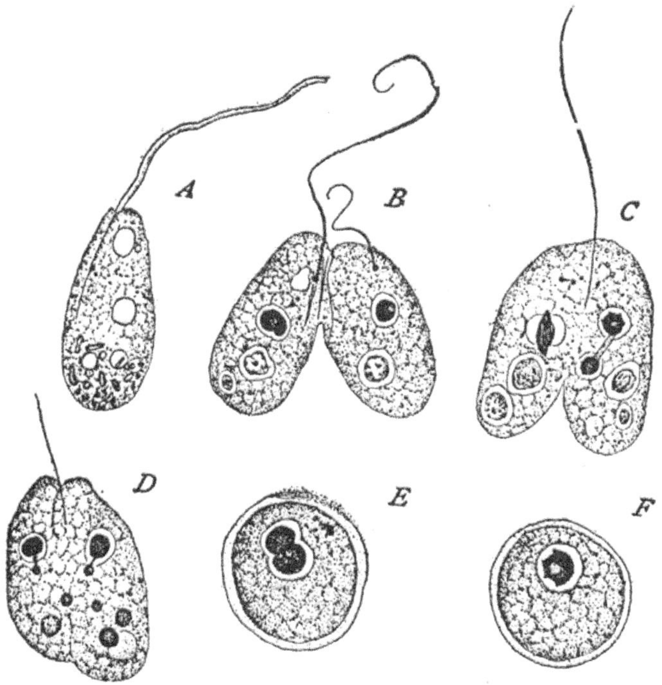

Fig. 100.—Conjugation in the Flagellate, *Copromonas subtilis.* From Calkins, "Protozoology," after Dobell. *A.* Vegetative form. *B.* Beginning of conjugation of two organisms; one flagellum withdrawn. *C.* Continued fusion. First stage in nuclear "reduction." *D.* Second "reducing" division (heteropolar). *E.* Conjugation completed; "reduced" nuclei fusing. *F.* Zygote within cyst.

and many other Rhizopoda; it is rare among Flagellates (*Stephanosphœra, Chlamydomonas*) and Ciliates. Finally, we find this reduction in size accompanied by a modification in form, as in other Rhizopoda whose gametes become flagellated, or where they are amœboid in forms usually flagellate or motionless.

We have then among isogamous organisms a series of forms, at one extreme of which the gametes are morphologically unmodified, at the other they are diminutive and structurally modified, usually in connection with the motor apparatus, in such a way as to render more likely the accident of their meeting, likelihood of which is largely increased

through the fact that reduction in size is usually the result of multiple
fission or brood formation, which increases the number as well as the
activity of the gametes. In all of these forms of isogamy the union
of the gametes is permanent, the conjugants fusing completely and
thereby losing their identity as individuals. It need hardly be added
that in these cell conjugations the essential step seems to be the fusion
of the gametic nuclei into a single zygote nucleus.

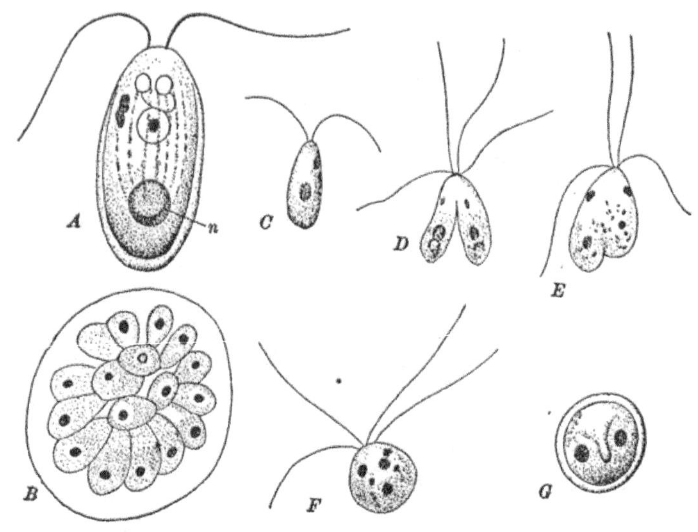

Fig. 101.—Gamete formation and fusion (isogamy) in the Flagellate, *Chlamy-
domonas steinii.* After Goroschankin. *A.* Vegetative form. *n,* nucleus.
B. Group of gametes formed by multiple fission. *C.* Single gamete. *D, E, F.*
Stages in the fusion of gametes (isogametes). *G.* Zygote.

A special form of isogamy needs particular notice on account of its
frequency among the most familiar Protozoa—the Ciliata. This is a
temporary form of isogamy which involves, not the fusion of two gametes
to form a zygote, but the *mutual* fertilization of the two gametes through
the *exchange* of nuclear substance and perhaps also a small amount of
cytoplasm. The details of nuclear behavior in the conjugation of *Paramœcium,* for example, are probably familiar but will bear brief restatement here. This outline refers particularly to that form of *Paramœcium*
having only a single micronucleus; one should recall that in these forms
which have both micronucleus and macronucleus, the former is, or
represents, the idiochromidia, the latter the vegetative nuclear structures. Two Paramœcia of normal vegetative size and external form
meet side by side, oral surfaces in contact, in a sort of plastogamic union,
The further course of events is exactly similar in each individual of

the pair (Fig. 102). In each cell the micronucleus divides and the
daughter micronuclei immediately divide again forming four. Of

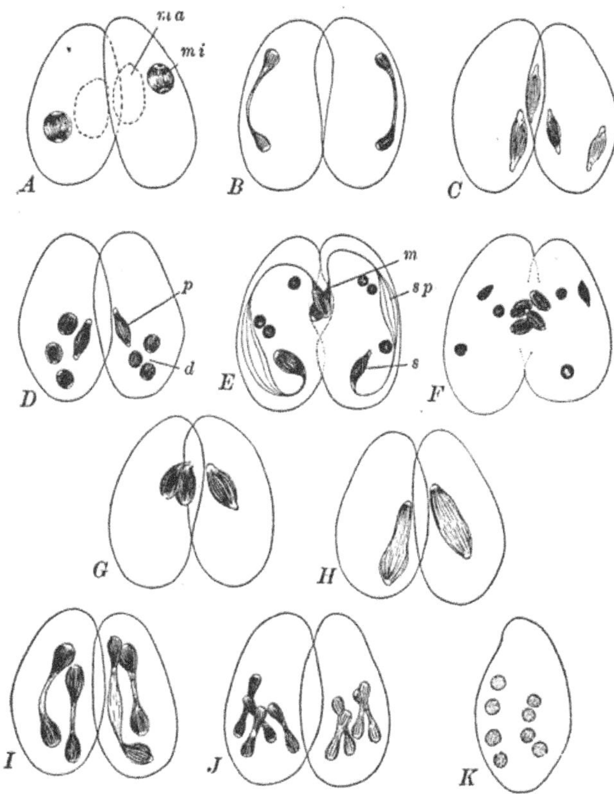

Fig. 102.—Nuclear history during conjugation in *Paramœcium putrinum*.
After Doflein. The macronuclear structures are omitted in all except *A*. *A*.
Formation of spindle for first division of micronucleus. *B*. Telophase of first
division. *C*. Second division of the micronucleus. *D*. Degeneration of three
of the micronuclei; the fourth, or permanent micronucleus, preparing for another
division. *E*. Division of the permanent micronucleus into the stationary and
migratory micronuclei. The spindle is greatly elongated and has a characteristic
mid-body. *F*. Grouping of micronuclei and exchange of migratory micronuclei.
G. Fertilization (mutual). Fusion of migratory with stationary micronuclei.
H. Formation of spindle for first division of the fusion nucleus in each conjugant.
I. Late phase of second division of fusion nucleus. *J*. Third division of fusion
nucleus. *K*. Exconjugant with eight micronuclei, derived from the fusion
nucleus. (The degenerating micronuclei are omitted from *G*–*K*.) *d*, degenerating micronuclei; *m*, migratory micronucleus; *ma*, macronucleus; *mi*, micronucleus; *p*, permanent micronucleus; *s*, stationary micronucleus; *sp*, spindle.

these, three degenerate (*cf*. maturation), while the one remaining
divides once more, this time unequally, forming a larger and a smaller

micronucleus in each organism. Each larger or "stationary" micronucleus remains passive, but the smaller or "migratory" nucleus becomes
active and moves through the bridge of fused cytoplasms to the stationary micronucleus of the other individual, with which it fuses forming
a single compound or zygotic nucleus in each individual. The two
Paramœcia now separate each with a nucleus of modified composition.
The macronucleus, which has taken no share in the events of fertilization,
now fragments and dissolves leaving the fusion nucleus as the only
nuclear structure present. Then by three successive divisions the
fusion nucleus gives rise to eight small nuclei; four of these in the posterior end of the cell, remain small, as micronuclei, while the other four,
in the anterior end, enlarge, forming macronuclei. During the first
fission of this cell each daughter cell receives two nuclei of each kind,
and at the next division each of the four granddaughters of the "zygote"
receives one micronucleus and one macronucleus, and the normal
vegetative condition is restored. This form of karyogamy is peculiar
for at least three reasons; the nuclei alone fuse, *both* of the gametes
undergo nuclear reconstruction, and the individuality of the gametes is
not lost.

The sessile Ciliates show an adaptive modification of this process which
is anisogamic in character. In the common *Vorticella*, for example,
while one of the conjugants or gametes retains its normal vegetative
form, the other is small and one of a brood of four, which become free-
swimming. A small individual upon meeting a large one, is actually
absorbed by it. The early nuclear history of each organism is much the
same as in *Paramcœium*, save that the final micronucleus of the megagamete is one of four, that of the microgamete one of eight, the remainder
in each gamete having degenerated. But after the equivalents of the
stationary and migratory micronuclei (idiochromidia) are formed
in each gamete, the process changes somewhat, for now one of the two
micronuclei (the equivalents of one stationary and one migratory body)
degenerates in each organism, while those remaining fuse together
forming thus only a single fusion nucleus in the single but duplex zygote.
Thus there is no mutual fertilization, and while strictly this is anisogamous, it is mentioned here because it is clearly derived from the more
typical Ciliate condition as an adaptation to the sessile life of the vegetative form.

Coming now to *anisogamous* karyogamy (Fig. 103), we should note
that transitional conditions between isogamy and anisogamy are not
infrequent. Thus in the Flagellate, *Bodo*, the conjugants may be
either of equal or unequal size, apparently in an accidental fashion.
The colonial *Pandorina* forms gametes of three sizes, small, medium,
and large, and conjugation may occur between any smaller and any
larger individuals anisogamically, or the small or the medium organisms

may conjugate together isogamically. Here then anisogamy is not obligatory, but facultative or accidental. In this case exogamy is the rule since a single colony forms gametes of one size only, though in isogamy endogamy may occur.

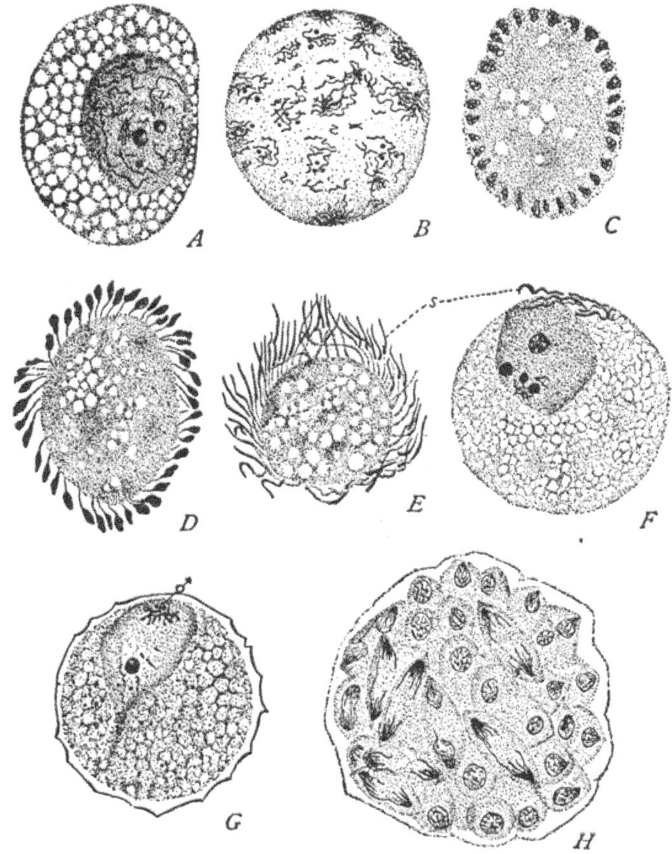

Fig. 103.—Formation of gametes and syngamy in the Sporozoan, *Klossia octopiana*. From Calkins, "Protozoa," after Siedlecki. The chromatin of the nucleus is distributed throughout the cell, *A, B*, finally forming nuclei of the future gametes, *C, D, E*. The mature gametes, *s*, swim about, and join a macrogamete, *F*. The nuclei mingle, *G*, and then the cleavage nucleus divides repeatedly by mitosis, to form the spores, *H*. ♂, microgametic nucleus.

In true anisogamy conjugation is practically always exogamous for as a rule a single organism forms gametes of only one size at a time. The essential difference between the gametes is probably that of behavior, *i.e.*, degree of activity, associated with which are constant differences in size. The larger gametes or *megagametes*, are less numerous and less active than the smaller *microgametes*, which may be formed in very considerable numbers. Occasionally the gametes differ in form only, as in *Dallingeria*, where one of the gametes has three flagella, like the vegetative cells, while the other gamete has but one flagellum. In a

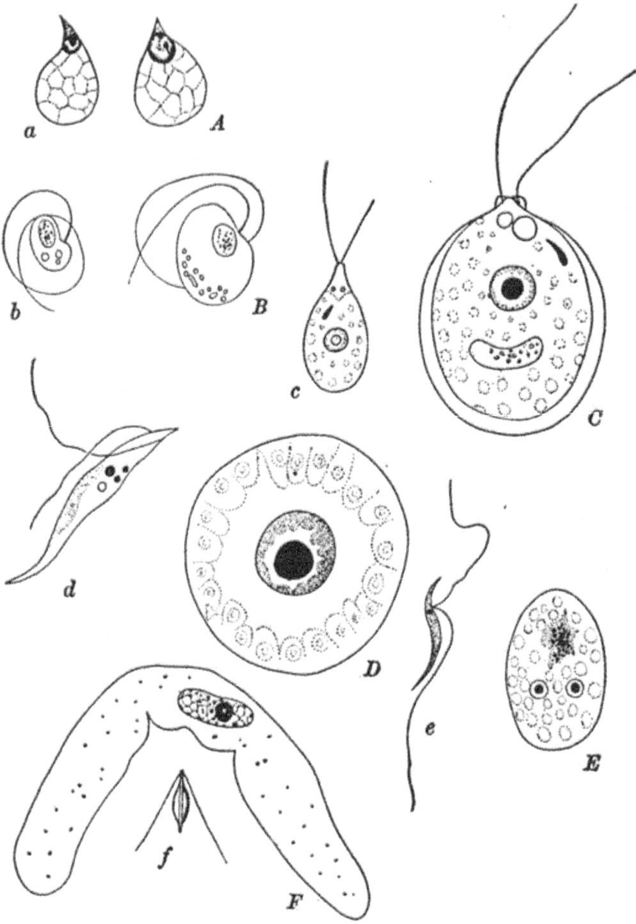

Fig. 104.—Micro- and macrogametes of various Protozoa, illustrating various degrees of differentiation. After Doflein, from various authors. *f. F*, × 562, others × 1125. *a, A. Urospora lagidis* (Brasil). *b, B. Collozoum inerme* (Brandt). *c, C. Chlamydomonas braunii* (Goroschankin). *d, D. Volvox aureus* (Klein). *e, E. Cyclosporia caryolytica* (with two polar bodies in cytoplasm of *E*) (Schaudinn). *f, F. Orcheobius herpobdella* (Kunze).

few forms such as *Monas* and many of the Gregarines, the only morpho-logical difference between the gametes is that of size, the microgamete being smaller and somewhat the more active, but not otherwise unlike the megagamete. In other forms the microgametes are considerably modified structurally, usually in connection with an increase in loco-motor activity. At the same time the megagamete may increase considerably beyond the ordinary vegetative size and may then lose motility more or less completely. So it finally comes about that gametes of two wholly different types are formed, both quite unlike vegetative cells, and the typical Metazoan condition is reached.

Several groups of Protozoa, *e.g.*, Gregarines, colonial Flagellates, afford interesting series showing stages in this differentiation of the gametes (Fig. 104). We may outline one such series selecting examples from the Volvocine group of Flagellates.

In *Stephanosphæra* all the individuals of the colony are, or may be, reproductive, and conjugation is isogamous and endogamous. There is no differentiation of gametes. In *Pandorina* (Fig. 8) all of the individuals may be reproductive, but some of the gametes may be differentiated in size, and conjugation may be either isogamous or anisogamous as described above. Here dissimilarity of the gametes is facultative. In *Eudorina* two kinds of colonies are found. In one, all the cells may become reproductive, the individuals forming megagametes only slightly larger than vegetative cells; in another only four cells of the colony are reproductive and each of these forms sixty-four very small and active microgametes. Fertilization is here strictly anisogamous and exogamous. The last step is represented by *Volvox* (Fig. 10), where the number of gamete forming cells is always limited. Here too differentiation of the gametes reaches its climax among the Protozoa, and the Metazoan condition is reached. The reproductive cells lose their motor organs and begin to enlarge. A few of them grow to a relatively enormous size and become the passive megagametes. The others grow to lesser extent and then divide rapidly, each forming, probably 128 microgametes. These are very small flagellated, extremely active cells, with an elongated rostrum or penetrating organ at one end (Fig. 104, *d*). The microgametes are liberated in large numbers and swim about until one reaches a megagamete which it then enters and their nuclei fuse forming a typical zygote which then reproduces a new colony. The resemblance to the gametes of the Metazoa is so complete that they are here termed the *oösphere*, or *ovum*, or *oögamete* (megagamete) and the *spermatozoön* or *spermagamete* (microgamete).

It should perhaps be noted here that the process of conjugation or fertilization is not always associated with the reproduction of the Protozoa mentioned above, not even in the colonial forms. For the usual reproductive processes are carried out by the simple fission of ordinary vegetative cells. In the simpler colonial forms, such as *Pandorina* and *Eudorina*, as many new colonies may be formed as there are individuals forming the original colony, in *Volvox*, however, the number of cells which may reproduce in this way is limited, and there seems to be a real distinction between soma and germ, much like that of the Metazoan organism, the mother cell which divides to form the multiple spermatozoöids, has even been compared with the testis of a Metazoan.

The principal forms of fertilization and gamete formation are summarized in the accompanying table.

Several facts of prime importance are to be drawn from this account. (a) Among the Protozoa as well as the Metazoa, the process of fertilization is widespread. (b) Out of a variety of forms comes that form of fertilization characteristic of the Metazoa, namely, karyogamy. (c) Accompanying this karyogamy is a gradual and finally complete differentiation of gametes, which differ morphologically and physiologically, both from vegetative cells and from each other. (d) The Metazoa show much less diversity than the Protozoa respecting the process of fertilization and the form of the gametes.

Such a series of stages as that outlined above, of the gradual differentiation and specialization of gametes, cannot fail to suggest the general subject of sex. It does indeed indicate the nature of the original distinction between the sexes. Among the Metazoa the primary and familiar facts upon which the definition of sex is based, are, that spermatozoa-producing individuals are males, ova-producers are females. In all cases of isogamic conjugation no distinction between the gametes, and therefore between sexes, obtains. In those instances transitional between isogamy and anisogamy, we may see the beginnings of sex distinction, often facultative. True anisogamy involves true sex distinction; at first relatively slight (*Pandorina*), in such forms as *Volvox* or *Coccidium* and many other Sporozoa, the fundamental differentiation of sex seems to be completely established, *i.e.*, the gametes are markedly unlike and conjugation occurs only between two dissimilar cells.

The essential processes of fertilization are entirely equivalent in isogamy and anisogamy, so that the fundamental distinction of sex is based only upon the external form and behavior of the *gametes*, not upon any differences in the nature of the conjugation processes in sexual and non-sexual forms, for none exist.

Most of the colonial Protozoa are monœcious or hermaphroditic, producing gametes of both kinds, but cross-fertilization (exogamy) is the rule here, as it is among the hermaphroditic Metazoa, and frequently for the same reason in both groups, namely, a difference in the times of ripening of the two forms of gametes of a single colony or individual. Many of the Sporozoa are clearly diœcious or unisexual, and in some of these there are also *secondary* sexual characters, usually size differences, such that macrogamete-forming or female individuals can be distin-

PRINCIPAL FORMS OF FERTILIZATION IN THE PROTOZOA

(Autogamy)
Endogamy — I. Cytotropy. To or more individuals. *Amœba.*
II. › My. To or more { Temporary. *Actinophrys,* mny Foraminifera.
                         { Permanent. Mycetozoa (Myxomycetes)

Autogamy. *Entamœba,* sne spes o *Amœba.*

Exogamy — III. Karyogamy {

Isogamy {
  ⸢ht { Full size { N mal vege-tative f on. } *Actinophrys,* mny Fla-gellates, *g.,* pdo-nas ‹ dg, *Noctiluca.*
                   { Mth ly. } Many Flagellates, *g., Cercomonas, Tetramitus.*
       { Reduced { dial vege-ite f on. } Mst Foraminifera, many Rhizopoda, a few bls, *e.g.,* ph-*anosphœra.*
                  { Modified. Rhizopoda ard a few Flagellates.
  Temporary. Exchange o ulei and tual r zin. Mst tG, *e.g.,* dB, lq.
}

(T ian) Anisogamy facultative. *Pandorina, Po* ı yd, oBo.

Anisogamy {
  Differ in form oly. *Dallingeria.*
  Differ in size only or slight y in bvior My Gregarines ard Flagellates, *g.,* M.
  Differ in af, zi, ard dbr. *Thalassicola, Coccidium,* mst Gregarines ard colonial Flagellates *e.g Eudorina* db. pd-ia. All M.
}

guished from microgamete-forming or male individuals, some time before
the gametes are actually formed; in a few rare instances this distinction
can be made throughout the life cycle, and individuals can be identified
at any time as males or females.

We now come to a consideration of the meaning and theoretical significance of these processes of fertilization or syngamy.
Probably there is, in the whole field of Biology, no process of
such widespread occurrence and obvious importance, where
the phenomena are so well known, which at the same time is so
little understood. Why fertilization should occur, what is
effected by it, and how syngamy brings about the results which
do follow it, are questions to which to-day, after decades of
speculation and research, no sure answers can be given.

Although we may be on uncertain ground, it will be profitable
to review some of the suggestions and hypotheses that have
been proposed in this connection, even if we accomplish little
more than to point out the possibilities and difficulties of this
fascinating subject. And furthermore, while no thoroughly
demonstrated solutions of the problems of fertilization have
been reached, there are several carefully worked out hypotheses
in the field, some of which are certainly to be regarded as close
approximations toward the correct explanation of some of the
problems of fertilization. We may conveniently arrange these
current ideas as to the results and primary "purpose" of fertilization in four groups. The results of fertilization may be
connected with (a) reproduction, (b) rejuvenation, (c) the
process of variation, (d) the process of heredity. In considering
each of these we shall state as briefly as possible the essentials of
the evidence for and against the central idea.

That fertilization is primarily a reproductive process was the
original view held by Harvey and his successors. There are
now two forms of the hypothesis. In one form we find the
idea that the ovum quite obviously seems to contain only a
part (*i.e.*, the cytoplasm and one-half a nucleus) of the mechanism necessary for development, and that the spermatozoön
brings into the ovum those parts (*i.e.*, one-half a nucleus and
the centrosome, or the stimulus to its formation) which com-

plete this mechanism and enable development to proceed.   In
its other form this idea is that the ovum is a quiescent, passive
body which needs to be stimulated to its normal activity
(development) by the entrance of the spermatozoön, the kinoplasm (centrosome) of which is the part chiefly acting as the
stimulus.   In short, fertilization is to be regarded normally as
the necessary antecedent, as the cause of development.

There are many facts opposed to this view.   Of these we shall
discuss two chief classes, first, those of parthenogenesis, both
normal and "artificial," and second, those drawn from the
relation between fertilization and reproduction among the
Protozoa.

For present purposes we may extend the definition of parthenogenesis to include all those cases where single cells, specialized
for the purpose, develop without undergoing syngamy.   In
the plant kingdom parthenogenesis, in this broad sense, is very
widespread.   Development from spores is very common, even
among the higher (vascular) plants, and in some instances
(some of the Fungi) reproduction by single unfertilized cells is
the exclusive method.   And the development of ova, typical
in every respect save that of needing to be fertilized, is not
uncommon.   Among the single-celled animals phenomena
equivalent to development from spores are frequent, and among
the multicellular animals normal parthenogenesis is known in
the Rotifera, some of the Crustacea, and in several orders of
Insecta.   In most of these forms fertilization does occur at
some period in the life cycle, after a widely variable number
of parthenogenetically produced generations, but there are a
few Metazoa, *e.g.*, the wasp, *Rhoditis*, and the Crustacea, *Cypris*,
Limnadia, and sometimes *Apus*, in which males never develop
and fertilization is therefore entirely unknown (Weismann).
In some of the parthenogenetic Crustacea the nucleus of one
of the polar bodies seems to act as a fertilizing nucleus (see
Chapter IV), so that a sort of autogamic process occurs,
recalling that of some of the Protozoa and perhaps analogous
with it.   In all of these Metazoa the form and history of the
parthenogenetic ova, the occasional presence of vestigial

spermathecæ, and other similar conditions, clearly indicate that this is a secondary or derived condition, a special adaptation, which therefore throws but little light upon the fundamental significance of the fertilization process; this proves only that fertilization is not in all cases a necessary antecedent to development, and that the ova may, in certain cases, contain in themselves complete developmental mechanisms, and need neither the addition of sperm structures nor the special form of stimulation afforded by the entrance of the sperm cell.

Instances of *merogony,* or "male parthenogenesis," where egg fragments containing no nuclear substance develop after penetration by a spermatozoön (Echinoderms), also show that the addition of egg and sperm structures, each incomplete in itself, is not a necessary feature of fertilization.

It may truly be said that the obviously secondary character of normal parthenogenesis renders the phenomenon of little value as evidence regarding the real meaning of fertilization in the vast majority of instances. But such an objection cannot be brought against the evidence from experimental parthenogenesis, and probably the clearest evidence upon this phase of the subject is that of "artificial" or induced parthenogenesis. In view of this fact, and of the great general importance of the subject, we may consider this matter rather fully.

The eggs of many animals, belonging to many different classes and phyla, normally requiring to be fertilized, may be stimulated to begin their development by chemical or physical treatment. Thus the ova of many Cœlenterates, Echinoderms, Annulates, Molluscs, and even some Chordata (Teleosts, Amphibia), may be induced to commence their development parthenogenetically by being subjected to the action of a great variety of organic and inorganic substances in solution, to unusually high or low temperatures, to physical shock, or to various other conditions.

In this process of artificial parthenogenesis two phases must be kept separate, first, the phenomenon of maturation, in those cases where this is not completed at the time fertilization normally occurs; and second, the phenomena of cleavage and differentiation, occurring subsequently to maturation. Certain acids and some other substances seem to have, according to Morse, a specific effect in bringing about maturation, either not producing cleavage or actually inhibiting it, in which latter case it may then be induced by other treatment. The precise actions upon the egg of those chemicals inaugurating cleavage are varied, but for the most part they appear to effect certain changes in the egg which are similar in nearly every instance. For example, according to Loeb, who is the pioneer in this important work, in the sea-urchin, the best result, that is, the closest imitation of natural fertilization, is secured by treating the eggs, first with a solution of a monobasic fatty acid, such as butyric acid, for one or two minutes. In many cases the butyric acid can be replaced by an alkaline solution of equivalent strength or by a solution of almost any fat solvent. This treatment results in the formation of an apparently typical fertilization membrane. Then second, the eggs are treated for some minutes or hours (sea-urchin eggs, thirty to fifty minutes at 15° C.) with a hypertonic sea water, that is, sea water whose osmotic pressure has been raised about 50 per cent. above normal by the addition of salts, such as sodium chloride. Finally the eggs are returned to normal sea water and cleavage then follows in quite the usual fashion.

What is actually accomplished within the egg by such treatment as this is largely conjectural. Loeb suggests that the process may be as follows. The mature ovum is surrounded by a relatively impermeable surface film which prevents the oxidations necessary to development. The butyric acid or similarly acting substance, by dissolving certain fatty constituents near the egg surface, frees from association with these, certain other osmotically active materials which then form the permeable fertilization membrane, and thus rapid oxidations are permitted. Loeb believes that the nuclear substance possesses a catalyzer which, in the presence of oxygen, brings about a synthesis of nuclein, one of the chief constituents of chromatin, and that this synthesis of nuclein is the chief chemical action of the segmenting ovum. This process of cleavage once started in the right direction, leads then in a perfectly natural and normal way to the later processes of development. Loeb suggests farther that the spermatozoön may contain certain substances, enzymes, which form within the egg, materials capable of producing effects similar to these, and that herein lies the natural stimulating effect of the spermatozoön.

Many facts regarding this hypothesis, both *pro* and *con*, have been forthcoming in recent years, but it is still too early to say how closely this approximates the truth. We might add, however, that Masing and others have not been able to detect any marked synthesis of nuclein such a Loeb describes during cleavage. And quite recently Conklin has determined that the synthesis of nuclear substance is, in *Crepidula*, at least no greater than the synthesis of cytoplasm.

The eggs of other forms are more successfully treated by other methods, and each may have a particular treatment which is most effective. But in very many of the instances of artificial parthenogenesis the essential result of the treatment seems to be a process of membrane formation accompanied by the withdrawal of fluids from the egg. This has led to the suggestion (Loeb) that the spermatozoön acts in this fashion, for it is relatively very deficient in fluids, and upon entering the egg reduces to some extent the relative fluidity of its cytoplasm, thus acting as a stimulus.

That the action of the spermatozoön is not specific, and that fusion of the two germ nuclei is really not necessary to inaugurate development, is clearly shown by the fact that almost any spermatozoön, of whatever species, that can gain entrance to an ovum, is capable of initiating development, and of effecting the apparently normal cleavage of the ovum; to what extent the internal processes of fertilization and cleavage are entirely normal in such a case, we shall see later; suffice it to say here that frequently a foreign sperm nucleus remains quiescent and takes no part in the formation of the mitotic figure.

It is true that the development of artificially fertilized ova seldom proceeds farther than the cleavage stages. As a rule the lower the organism in the evolutionary series, the farther its development may proceed. And while some artificially fertilized Echinoderm eggs have been carried past the larval stage (Delage), the Chordate ovum (Teleost, Cyclostome, Urodele) will cleave only a few times. This indicates clearly that the parallelism between natural and artificial fertilization is not complete, although it is not unlikely that ultimately a form of treatment may be found which will produce just the same result as normal fertilization, save in so far as this is concerned in the inheritance of individual characteristics.

In all cases of artificial parthenogenesis the cleavage figures are essentially normal except that the reduced or $\frac{s}{2}$ number of chromosomes is present (exceptions to this have been reported by Tennent and Hogue, and others); the poles of the spindle are occupied by typical centrosomes, formed anew by the substance of the egg after the disappearance of the oöcyte centrosomes of the maturation spindle. This is also true regarding the centrosomes of normally parthenogenetic eggs. And there are several instances known where the specific effect of certain reagents or external conditions is the formation, out of the cytoplasm of the ovum, of numerous centrosomes, apparently of normal structure and each with a small aster (Yatsu).

To summarize the evidence from parthenogenesis, both normal and artificial, we may say that, among the Metazoa, the ovum contains within itself a mechanism sufficiently complete to function for a time at least, although the spermatozoön, when it enters, does add to this mechanism and supplies some parts not present in the egg; these parts either are not absolutely necessary or they may, under certain conditions, be supplied from the structure of the ovum in the absence of the sperm. And further, while the egg may be stimulated to develop by means other than the entrance of the sperm, this is normally the form of stimulus which inaugurates the series of reactions we call development. Taking this view of fertilization the formation of the spermatozoön is a means of insuring the properly effective form of stimulus, which might otherwise be lacking in the environment of the egg.

Turning now to the evidence which the Protozoa offer regarding the relation of fertilization and reproduction, we may approach more closely the problem of the fundamental significance of syngamy. Nearly all the known life histories of Protozoa are cyclic in character. The process of reproduction by simple fission may proceed uninterruptedly for a longer or shorter period, but finally this is interrupted by some form of syngamy. Formerly this seemed obviously to mean that the ordinary reproductive processes depended ultimately upon a process of conjugation or fertilization, and that the life cycle in the Protozoa was essentially similar to that of the Metazoa, the divisions of the somatic cells of the latter being equivalent to the simple fissions of the former, and the fertilization processes of both being essentially similar (homologous). It was overlooked at first that the process of fertilization might just as well be considered the result of vegetative divisions as the cause of them; to this phase of the relation we shall return shortly.

It is true that in some Protozoa, *e.g.*, *Noctiluca*, *Trichosphærium*, some Gregarines, fertilization is really followed by a marked increase in reproductive activity. And it is often true that multiple fission tends to follow conjugation. But in other cases reproductive activity seems not to be affected by fertilization. And in many, probably most Protozoa, fertilization tends to inhibit reproduction. In many Rhizopods, Flagellates, and Ciliates, a pause in the succession of fissions may be quite marked after conjugation. In many of the Sporozoa a period of encystment follows, and the same is true of many Algæ. In such cases, therefore, fertilization seems opposed to reproduction, or at least to any immediately ensuing processes of multiplication; it may still be true that the *ultimate* effect of fertilization may be increased rate or duration of fission. Conjugation may occur without reproduction; reproduction may occur without conjugation. And that conjugation is frequently to be regarded as determined by external rather than internal conditions is indicated by the occurrence of so-called "epidemics" of conjugation which may often be observed in Protozoan cultures. In such cases conjugation may often be artificially induced by regulating the character and amount of the food supply (Jennings).

It may be concluded, therefore, that among the Protozoa the processes of reproduction and fertilization are not fundamentally related, and the primary significance of fertilization must be sought in some other relation. This view is widely accepted to-day and it consequently becomes necessary to explain the practically universal association of the two processes among the Metazoa, the only exceptions being, as we have seen above, the secondary and obviously derived instances of normal parthenogenesis. The commonly suggested explanation is the following.

Whatever the real significance of fertilization may be, it seems, for reasons which will appear later, a condition for continued existence of specific forms of protoplasm that occasionally some disturbance of its inner structure should occur, such as would result from the mingling of the substances of two distinct individuals. Among the single-celled organisms this may occur at any time, whenever that action would form a natural response to internal conditions of the organisms. Among the Metazoa, on the other hand, such a complete fusion of cells, and particularly of nuclei, can occur only when the organisms are in the form of single cells, *i.e.*, gametes, and differentiation of the organism is at a minimum. Thus whereas the two processes are originally distinct and unrelated in their origin in the Metazoa, they have come to be related, and now fertilization appears as the first step in reproduction.

Another general hypothesis regarding the function and significance of fertilization is the *rejuvenation* hypothesis, associated chiefly with the names of Bütschli, Maupas, and Richard Hertwig. This is based to a large extent upon the phenomena of the Protozoan life cycle. It involves as a starting point the assumption, partly based upon observation, that protoplasmic activity tends gradually to diminish in intensity, and that associated with this diminution are certain morphological alterations in the structure and composition of the cell. Altogether these modifications are known as *senescence*. A frequent characteristic of a senescent cell, in both Protozoa and Metazoa, is the relatively large proportion of cytoplasm as compared with nuclear substance. It is further assumed that syngamy and the consequent admixture of nuclear and cytoplasmic materials of two individuals, perhaps representing different races, causes the restoration of the senescent protoplasm to a condition of vigor, in a word, brings about rejuvenation. It would follow from this, that protoplasmic activity is cyclic, and that periods of senescence would be followed by death unless fertilization, or an equivalent process, should occur.

Precisely what is involved in the process of rejuvenation cannot be stated definitely. Richard Hertwig suggests that senescence is due chiefly to changed nuclear-cytoplasmic relations resulting from repeated cell-fissions, and that in rejuvenation there is essentially a restoration of the normal nuclear-cytoplasmic ratio, as well as a certain chemical and physical reorganization of the protoplasm through the combination of materials from two more or less unlike individuals. Loeb and others, as we have seen, also regard the rapid synthesis of nuclein as the most important consequence of fertilization. Still others (Minot, Bernstein) believe that rejuvenation is not only a nuclear-cytoplasmic phenomenon involving or resulting from an increase in the relative amount of nuclear substance, but that it further includes an increase in the property of growth, *i.e.*, the formation of new protoplasm, both nuclear and cytoplasmic, out of non-living substance.

The real evidence for the cyclic character of the life processes of the Protozoa is chiefly that of Maupas and Calkins, who showed that in *Paramœcium* and some other Ciliates, when conjugation is prevented, there occur, under laboratory conditions, periods of depression in vital activity, accompanied by changes in structure, *i.e.*, periods of senescence. This depression leads finally to death unless conjugation is permitted, or unless the organisms are subjected to some form of stimulus. If stimulated by chemical or physical means, or naturally through conjugation, the organisms may in some cases recover their original vigor and begin a new cycle with youth renewed.

But rejuvenation is by no means always the result of conjugation, for frequently the senescent organisms perish in spite of conjugation; and it may even be the case that the descendants of cells which have conjugated before the signs of senescence have appeared, perish sooner than their immediate relatives of approximately the same age, which have been prevented from conjugating. Moreover, Jennings has shown that in certain races of *Paramœcium aurelia-caudatum* conjugation may occur at intervals of only one or two weeks, while in other races of the same species conjugation occurs only at intervals of a year or longer, and in still a third race no conjugation was observed during a period of three years, although during this time observation was not so continuous as to preclude the possibility of conjugation having occurred.

Valuable evidence upon the question of the cyclic character of the Protozoan life history is afforded by the work of Woodruff, who has shown that if more natural conditions are substituted for the artificial and more uniform conditions of the laboratory, no cyclic relation appears, in some strains of *Paramœcium* at least. By continually altering the character of the food, and by imitating in other ways the naturally variable conditions of pond life, he has been able to continue a single race of *Paramœcium* for over five years. During this period more than 3000 generations were formed by simple fission, and in all this time conjugation did not occur, and no periods of depression or signs of structural modification could be ob served.[1] Finally, Woodruff has been able to carry a culture of *Paramœcium* on a uniform diet of beef extract, which is sup posed to contain all of the materials necessary for their life, for ten months (about 450 generations) without any indication of senescence.

Such facts as the foregoing show, first, that protoplasmic activity among the Ciliates may not be cyclic in character under

[1] On Sept. 27th, 1912, Professor Woodruff writes that this culture is in its 3265th generation, and still normal.

certain conditions, and second, that when cyclic periods of protoplasmic depression do occur the protoplasm may be restored to a condition of normal vigor, either by physical or chemical stimuli, or by fertilization. Supposing, and the supposition is highly probable though not completely demonstrated, as a fact, that the living processes do tend, in the absence of continued stimulation, to diminish in intensity or otherwise to deviate from the normal, then we find in the process of fertilization a natural means of *insuring* the receipt of stimuli which might otherwise be lacking. The onset of those structural and physiological modifications called senescence, leads to a modification in the behavior of the organisms, *i.e.*, they form gametes and conjugate.

This becomes clearer when we recall that life itself is response—reaction to the stimuli resulting from changed relations. Such a changed relation may result (*a*) from changes in both the environment and the cell or organism, or (*b*) from changes in the environment alone while conditions within the organism remain comparatively uniform, or (*c*) from changes within the organism while the external conditions remain comparatively uniform. Of these three possibilities the first two are certainly the most frequent in the lives of most free-living Protozoa. But we may interpret fertilization as fundamentally a means of ensuring a changed relation through the realization of the third possibility in the absence of the other two. In a way the Ciliates act so as to ensure automatically a changed relation between organism and environment; when external conditions become too uniform to bring forth the normal vegetative activities, the form of reaction actually changes and is modified into gamete formation and fertilization, which immediately leads to an internal disturbance and the condition of uniformity is corrected, whenever it may occur.

Among the Protozoa we find this division of labor between vegetative and conjugative or fertilizing cells occurring whenever internal-external relations demand it. Among the Metazoa, however, such a division of labor must occur at a certain period in the life history, on account of the impossibility of the complete fusion of two whole organisms at any time other than when they are in the form of single cells; consequently we find vegetative and gamete-forming tissues differentiated side by side, and since these are, as components of a single organism, in a fairly constant environment removed from continuously rejuvenating stimuli, it is the function of the gamete-forming tissues to form single cells which can fuse with cells of other individuals and thus, by altering the composition of the organism, alter its relation to external conditions. And the almost universal association of reproduction and fertilization considered as a rejuvenative process among the Metazoa may have an added significance; the two occur together, not because they are directly related to one another, but because they are both occasioned by the same condition, or rather by the same limitation of opportunity, for the complete fusion of multicellular organisms can occur only when these are in the form of single cells, or gametes.

Reference to the third possible significance of fertilization may be more brief because of its extremely hypothetical character. The idea that the process of fertilization is primarily related to the phenomena of variation is associated chiefly with the names of Weismann and Oscar Hertwig. The scanty and uncertain character of the evidence here is indicated by the fact that there are two exactly opposed views as to the nature of the relation. Hertwig maintains that the effect of fertilization is to limit variation within a species, by tending to bring back to the normal, through the process of heredity, the progeny of extreme fluctuations and unusual or abnormal variations (mutations), because the likelihood of their mating with the much more common mediocre or average individuals is so much greater than that of their mating with their likes. Weismann, on the other hand, maintains that the effect of syngamy or "amphimixis" is to cause or promote variation, which would result from the new organic combinations in the continued admixture of the gametic nuclei of different individuals. In this way the process of fertilization becomes of great evolutionary significance in that it accounts, in part at any rate, for the presence, indeed the origin, of variations and fluctuations, the "raw materials" of evolution.

Both of these views are based upon the more fundamental and underlying hypothesis of the representative particle nature of the elements of the chromosomes, or perhaps of other portions of the germ cells, which themselves vary in their structure or their combinations. Here again the occurrence, among the Metazoa, of fertilization only when the organisms are in the form of single cells, grows out of the fact that complete nuclear fusion can occur only when in this state.

There is little direct factual evidence for or against these views, either one of which can be maintained upon theoretical grounds. In a few cases it is known that the amount of variability is not significantly different among sexually (gametically) and asexually (parthenogenetically) produced individuals of the same species. And from the standpoint of more recent studies upon heredity and variation the evidence is chiefly either negative or opposed to the idea that this relation constitutes an important element in the origin or present function of fertilization. The present aspects of this relation between fertilization and variation merge in the larger question of the relation with heredity which we may refer to next.

Whatever the significance of fertilization may prove to have been originally, its relation to the phenomena of heredity is to-day undoubtedly its most important aspect, at any rate among the Metazoa. The general subject of the relation of the structure of the germ cells to the main facts of heredity is reserved for consideration in Chapter VII, but we should point out here some of the underlying conditions involved in the fundamental fact of the union of the two germ cells derived in nearly all cases from two different individuals of the same group or species.

As pointed out in the introductory chapter, the germ cells are not to be regarded as the material links between successive generations of specific organisms, for organismal specificity is not discontinuous, but continuous, and the germ cells are no less specific, no less the organism, than are the mature individuals producing the germ cells or produced by them. From the standpoint of the fertilization process and of heredity, the essential fact is not that the zygote develops into an individual of the same species to which belonged the organisms producing the gametes, for in parthenogenesis, for example, specific organisms are produced in the absence of fertilization. The significant fact here is that offspring may possess some of those characteristics which are the *individual* possessions of either of the parents. On the whole, offspring inherit, or may inherit, equally from both parents, and such a possibility must depend upon the fact that the zygote is composed of substances or structures derived from both the parent organisms.

Of course the only substance of the zygote which is derived in equal or approximately equal parts from the two parents is the chromatic portion of its nucleus, and it is frequently said that therefore it must be the nuclear structures of the germ cells which are involved in this fact of equal biparental inheritance. And yet the fact should not be disregarded that the sperm does bring into the egg a certain though indeed a small amount of cytoplasm. The fact that the individual parental characters are inherited equally does not necessarily mean that all non-individual characteristics are thus inherited, for all of the more general species characters are common to both parents, and the offspring might conceivably inherit these wholly from either parent. From this point of view the contribution by the ovum of practically the whole of the cytoplasm of the zygote might have at least two meanings. It might mean that the general species characters of the offspring are determined by the structure of the cytoplasm, and only the individual traits by the nuclear structures; it would follow from this that the spermatozoön takes a relatively subsidiary part in heredity. Or it might mean that the two gametic nuclei are from the beginning equally involved in the determination or direction of development, while the cytoplasm of the ovum merely affords the great bulk of the material basis for this development, and is itself in no wise involved in the qualitative determination of either specific or individual characters of the offspring. It would follow from this that the two germ cells are of more nearly equal importance in the process of heredity.

We may finally conclude from all of the foregoing discussion that little can be very definitely asserted regarding the real function of fertilization now, and still less regarding the original significance of the process. Some of the questions involved here are to-day the most interesting and important of the unsolved problems in the fields of Embryology and Biology.

It is reasonably clear that fertilization is not, at the present time, a simple process, although it may have been so originally. Doubtless there has been an evolution both of the process and of the consequences of fertilization, just as there has been of all organ structure and organ physiology. Furthermore, it seems clear that the various possibilities described above, as to the significance of fertilization are not mutually exclusive: fertilization may be important for several of these reasons, even in a single case, and probably it has no one meaning that is exclusively true. It is quite possible that normally, among the Metazoa to-day, the spermatozoön may bring about in the ovum the formation of centrosomes which do, as a matter of fact, take an important part in the succeeding cleavages of the zygote, it may also chemically and physically stimulate the ovum to develop by bringing about initial changes in its chemical or physical structure or organization, it may at the same time introduce substances, the effect of which is "rejuvenation" of the specific protoplasm apart from the reproductive phenomena, and finally the structure of the spermatozoön which does these things may also affect the course of development so that individual characteristics of the male parent, as well as of the female parent, may appear. And to say that the result of fertilization is, for example, rejuvenation, need not mean that it is not also a stimulus to reproduction, a controlling factor in variation, and a means of heredity.

REFERENCES TO LITERATURE

ADOLPHI, H., Ueber das Verhalten von Wirbeltierspermatozoen in strömenden Flüssigkeiten. Anat. Anz. **28.** 1906.

VAN BENEDEN, E., (Ref. in Ch. II.)

BOVERI, T., Zellenstudien. I, II, III. (Ref. in Ch. II, III.) Zellen-studien. IV. Ueber die Natur der Centrosomen. Jena. Zeit. **35 (28).** 1901. Die Polarität von Ovocyte, Ei und Larve des *Strongylocentrotus lividus.* Zool. Jahrb. **14.** 1901. Das Problem der Befruchtung. Jena. 1902.

BULLER, A. H. R., Is Chemotaxis a Factor in the Fertilisation of the Eggs of Animals? Q. J. M. S. **46.** 1902.

CALKINS, G. N., Studies of the Life-history of Protozoa. I. The Life-cycle of *Paramœcium caudatum.* Arch. Entw.-Mech. **15.** 1902. IV. Death of the A Series. Conclusions. Jour. Exp. Zool. **1.** 1904. (See also, ref. Ch. I, II.)

COE, W. R., (See ref. Ch. IV.)

CONKLIN, E. G., (See ref. Ch. II, III.)

DELAGE, Y., Études sur la Mérogonie. Arch. Zool. Exp. (III), **7.** 1899. Élevage des larves parthénogénétiques d' *Asterias glacialis.* Arch. Zool. Exp. (IV), **2.** 1904.

DOBELL, C. C., (Ref. in Ch. II.)

DOFLEIN, F., (Ref. in Ch. I.)

DRAGO, U., Nuove ricerche sull' "attrazione" delle cellule sessuali. Arch. Entw.-Mech. **26.** 1908.

ELPATIEWSKY, W., Zur Fortpflanzung von *Arcella vulgaris,* Ehrb. Arch. Protist. **10.** 1907.

GODLEWSKI, E., Untersuchungen über die Bastardierung der Echiniden- und Crinoidenfamilie. Arch. Entw.-Mech. **20.** 1906.

HARTMANN, M., (Ref. in Ch. II.)

HARVEY, E. N., Methods of Artificial Parthenogenesis. Biol. Bull. **18.** 1910.

HERTWIG, O., Beiträge zur Kenntniss der Bildung, Befruchtung und Theilung des thierischen Eis. Morph. Jahrb. **1-4.** 1875–1878. (See also ref. Ch. III.)

HERTWIG, R., Ueber Kerntheilung, Richtungskörperbildung und Befruchtung von *Actinosphærium Eichhorni.* Abh. Acad. München (II Cl.) **19.** 1899. Was veranlasst die Befruchtung der Protozoen? Sitz.-Ber. Ges. Morph. Phys. München. **15.** 1899. Eireife und Befruchtung. Hertwig's Handbuch, *etc.* I, 1, 1. 1903. (1906). Ueber den Chromidialapparat und der Dualismus der Kernsub-stanzen. Sitz.-Ber. Ges. Naturf. München. 1907.

JENNINGS, H. S., What Conditions induce Conjugation in Paramecium? Jour. Exp. Zool. **9.** 1910.

KORSCHELT UND HEIDER, Lehrb., *etc.* (See ref. Ch. III.)

KOSTANECKI, K., Ueber parthenogenetische Entwicklung der Eier von *Mactra* mit vorausgegangener oder unterbleibener Ausstossung der Richtungskörper. Arch. mikr. Anat. **78, II.** 1911

KUPELWIESER, H., Entwicklungserregung bei Seeigeleiern durch Molluskensperma. Arch. Entw.-Mech. **27.** 1909.

Lillie, F. R., Studies of fertilization in *Nereis*. III. The morphology of the normal fertilization of *Nereis*. IV. The fertilizing power of portions of the spermatozoön. Jour. Exp. Zool. **12**. 1912. (See also ref. Ch. III.)

Loeb, J., Ueber die Befruchtung von Seeigeleiern durch Seesternsamen. 2 Mitth. Arch. gesamte Physiol. **99**. 1903. Artificial Membrane-formation and Chemical Fertilization in the Starfish (*Asterina*). Also, On an Improved Method of Artificial Parthenogenesis. Univ. Calif. Publ. Physiol. **2**. 1905. Die Chemische Entwicklungserregung des thierischen Eies. Berlin. 1909.

Mark, E. L., Maturation, Fecundation, and Segmentation of *Limax campestris*. Bull. Mus. Comp. Zool. Cambridge. **6**. 1881.

Masing, E., Ueber das Verhalten Nucleinsäure bei der Seeigeleies. Zeit. Physiol. (Hoppe-Seyler.) **67**. 1910. See Arch. Entw.-Mech. **31**. 1911.

Maupas, E., Le rejeunissement karyogamique chez les Ciliés. Arch. Zool. Exp. (II), **7**. 1889.

Minot, C. S., The Problem of Age, Growth, and Death. New York. 1907.

Morse, M. W., Maturing Reagents and those inducing Segmentation in Artificial Parthenogenesis. Science. (Proc. Amer. Soc. Zoologists.) **33**. 1911.

Němec, B., Das Problem der Befruchtungsvorgänge und andere cytologische Frage. Berlin. 1910.

Phillips, E. F., A Review of Parthenogenesis. Proc. Am. Phil. Soc. **42**. 1903.

Rückert, J., Ueber Polyspermie. Anat. Anz. **37**. 1910.

Schaudinn, F., Ueber die Copulation von *Actinophrys sol*. Sitz.-Ber. Acad. Wiss. Berlin. **5**. 1896.

Silvestri, F., Ricerche sulle fecondazione di un animale a spermatozoi immobili. Rich. Lab. Anat. Roma e altri Lab. Biol. **6**. 1902.

Sobotta, J., Die Reifung und Befruchtung des Eis von *Amphioxus lanceolatus*. Arch. mikr. Anat. **50**. 1897.

Tennent, D. H., and Hogue, M. J., Studies on the Development of the Starfish Egg. Jour. Exp. Zool. **3**. 1906.

Weismann, A., (See ref. Ch. I.)

Wilson, E. B., (See ref. Ch. II.)

Woodruff, L. L., The Life Cycle of *Paramecium* when subjected to a Varied Environment. Amer. Nat. **42**. 1908. Two Thousand Generations of *Paramœcium*. Arch. Protist. **21**. 1911. A Summary of the Results of certain Physiological Studies on a Pedigreed Race of *Paramœcium*. Biochem. Bull. **1**. 1912.

Yatsu, N., The Formation of Centrosomes in Enucleated Egg-Fragments. Jour. Exp. Zool. **2**. 1905.

CHAPTER VI

CLEAVAGE

THE grosser and externally visible processes of development begin with the cleavage of the fertilized ovum, or zygote. The period of cleavage therefore may be regarded as the second of the "grand periods" in individual history. During the first general period occur all the events leading up to and including the final establishment of the zygote, a single cell, but a new organism. During this second period the Metazoan really becomes made up of many cells.

The essential process underlying many of the varied phenomena of cleavage is a process already familiar, mitotic cell division; but it is true that cell division continues long after the cleavage period proper is terminated, in some tissues throughout the life of the organism. And as we shall soon see, the process of cleavage involves a great deal more than merely a succession of cell divisions.

Certain general characteristics of the mitoses of the period of cleavage, or *segmentation,* of the zygote, may be observed, but it is difficult to state precisely wherein these cell divisions differ from those of later development. Probably the most significant characteristic of the divisions of this period is that they are rarely at random, but nearly always occur in an orderly fashion, according to a definite schema or plan, which is quite fixed for each species or larger group, and which involves the entire cell community. The mitoses of cleavage are frequently very unequal and the daughter cells may be very unlike, not only in size, but further as regards cytoplasmic character and the nature of various cell inclusions, which may be distributed dissimilarly during these divisions. After the first few mitoses the blastomeres may not divide synchronously, so that the regular and rhythmic geometric increase in their number to 2, 4, 8, 16, *etc.*, is very rarely continued after eight or sixteen cells have been formed; the regularity of division may be disturbed as early as the second cleavage. In some forms (*e.g.*, Echinoderms, Godlewski) the nuclei of the daughter cells enlarge considerably after each division, in some cases perhaps even to the original size: the cell bodies fail to do so. The result is the constant increase in the relative size of the cell nuclei; in other words, while the amount of cytoplasm increases only slightly or not at all during cleavage, the amount of nucleo-

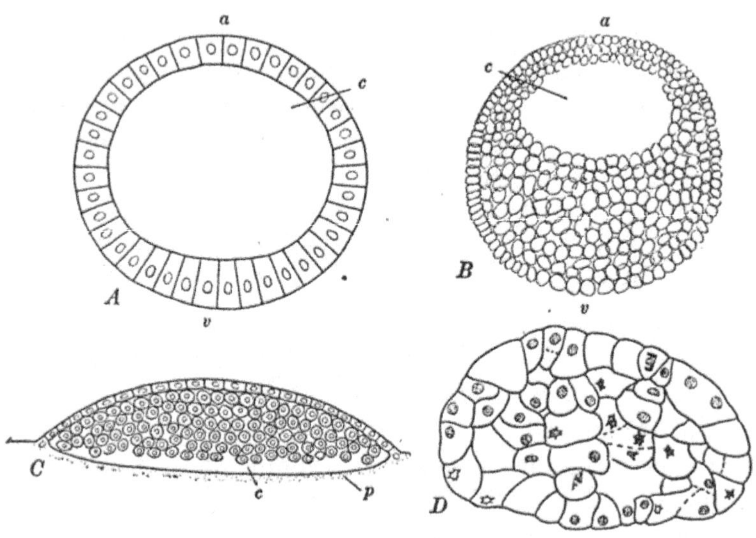

Fɪɢ. 105.—Types of blastulæ. *A.* Amphioxus (cœloblastula). *B. Petro. myzon.* After von Kupffer. *C. Noturus* (Teleost) (discoblastula). *D. Clava* (Hydroid). After Hargitt. (Solid type.) *a*, animal pole; *c*, segmentation cavity or blastocœl; *p*, periblast (a non-cellular protoplasmic layer resting upon the yolk mass); *v*, vegetal pole.

plasm increases considerably, so that at the close of this period the organism contains an appreciably greater proportion of nuclear material than did the zygote. This, however, may not be regarded as a general characteristic of the cleavage mitoses in all organisms (Conklin).

After a number of cells, varying in different species, have been formed they become arranged so as to limit an internal cavity filled with a fluid. In its simplest and apparently most primitive or typical form, the figure thus established is a hollow sphere, the wall of which is composed of a single layer of cells or blastomeres. This structure, however its actual form may deviate from this type, is termed the *blastula*, and the cavity within is the *blastocœl*, or *segmentation cavity* (Fig. 105). The diverse forms of the blastula depend immediately upon the arrangement of the preceding cell divisions; the blastula may in some cases be almost or quite solid, so that the blastocœl exists only virtually.

About the time the blastula is formed the successive cleavages have reduced the cell size to a physiological minimum and thereafter the daughter cells increase in size subsequently to each division, and there is no further reduction in the size of the blastomeres; the volume of the cytoplasm, as well as of the nucleoplasm, commences to increase, in other words the organism begins to grow. The relative time of the appearance of this growth phase is widely different in different forms; in the Echinoids it appears when about sixty-four cells have been formed (Godlewski). While there is no general and externally visible indication marking a definite close of the cleavage period, the formation of some type of blastula, or the initiation of cytoplasmic growth, is more or less arbitrarily assumed to mark its termination, although many of the processes characteristic of this phase of development, including of course cell division, may continue for some time longer. We may now define cleavage as that early period of development characterized externally by a rapid and orderly succession of mitoses, which results in the formation, from the zygote, of a regularly arranged group of small blastomeres possessing relatively large nuclei.

In most cases there is a marked tendency for the blastomeres to assume a spheroidal form, more or less modified by the tension with which the cells are held together, by the pressure of the egg membranes, *etc*. Sometimes the cells round up so as to become practically spheres, in contact with one another by greatly restricted surfaces (Amphioxus, Echinoderms, Cœlenterates). In other instances (frog, chick, Arthropods) the separations between the blastomeres are mere grooves or furrows on the surface of the mass; here the cells are broadly in contact and in some instances remain connected by very delicate protoplasmic bridges similar to those connecting tissue cells previously mentioned. In a few instances (some Arthropods, a few Cœlenterates) these early cell divisions may be imperfect, the nuclei alone dividing and forming a syncytium; later the cytoplasm also divides simultaneously into a number of complete cells. Or cells once formed may fuse into syncytial masses, as in some Crustacea.

The internal processes of development occurring during this period are of greater importance than the external phenomena. As stated above (Chapter V) one of the underlying processes of great importance seems to be the synthesis of chromatin which occurs at this time.

In the opinion of many this is a highly characteristic chemical process of early development. It results from rapid oxidations within the egg which were made possible by the transformation of the egg membrane from a condition of relative impermeability, to a state of high permeability, oxygen thus being readily admitted from without. This transformation is thought to result from the chemical reactions of the cytoplasm following fertilization, during which there occurs also the activation, or perhaps the introduction, of specific enzymes which bring about this characteristic oxidative synthesis.

This view as to the chemical process most essential in cleavage agrees well with the "kern-plasma" theory of Richard Hertwig, according to which, as already mentioned, the ovum and zygote are to be regarded as instancing abnormal or especially adapted relations between nucleus and cytoplasm; for here the relative amount of cytoplasm is far in excess of the common proportion. In cleavage, with its proportional increase in nuclear substance, we should see a restoration to or toward the normal of the kern-plasma ratio $\dfrac{\text{Volume nucleus}}{\text{Volume cytoplasm}}\left(\dfrac{V_n}{V_c}\right)$

Another internal process is of prime importance. We have already become familiar with certain facts—that one result of maturation and fertilization is the presence in the zygote of two similar chromosome groups, derived respectively from the male and female parents; that while the egg nucleus nearly always returns to a "resting" stage before its fusion with the sperm nucleus, the latter may or may not do likewise before the fusion of the two germ nuclei into the cleavage nucleus; that the egg and sperm nuclei may or may not form separate spiremes each with $\frac{s}{2}$ chromosomes during the prophase of the first cleavage figure of the zygote; and finally that whatever the preliminary details may have been, the constant and essential facts regarding this first cleavage figure are, (1) that the chromosome group now consists of the full somatic number of univalent elements, (2) that these are present in pairs, and (3) that they have been derived equally from the two parents.

As cleavage begins the first important step is the longitudinal division of each chromosome; the halves diverge during the anaphase of the first mitosis, and into the nucleus of each daughter cell there passes a precisely similar group of s chromosomes, paired as before and *derived in equal numbers from each of the two parents*. In each succeeding mitosis the same thing happens. So that in every cell of the blastula, and probably even of the fully matured organism, the nucleus is composed of substance derived in equal parts from the male and the female parents (Fig. 106). In some forms (Copepods, *Ascaris*) the two parental chromosome groups appear to remain fairly distinct up to a late stage in cleavage (Fig. 107). And in some cases of hybridization, when the chromosomes of the two parents are easily distinguished by differences in size and form (*e.g.*, the hybrids of *Fundulus* and *Menidia* described by Moenkhaus), such a process of equal distribution of the chromosomes can be clearly followed into the blastula stage (Fig. 39). It may fairly be assumed, therefore, that Huxley's comparison of the body of an organism with a web, of which the warp comes from one parent, the woof from another, has been justified by the subsequently discovered facts of development. This idea has been spoken of as the "autonomy of the male and female chromosome groups." It follows from this that the parental chromosome groups of the primordial germ cells of the new

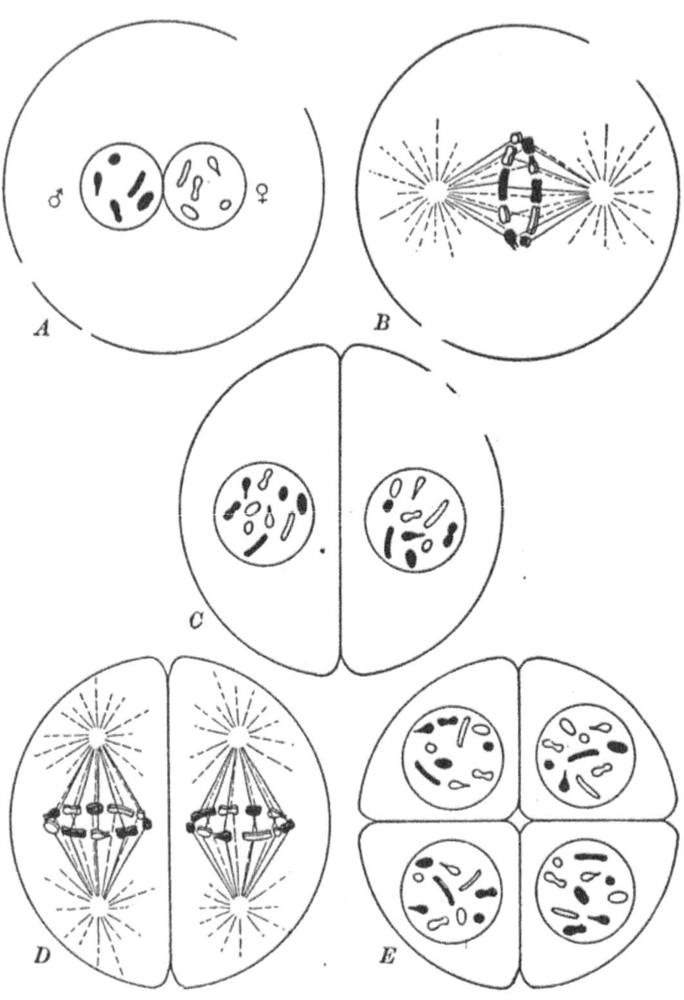

Fig. 106.—Diagrams illustrating the distribution of the paternal and maternal chromosomes during cleavage. *A*. Zygote containing sperm, ♂, and egg, ♀, pronuclei, with similar chromosome groups. *B*. First cleavage figure. All the chromosomes on the spindle, and each divided. *C*. Two-cell stage, each nucleus containing equivalent chromosome groups of paternal and maternal origin. *D*. Second cleavage figure; the first figure is repeated in each cell. *E*. Four-cell stage. Nuclei all alike, and each composed of similar contributions from each parent.

organism are also separate and remain so through their descendants, the oögonia and oöcytes, or spermatogonia and spermatocytes, of the mature individuals, until their period of synapsis, when the members of each pair of chromosomes, similar but of diverse ancestry, unite forming a single bivalent chromosome which is represented in the mature ovum or spermatozoön finally formed. It has already been suggested that this process of synapsis, the ultimate fusion of paternal and maternal chro-

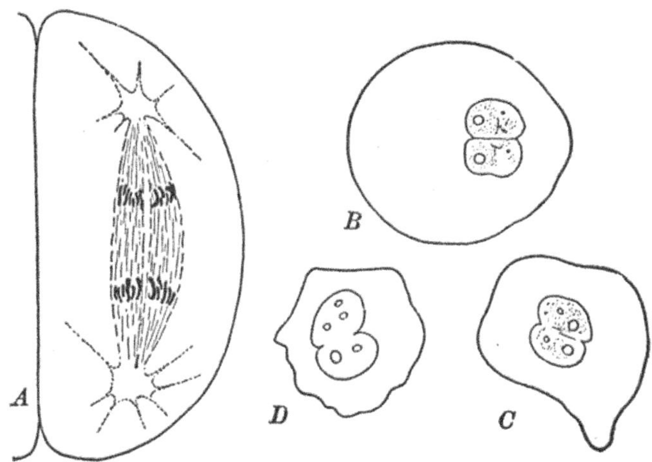

Fig. 107.—*A*. Cleavage figure in one of the first two blastomeres of the egg of the Crustacean, *Cyclops strenuus*, showing the independence of the paternal and maternal chromosome groups. After Rückert. *B, C, D*. Primitive germ cells from embryos of the skate, *Raja*, showing the duplex character of the nuclei. *B* and *C* are from a stage about the close of gastrulation; *D* from a larva of 10 mm. After Beard.

mosomes, may be regarded as the final step in syngamy, and that it is at the same time the first step in the beginning of a new organism.

With these briefly stated introductory facts in mind we may proceed to a more precise description of the events of the cleavage period. The process of cleavage may be described in several different ways, or rather from several different viewpoints. We may group these all under two heads and consider cleavage, first, as a morphological process, emphasizing primarily the *forms* of cleavage and describing the relation of the cleavage planes and the blastomeres (*a*) to the entire zygote, and (*b*) to each other. Then second, we may emphasize chiefly the physiological aspects of cleavage describing the relation of the cleavage *processes* (*a*) to the structure or organization of the ovum and zygote, (*b*) to the later stages in the development of the mature organism. As a matter of fact these aspects of cleavage are not really separate, for all the particular phenomena of cleavage, as of development in general, are to be referred to a single fundamental condition, namely, the organization of the ovum or zygote as it is related to external conditions; and if our knowledge were complete here we should be able to describe all the phenomena of cleavage from a single viewpoint. But for the present we shall find it more convenient as well as more instructive to separate more or less arbitrarily and to consider apart, the chiefly morphological and the chiefly physiological aspects of this process.

In considering the relation of cleavage to the grosser structure of the zygote we find that one of the primary factors in determining the form of cleavage is the relative amount and the form of distribution of the yolk and other deutoplasmic substances contained in the ovum. In Chapter III, three types of eggs were described on this basis: (1) *homolecithal* or *isolecithal* (*alecithal*), containing little deutoplasm, distributed with considerable uniformity throughout nearly the entire ovum: (2) *telolecithal*, containing varying, often considerable amounts of deutoplasm chiefly localized toward the vegetative pole of the ovum; (3) *centrolecithal*, really a form of telolecithal ova in which the deutoplasm has a central rather than a polar localization.

Corresponding in a general way with these variations in yolk distribution we may distinguish certain types of cleavage, each however with certain variations which may sometimes appear as connecting intergradations. First we may distinguish *complete* and *incomplete* cleavage. When the eggs are comparatively small and of the homolecithal type, the earlier cleavage planes pass completely through them in meridional and latitudinal planes. Theoretically the simplest form of complete cleavage is that known as *equal* cleavage, where the egg and the blastomeres formed from it are always divided equally, so that the constant result is a group of similar cells. This is rarely if ever completely realized; the nearest approach to it is seen in the Holothurian, *Synapta* (Fig. 108). In most examples of so-called equal cleavage slight inequalities may be detected even as early as the two-cell stage (Amphioxus), and quite frequently in the four-, or eight-cell stages. This modification of equal cleavage is known as *adequal*. Amphioxus and some of the Echinoderms (Fig. 109) illustrate the fact that no sharp distinction can be drawn between equal and unequal cleavage, for equal cleavage soon becomes unequal, and the transition appears gradually.

More frequently the cleavage though still total is distinctly *unequal*, at least by the time eight cells are formed, and often from the very beginning of cleavage. This is characteristic of those telolecithal eggs in which the accumulation of yolk is slightly or moderately marked, as in most of the Platyhelminthes, Nemathelminthes, Annulata, Trochelminthes, Mollusca, Ganoids, and Amphibia (Figs. 110, 111). This leads to a second general type of cleavage, the *incomplete* type, where a portion of the ovum remains uncut by the cleavage planes (Fig. 113). Such eggs are known in general as *meroblastic*, in distinction from the *holoblastic* ova whose cleavage is complete. In telolecithal eggs with very large accumulations of deutoplasm, the cleavage planes are nearly restricted to the protoplasmic region and extend only a short distance out into the yolk; this is known as *partial* cleavage. When the protoplasmic part is quite definitely restricted to the animal pole, cleavage is of an extremely incomplete type known as *discoid* (Fig. 116), and the result is the formation of a small cap or disc of cells on the surface of the yolk mass (Teleosts, Reptiles, Birds). Or, if the egg is of the *centrolecithal* type, cleavage is limited to the peripheral protoplasmic layer and is known as *superficial* (most Arthropods) (Figs. 117, 118).

Since there are all intermediate conditions between homo-, telo-, and centrolecithal types of ova, we find, as we should expect, all corresponding intermediate conditions between these various forms of cleavage; the details are not particularly instructive and may be omitted.

We may now turn to the morphological description of the relations of the blastomeres among themselves. Before describing the various relations which these may exhibit, it will be useful to describe briefly a simple form of total and equal cleavage, which may be regarded as a typical form. Such a form of cleavage is indeed rare but it is found in the homolecithal and holoblastic egg of the sea-cucumber, *Synapta*, as

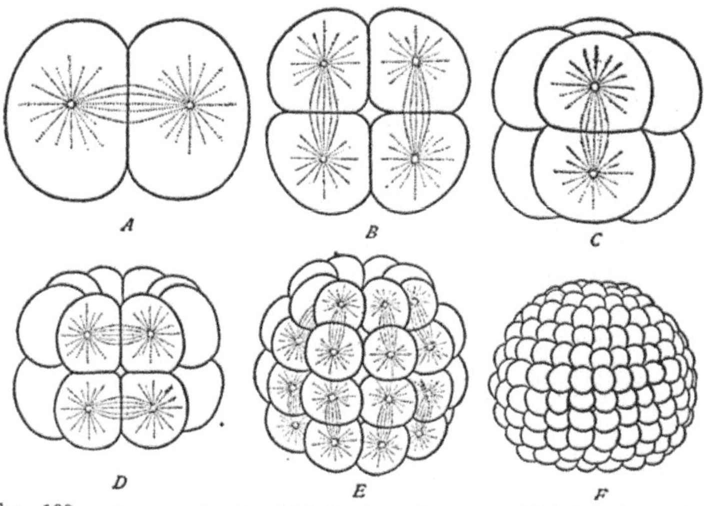

Fig. 108.—Cleavage in the Holothurian, *Synapta*, Slightly schematized. From Wilson, "Cell," after Selenka. *A–E*. Two-, four-, eight-, sixteen-, and thirty-two cell stages. *F*. Blastula of 128 cells. *B*, in polar view, others in side view.

described by Selenka (Fig. 108). The earlier cleavage planes always appear in a definite relation to the polar structure of the ovum, and they are described as if the main axis of the egg were in a vertical position. The first cleavage plane passes through both poles and the chief axis of the egg, dividing it equally and appearing on the surface as a complete meridian. The second plane is similarly *meridional* or *vertical*, is at right angles to the first, and divides the egg into four equal quadrants. The third division plane is at right angles to the first two and is therefore *horizontal*. In *Synapta* it is practically midway between the poles of the egg and is therefore described as *equatorial*. In most ova cleaving according to this general rule, the third plane is displaced a variable distance above the equator and is then termed *latitudinal*. When equatorial this cleavage divides the four equal cells into eight, again equal, arranged in upper and lower groups of four, known as the upper and lower *quartets*. The fourth cleavage is again meridional and is really double, for two planes appear simultaneously dividing each pair of opposites in each quartet similarly; this results in the formation of upper and lower octets. The fifth cleavage is horizontal or latitudinal, and is again double for it divides simultaneously the upper and lower octets each into two horizontal groups of eight cells, so that the ovum is now divided into eight vertical rows of four cells each. The cleavages continue to alternate meridionally and vertically, until about the ninth cleavage when 512 cells are formed. After this, and in fact usually before this time, the synchronism of cleavage begins to be disturbed, some of the cells dividing more rapidly.

One of the very frequent causes of departure from this simple schema is the telolecithal character of the ovum. Here the upper quartet is usually smaller than the lower and the fourth cleavage appears earlier in the upper quartet or cells of the animal pole. This leads very soon to an irregularity in the rhythm of cleavage, which may be entirely lost after eight or sixteen cells are formed.

This typical outline of cleavage serves as a basis to illustrate certain "laws" of cleavage which may be referred to briefly at this point, although their applicability is now known to be very limited. The first of these is the Sachs-Hertwig law describing the geometric relations of the successive cleavage planes. This law really consists of two parts which may be stated as follows: (1) The nucleus of a blastomere (or of any cell) tends to assume a position near the center of the *protoplasmic* mass. From this results the equal division of the cell, provided it is free from deutoplasm, or its unequal division if the cell contains deutoplasm not uniformly distributed, for in the latter case the center of the protoplasmic mass does not correspond with the center of the entire cell. (2) The chief axis of the mitotic figure tends to lie in the longest axis of the protoplasmic mass. The result of this is that in cells that are approximately spherical and homogeneous with respect to yolk content, successive cleavage planes tend to alternate at right angles with one another, for it would always be the longest axis of the cell that is divided, and in most cases any other axis would be greater than one-half the longest and no two successive spindles would be parallel. The regular alternation of cleavage planes probably depends, as a matter of fact, upon a more fundamental relation, namely, the position of the centrosome. At the conclusion of a mitosis the centrosome lies at one end of the axis passing perpendicularly to the plane of division; when the centrosome divides, its halves usually migrate symmetrically to opposite sides of the nucleus, occupying the poles of an axis lying parallel with the plane of the preceding division, and since division always occurs at right angles to the axis connecting the centrosomes, the plane of one division will be at right angles to that of the preceding or succeeding cleavage (Fig. 24). Any other relation between successive cleavage planes involves either a change in the relative position of the centrosomes, or a rotation of the cleavage spindle after its formation. Thus in the formation of a simple epithelium, where successive cleavages are nearly parallel, the centrosomes migrate through approximately 90° at some time during the interkinesis; and in the cleavage of some ova encased in comparatively rigid shells, the position of the spindle may change (*Lepas*, Bigelow).

Balfour's law of cleavage, which is really a corollary of the first part of the Sachs-Hertwig laws, concerns the rate rather than the geometrical relations of cleavage. This law states that the *rate* of cleavage is inversely proportional to the amount of deutoplasm contained within the cell. It follows from the fact that the nucleus tends to lie in the center of the protoplasmic mass, that in the unequal division of cells containing localized deutoplasm, the smaller cell will contain relatively a smaller proportion of yolk than the larger cell, and consequently, being free from the influence of the dead and inert deutoplasm, will be able to divide sooner.

While these laws are often applicable to the processes of cleavage in a general way, the exact study of cleavage in a great variety of forms has disclosed very numerous exceptions and contradictions. On the whole we may say that such laws, though still retaining a limited applicability, are chiefly interesting as indicating the attempt to refer the phenomena of cleavage to the grosser mechanical relations of cell structures. It is now clear, as we shall see later, that other factors are of greater importance in determining the form and rhythms of cleavage. The fundamental "organization" of the ovum, which is not only morphological but physiological as well, is the primary factor in determining the characteristics of cleavage. The numerous "exceptions" to these laws of cleavage are definitely related to both the organization of the ovum and also to the structural and functional characters of the later stages of development, since these too are primarily determined by the same organization factor.

Most of the conditions which form exceptions to these rules, and are therefore deviations from the simple and regular form of cleavage like that of *Synapta*, may, following Wilson ("The Cell," *etc.*), be grouped under three heads. (1) *Unequal Division*. While this is usually related to differences in deutoplasmic content, there are many instances where no such relation can be made out and the inequalities must be explained upon other grounds (*e.g.*, the micromeres of the Echinoids, the teloblasts of certain Annelids and Molluscs). (2) *Cell Displacement*. This may result from the atypical position of the spindle or from the shifting of blastomeres after they have been formed. Often the individual blastomeres are only lightly held together since they normally show a tendency, often very marked, to assume a spherical form. Under these conditions they might tend to assume a position described by the law (Plateau's) of "least surfaces" or "minimal contact," according to which a group of elastic spheres, like bubbles, held together and yet free to move, tend to become arranged in such a way as to reduce their exposed surfaces to a minimum. But there are frequent exceptions here as in the case of the other "laws" of cleavage. Furthermore, the active migration of blastomeres is not infrequent, so that cells may ultimately be found in regions considerably removed from the place of their formation (Rotifers, Molluscs). (3) *Rhythm*. The rate of division frequently does not correspond with the relative amount of deutoplasm. The factors regulating the rhythm of division still remain largely unknown. It is true here as in many other "exceptional" cleavage phenomena, that the deviation is related to the future morphological or functional character of the developing organism or of parts of it. We shall return to this aspect of cleavage later.

With these general considerations in mind we may proceed now to a more exact description and classification of the geometric forms of cleavage. Here we shall find illustrations of many of the preceding statements. Considering first the various forms of *complete* cleavage (holoblastic ova), we may distinguish rather roughly, four types, *radial, spiral, bilateral, irregular*.

(1) *Radial*.—This is the form exemplified by *Synapta* (Fig. 108), already described as being geometrically the simplest. This should perhaps better be termed *rotatorial* than radial, for while the blastomeres are arranged in symmetrical fashion in any single plane perpendicular to the main axis of the egg, there are usually considerable differences in size between the cells of the animal and vegetal poles. Cleavage of this type is found in the sponges, jelly-fishes, and many Echinoderms, in some Nematodes and Rotifers. In the sea-urchins (Fig. 109) the third cleavage is meridional in the upper quartet, in the lower latitudinal and very unequal, cutting off a quartet of very small cells or micromeres which curiously are found at the lower or vegetative pole.

(2) *Spiral*.—This may be regarded as a modification of the radial type resulting from the displacement of cells so that the blastomeres above and below any horizontal cleavage furrow tend to alternate with one another in a vertical direction, some-

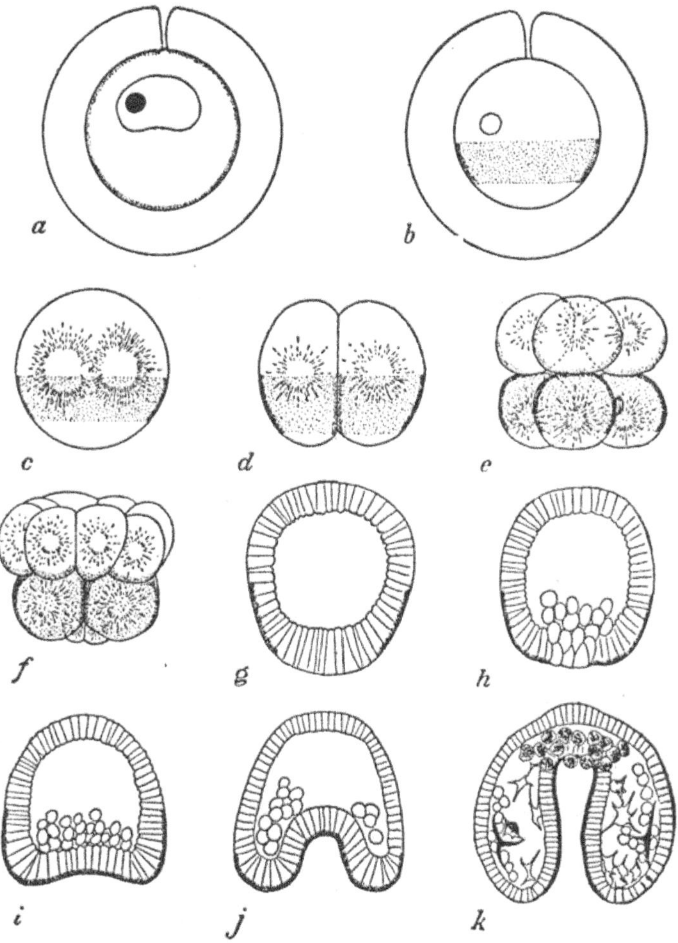

Fig. 109.—Cleavage in the sea-urchin, *Strongylocentrotus lividus*. From Jenkinson, after Boveri. Animal pole uppermost in all cases. *a*. Primary oöcyte surrounded by jelly, and containing large germinal vesicle with nucleolus. Pigment uniformly distributed over surface. *b*. Ovum after formation of polar bodies. Pigment forms a band below the equator. *c, d*. First cleavage. *e*. Eight-cells. Pigment almost wholly in lower quartet (vegetative blastomeres). *f*. Sixteen-cells. The lower quartet has divided latitudinally and unequally, forming four micromeres at the vegetal pole; the upper quartet has divided meridionally forming a plate of eight cells. *g*. Section through blastula. *h*. Later blastula, showing formation of mesenchyme at lower pole. *i, j, k*. Three stages in gastrulation, showing the infolding of the pigmented cells to form the endoderm (archenteron). In *j* the primary mesenchyme is separated into two masses, in each of which a spicule is formed (*k*). In *k* the secondary, or pigmented, mesenchyme is being budded off from the inner end of the archenteron.

what like the bricks in a wall or the bones of the wrist. They may be cut off in this way on account of the obliquity of the spindle in the parent cell, or they may shift to this position after having been formed according to the radial plan. This spiral arrangement may be foreshadowed in the four-cell stage by the meeting of the first two cleavage planes at the poles of the egg in the form of a zig-zag line instead of at a common point.

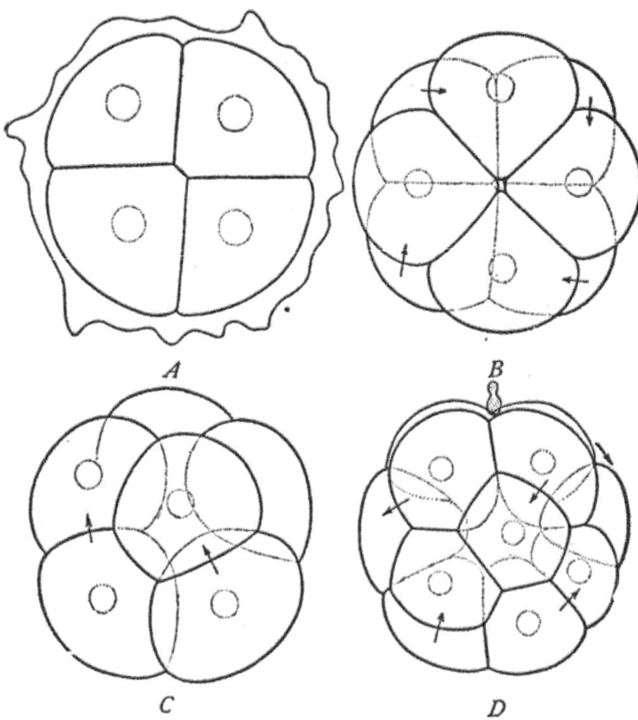

FIG. 110.—Cleavage in the Annulate, *Polygordius*. From Wilson, "Cell." *A, B.* Four- and eight-cell stages, from the animal pole. *C.* Side view of eight-cell stage. *D.* Side view of sixteen-cell stage.

One of the simplest illustrations of this type is the adequal cleavage of *Polygordius* (Fig. 110) but it is also well represented by the markedly unequal cleavage of many Platyhelminthes, Nemertines, Annelids, and Molluscs (Figs. 111, 112, 119). These forms illustrate at the same time a graduated series in the inequality of the blastomeres. This inequality may appear in

different stages; in the very first division of the egg (*Nereis*);
at the second (*Clavelina*); third (*Cerebratulus*, Fig. 112); fourth
(sea-urchin), or still later (*Synapta*).   The direction which the
spiral takes is fixed in each species; it is described as *dextral*

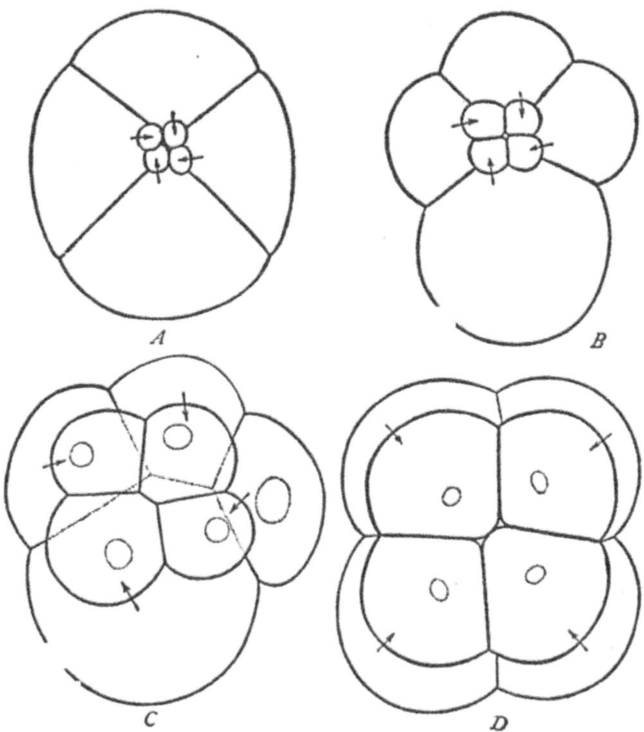

Fig. 111.—The eight-cell stage of four animals showing gradations in the
inequality of the third cleavage, and in the extent of the spiral rotation of the
micromeres.   From Wilson, "Cell."   All viewed from the animal pole.   *A*. The
leech *Clepsine* (Whitman).   *B*. The chætopod *Rhynchelmis* (Vejdovský).
C. The lamellibranch *Unio* (Lillie).   *D*. *Amphioxus*.

(*dexiotropic*) or *sinistral* (*læotropic*) when the upper cells are
rotated clockwise or counter-clockwise respectively, as viewed
from the animal pole.

(3) *Bilateral.*—In this form we see a second modification
of the radial type, which is first indicated by the fact that the
third of the meridional cleavages fails to reach precisely the
poles of the egg but meets either the first or second plane at
some distance from the pole.   In this way it comes about that

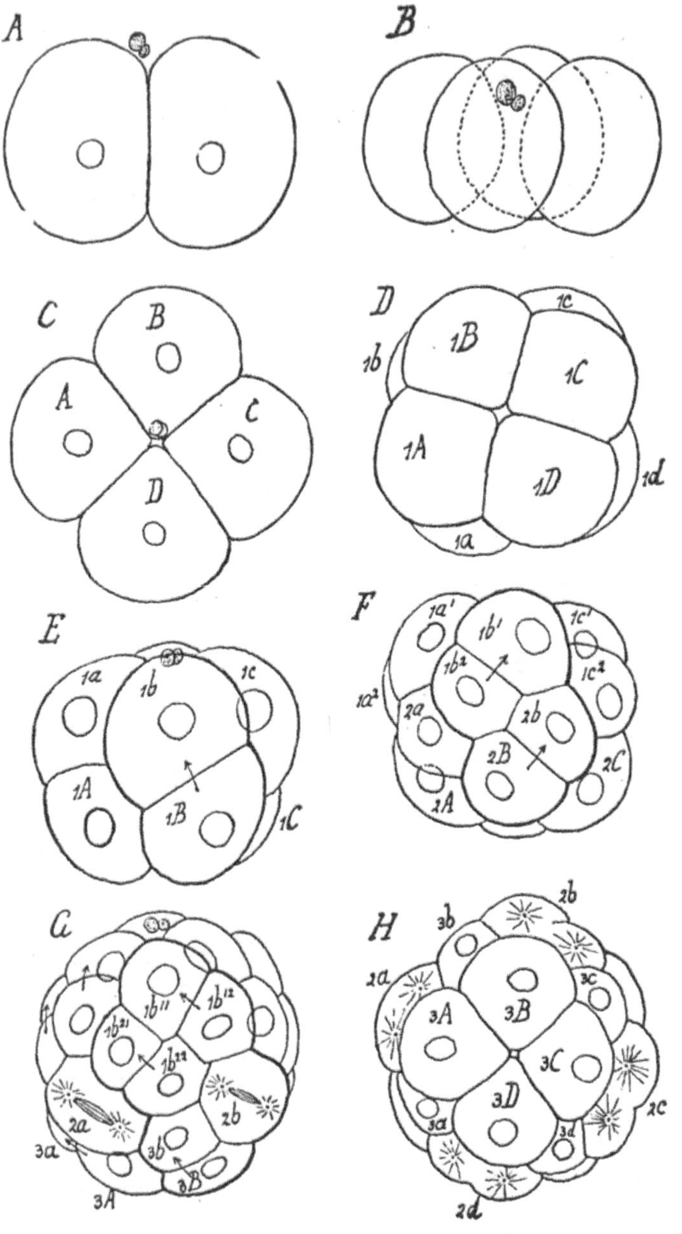

Fig. 112.—Cleavage in the Nemertean, *Cerebratulus marginatus*. From Korschelt and Heider, after Zeleny. × 216. *A, B.* Two- and four-cell stages in side view. *C.* Four-cells, from animal pole. *D.* Eight-cells from vegetal pole. *E.* Eight-cells, side view. *F.* Sixteen-cells, side view. *G.* Twenty-eight cells, side view. *H.* Twenty-eight-cells from vegetal pole. For explanation of lettering, see p. 248.

one of the first two cleavage planes, usually the first, becomes the plane of a bilateral symmetry which may remain quite pronounced for some time, and indeed corresponds with the median plane of the bilaterally symmetrical adult. The bilateral type

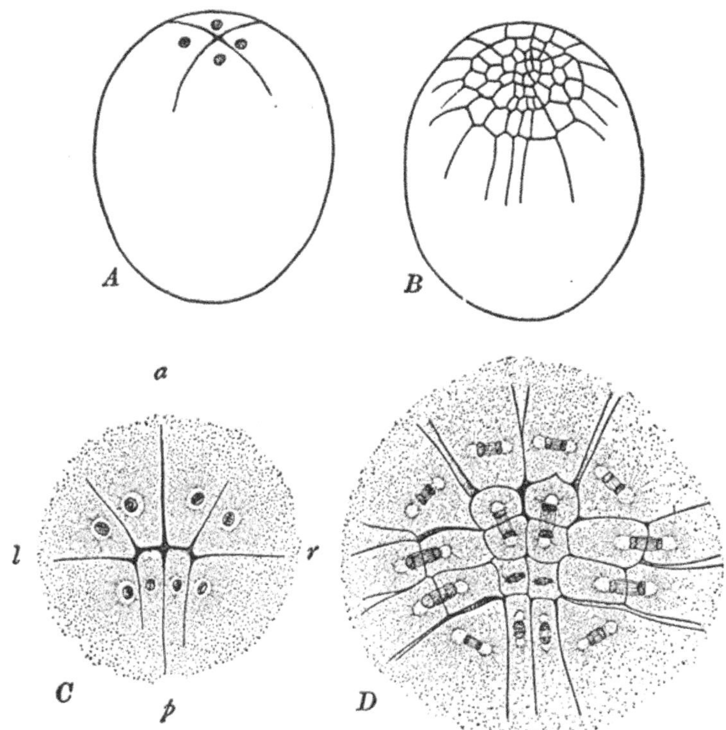

Fig. 113.—Meroblastic cleavage in the squid, *Loligo pealii*. *A, B*. Egg viewed obliquely, showing animal pole. × 45. After Watasé. *C, D*. Surface views of animal pole, more highly magnified, to show bilateral arrangement of blastomeres. From Wilson, "Cell," after Watasé. *A*. Four-cell stage. *B*. About sixty-cells. Cells at animal pole very small, lowermost cells incomplete, cell walls extending down toward the uncleaved lower pole. *C*. Eight-cell stage. *D*. The fifth cleavage (sixteen to thirty-two cells). *a–p*, marks the plane of the first cleavage and the median plane of the organism; *l–r*, marks the second cleavage, and the transverse plane of the organism.

of cleavage is found in the Cephalopods (Fig. 113), a few Rotifers and Nematodes, in Amphioxus and the Ascidians, and perhaps in most of the Craniates, but in these last forms variations are more frequent, especially in those forms with discoid cleavage.

A rather special form of bilateral cleavage known as the *disymmetrical* type is found in the Ctenophores. In these Cœlenterates the first and second furrows are meridional, the planes are complete, and divide the egg adequally. The third, a double cleavage, is oblique to the first, passing in the same general direction as this, on each side of it, but approach-

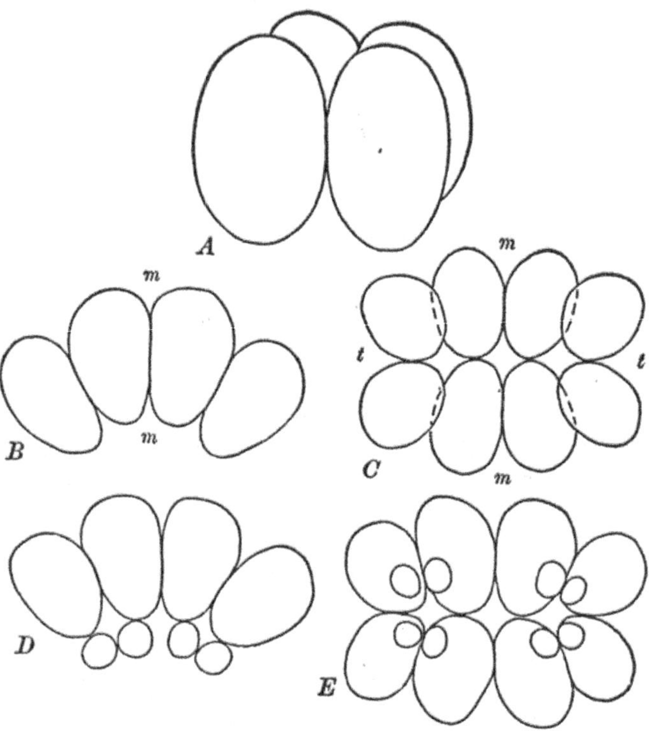

Fig. 114.—Diagrammatic representation of the cleavage in the Ctenophore (based upon *Beroë*). After Ziegler. *A.* Four-cells, from side and above. *B.* Eight-cells in side view (the animal pole is downward here, and in *D*). *C.* Eight-cells, from animal pole. *D.* Sixteen-cells, from side. *E.* Sixteen-cells, from animal pole. *m, m.* Median plane; *t, t,* transverse plane.

ing the vegetal pole more closely than the animal. The descendants of each of the first two cells then become symmetrically arranged about the second plane so that two similar groups of cells are formed, each group bilaterally symmetrical with reference to a plane perpendicular to the plane of symmetry of the entire cell group (Fig. 114).

(4) *Irregular.*—In many phyla, scattered forms are known
in which cleavage adheres to no single or simple type and may
truly be said to be irregular. This does not mean that no
definite plan is followed, for each species follows a fixed rule;
these cleavage forms are in this sense regular, but cannot

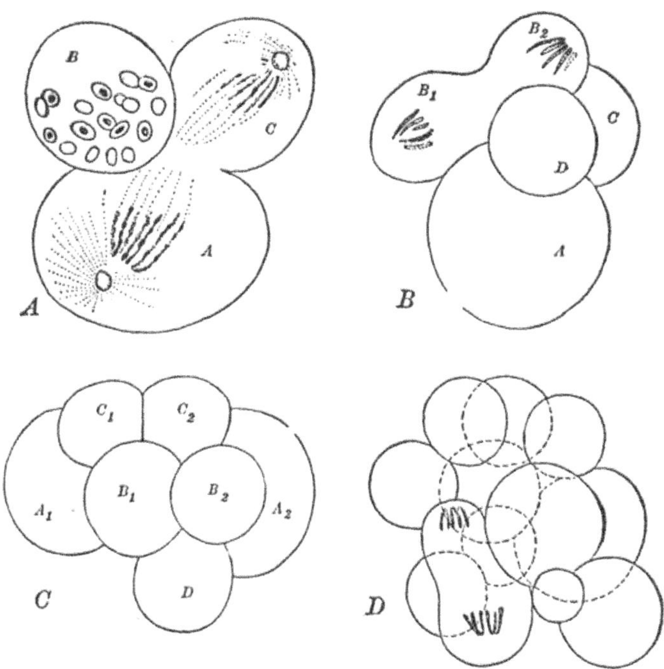

Fig. 115.—Irregular cleavage in the Turbellarian, *Mesostomum ehrenbergi.*
After Bresslau.   × 700.   *A.* Three-cell stage, in section.   *B.* Four-cells
becoming five.   Side view.   *C.* Seven-cell stage, from animal pole.   *D.* Twelve-
cells.   *A.* Macromere, giving rise to A_1 and A_2 in the seven-cell stage.   *B.* First
micromere, forming B_1 and B_2 in the five-cell stage.   *C.* Second micromere
formed in three-cell stage, and giving rise to C_1 and C_2 in the seven-cell stage.

be described in general terms. We cannot stop to describe
any of these instances in detail.   Irregular cleavage may be
found among the Porifera, Cœlenterates, Platyhelminthes,
Molluscoids, Enteropneusta, and Teleosts; a typical example
is illustrated in Fig. 115.

The remaining forms of cleavage are grouped as *incomplete,*
and are found among those species with markedly telolecithal

or centrolecithal ova (meroblastic).   Here little or no geometric
regularity of cleavage pattern can be made out.   We may add
a few details concerning discoid and superficial cleavage to the
brief statements made on a preceding page.

Discoid.—This is chiefly characteristic of the Craniata but it
is found occasionally in the Arthropods (Scorpions).   In the

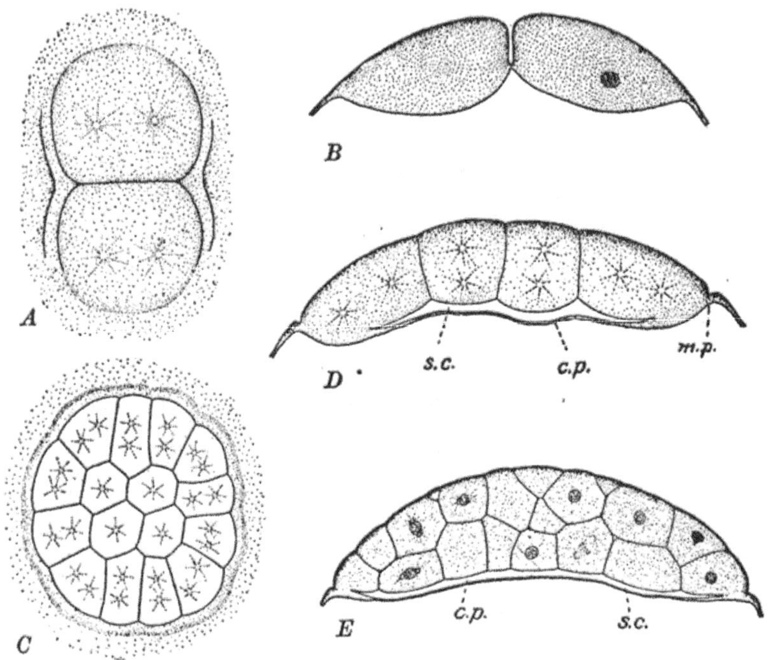

Fig. 116.—Cleavage in the sea-bass, *Serranus atrarius*.   From H. V. Wilson.
A. Surface view of blastodisc in two-cell stage.   *B.* Vertical section through
four-cell stage.   *C.* Surface view of blastodisc of sixteen cells.   *D.* Vertical
section through sixteen-cell stage.   *E.* Vertical section through late cleavage
stage.   *c.p.*, central periblast; *m.p.*, marginal periblast; *s.c.*, segmentation cavity
(blastocœl).

ova of these forms there is often a fairly definite demarcation
between the protoplasmic and deutoplasmic portions (Elasmobranchs, Teleosts, Reptiles, Birds) and the cleavage planes are
practically limited to the former region, known as the *blastodisc* (Figs. 48, 157–159).   The early cleavages may be fairly
regular, and approximate either radial or bilateral arrangements, and as far as the protoplasm alone is concerned, the

cleavage is frequently equal. When the protoplasm forms merely a disc resting upon a large mass of yolk, it is obviously impossible to speak of meridional or latitudinal cleavages; hence the cleavages are described as vertical, either radial or circular, and horizontal. The vertical cleavages soon become connected below the surface of the ovum by horizontal planes, separating the lower surface of the protoplasm from the underlying yolk, and the peripheral circular cleavages similarly separate the protoplasm from the outlying yolk (Figs. 116, 158, *A*). This is seen in the Teleosts, and in many Elasmobranchs, Reptiles, and Birds. After the protoplasmic blastodisc is divided into a number of cells, that is after it becomes a *blastoderm*, other cleavages may occur parallel with the surface, forming internal cells not visible upon the surface (Fig. 116), and the blastoderm may thus come to be many cells in thickness (Figs. 158, 105, *D*).

Some interesting transitions are to be found between total unequal cleavage and discoid cleavage, in those telolecithal eggs where the accumulation of yolk is not as great as it is in the Elasmobranchs, Teleosts, and some of the higher Craniates. Thus in the ganoid, *Amia*, and some of the Urodeles, as well as in the squid (*Loligo*), while cleavage is at first limited to the upper or animal pole, the earlier cleavages gradually extend down through the yolk mass and may finally divide it into a few large cells. Here the more peripheral circular cleavages (latitudinal) do not form any sharp separation between protoplasm and deutoplasm, and the yolk mass is for a long time only partially divided by meridional cleavages alone (Fig. 113).

Superficial.—This type of cleavage is characteristic of Arthropods in general and occurs elsewhere only in a few Cœlenterates. The central accumulation of the yolk, as it occurs in these forms, is an unusual condition, and correlated with this we find several unusual features in cleavage. Of course in such an arrangement the most obvious distinction between animal and vegetal poles is entirely lacking, and usually the position of the polar bodies and the external form of the egg are the only outward indications of the polarity of the ovum. Before cleavage begins the nuclear structures are located centrally, together with a small amount of protoplasm, and surrounding this is a dense mass of yolk, interpenetrated by a very fine network of protoplasm. At first nuclear division is not followed by division of the inert remainder of the egg mass. But after a varying number of nuclear divisions, the daughter nuclei separate, each accompanied by a small mass of protoplasm, the "*protoplasmic island*," and migrate to the surface of the egg, continuing to multiply as they go (Figs. 117, 118). In

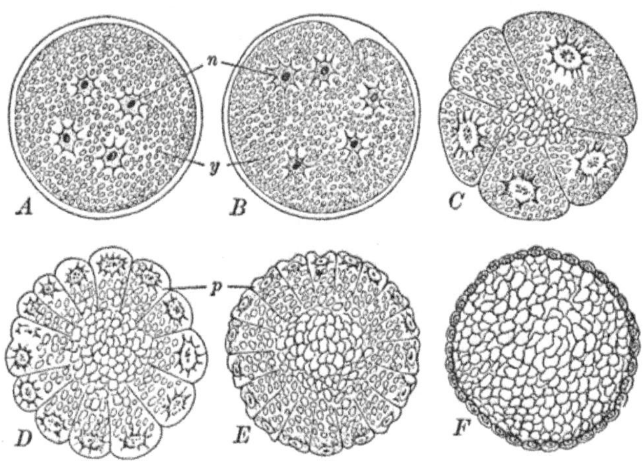

Fig. 117.—Superficial cleavage in the Decapod, *Dromia*. (Sections.) After Cano. *A, B.* Intravitelline divisions of the nucleus. *C, D, E.* Formation of yolk-pyramids. *F.* Blastula; a superficial layer of cells enclosing a mass of yolk. *n,* nuclei; *p,* yolk pyramids; *y,* yolk bodies.

this way the nuclear and cytoplasmic portions form a kind of superficial syncytium leaving the condensed and undivided yolk centrally.

Now either of two things may occur. In some forms (Decapods, Copepods, Ostracods, Amphipods) cleavages appear almost simultaneously, dividing the egg completely into a number of cone-shaped cells with the apices directed centrally; these cells are known as *yolk pyramids* (Fig. 117). In some cases the formation of the yolk pyramids does not occur simultaneously throughout the egg, but occurs earlier on that side of the egg which corresponds with the ventral surface of the embryo and larva. After a variable but considerable number of yolk pyramids are formed the planes of separation gradually disappear except in the superficial protoplasmic layer, which alone remains cellular, while the yolk again becomes a solid mass. Such a process as this is taken to mean that this type of cleavage may have been derived from the total adequal type.

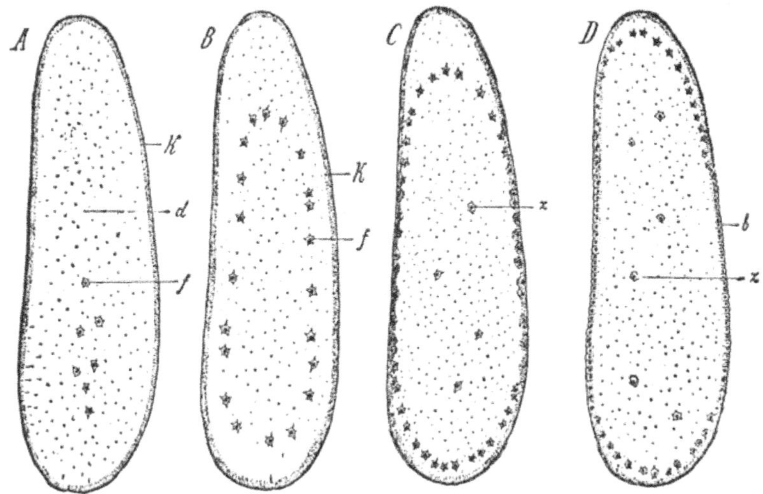

Fig. 118.—Cleavage in the beetle, *Hydrophilus*. From Korschelt and Heider, after Heider. *A, B.* Intravitelline divisions of the nucleus. *C.* Beginning of the formation of a superficial layer of cells. *D.* Later stage in formation of "blastoderm," or superficial cell layer. *b*, blastoderm, or superficial layer of cells; *d*, yolk; *f*, nuclei surrounded by protoplasm (protoplasmic islands); *z*, nuclei remaining in yolk (merocytes).

In other examples of nearly all groups of Arthropods the yolk pyramids are incompletely developed or even entirely absent and the cleavage is strictly superficial (Fig. 118). Here, as in the preceding, blastomeres may be formed either wholly or only partially around the ovum. The preliminary divisions of the nucleus and the formation of protoplasmic islands occur as described above. In many of the Insects, whose cleavage is typically superficial, the substance of the ovum ultimately becomes completely divided internally.

The preceding classifications and descriptions of cleavage have been almost wholly morphological in their basis. But development is a process, not a succession of morphological stages, and there remains to be described the most important aspect of cleavage as a developmental process. We come then to still another classification of cleavage types as (1) *determinate* and (2) *indeterminate*.

Cleavage is said to be *determinate* when exact morphological and physiological relations exist between the individual blastomeres and, (*a*) specific structures in the embryo and fully developed organism, and also (*b*) specific regions or substances in the ovum. Each of the products of cleavage is here a true organ, of particular and known value in development: blastomeres are not interchangeable, and their removal or destruction may lead to specific, related defects or abnormalities in later development (*e.g.*, the Ascidians). Cleavage is described as *indeterminate* when the blastomeres seem to have no specific relation to the structure of either the egg or embryo and adult. Here all the blastomeres may have equal value as factors in development; they are more or less interchangeable, and removal or destruction leads only to the loss of a corresponding amount of substance, not to the absence of any specific or related parts (*e.g.*, the Echinoderms).

In order that this classification should not be misleading, it should be said at once that this is a purely artificial distinction, based rather upon historical grounds than upon the facts of development, for these two types are completely connected by transitional conditions, and soon or late in development, all cells come to have specific, determined values.

Such a grouping as this, of the varieties of cleavage, obviously rests primarily upon a physiological rather than a morphological basis, for cleavage here is regarded as a *process* of development. In all cases of determinate cleavage the essential fact is that cleavage is not the mere division of the zygote into separate masses and units which can be moved about and moulded into the form of an embryo; cleavage is not merely a series of cell divisions, not the mere "vegetative reduplication of parts" occurring in accordance with certain mechanical rules like those of Balfour, Sachs-Hertwig, or Plateau, mentioned above.

We have already seen that the exceptions to these rules are so numerous and so fundamental that they must have some real significance. It is now clear that in determinate cleavage all the details have a significance that is prospective, looking toward the structural and physiological characteristics of the larva or fully formed organism. It has been said regarding determinate cleavage that "One can . . . go over every detail of cleavage, and knowing the fate of the cells, can explain all the irregularities and peculiarities exhibited" (Lillie).

Why this should be true is partly explained when we remember that the characters of cleavage and of the fully developed organism are *both* the primary result of the underlying structure of the ovum. Cleavage stands as an intermediate process between egg organization and adult structure; it is one of the processes through which the primary organization of the ovum gains expression in adult form.

This view of the cleavage process is by no means the only, or the original view, but it serves to bring out clearly the fact that the problems as to the nature and causes of the differentiations occurring during the cleavage process are related to the problems of the nature and causes of the differentiations of adult structure. Indeed these differentiations have a common cause in the structure and reactions of the ovum, and are therefore fundamentally equivalent. In our introductory chapter we said that the organism is specific at every stage, the zygote, the group of blastomeres, the embryo, the adult, are all the same specific organism, and the question why the cleavage group is what it is, is the same as the question why the mature organism has its own specific and individual characteristics. The problems and processes of development are fundamentally alike throughout.

In continuing our discussion of this determinative aspect of cleavage we shall make little further attempt to distinguish the determinate and indeterminate forms since this separation is clearly artificial. The apparent differences between determinate and indeterminate cleavage may arise from the fact that one of the determining factors contains a variable. That is, the time at which the organization of the egg becomes sufficiently complete to be effective in determining the course of differentiation of the blastomeres, may vary. Thus cleavage would be completely determinate if the egg organization were entirely or largely completed before the cleavage process begins, incompletely determinate if the organization is only partial, and indeterminate if the organization is only slightly marked during the earlier cleavages. For egg organization is progressive, it develops. So the determinate or indeterminate character of *cleavage* may depend, partly at least, upon the relative *time* during cleavage at which the organization becomes marked to such an extent as to determine the fate of particular blastomeres. Other factors obviously enter into the process and we shall review the subject from another point of view in the next chapter.

We have seen above that in nearly all species the earliest cleavage planes are definitely related to the polar axis of the ovum. The polarity of the egg is one of the fundamental aspects of its organization. We have seen also that the ovum often contains formed substances of various kinds, both protoplasmic and deutoplasmic, distributed in the cytoplasm in a definite and usually specific manner. It is a common feature of cleavage that the first plane symmetrically divides the egg or that part of it which takes part in the process of cleavage. And furthermore, with very few exceptions, this first cleavage plane coincides either precisely or approximately with the median plane of the embryo and adult. *Nereis* is one of the few exceptions to this rule; in this Annelid, as in some of the Urodeles, the second plane marks the future median plane.

The factors which appear immediately to determine the location of the first cleavage plane are mainly two. First and most important is the structure of the ovum itself, which in many cases, even in the unfertilized condition, is obviously bilaterally symmetrical; the first cleavage plane corresponds closely with this plane of symmetry, which is therefore determined by the same "organizational" factor that determines the polarity and other structural features of the ovum. In other cases the symmetry of the egg appears to be radial or rotatorial before fertilization and is converted into a bilaterally symmetrical structure by the entrance of the spermatozoön, the entrance path of which marks the plane of symmetry of the egg and developing organism, and determines the location of the first cleavage. It is quite possible, though hardly demonstrated as yet, that even in such cases there is really an invisible bilateral structure of the ovum which underlies the radial symmetry and really determines the point at which the sperm shall enter. In such a case the entrance path of the spermatozoön would itself be predetermined and could not be regarded as a primary factor in fixing the position of the first cleavage. This would obviously be the case in many of those eggs possessing micropyles. But in some eggs whose cleavage is indeterminate, even though they possess micropyles (Teleosts), there seems to be no regularity in the position of the first cleavage plane and no correspondence between this and any morphological characteristic of either ovum or adult.

The second cleavage, usually at right angles to the first, ordinarily corresponds with the median transverse axis of the egg, embryo, and adult. The third cleavage is usually horizontal and separates animal and vegetal poles and corresponds most frequently with the separation, in the embryo and adult, of the more active animal, and less active vegetative tissues.

The facts that in all eggs of a given species or genus cleavage occurs according to a definite pattern, and that there may be an exact relation between the individual blastomeres of the cleaving ovum and the tissues and organs of the later organism, make it possible to speak of the "cell lineage" (Wilson) of an organism. In forms with determinate cleavage it becomes possible to identify, even in a comparatively late embryonic stage, various groups of cells as the real lineal descendants of certain individual cells of the earlier cleavage group. In other words, it is in such cases possible to trace the structures of the embryo and adult back to single cells or parts of cells.

In order to illustrate the nature of the facts of cell lineage, and the completeness and exactness of the correspondence between blastomeres and differentiated groups of cells in the later embryo, we may describe briefly a typical instance, using as the subject one of the simpler and more regularly cleaving types, the Turbellarian, *Planocera*, as described by Surface. This account of the cleavage of this form should be read with the expectation of finding frequent "exceptions" to the laws of cleavage mentioned above.

In *Planocera* (Figs. 119, 120) cleavage is total, unequal, and spiral (dexiotropic). The first plane is meridional and divides the egg into two adequal blastomeres known as AB and CD. The division of each of these is also meridional but is unequal, each forming a smaller and a larger cell, and dividing the entire ovum into two larger, and slightly unequal, cells known as B and D, and two smaller cells, also slightly unequal, known as A and C. Of these D is the largest, and from later development is known to be posterior in position; B is anterior, A on the left, and C on the right, as viewed from the animal pole and with reference to later structure.

The third cleavage is horizontal, unequal, and strongly spiral (dexiotropic). As usual among the Turbellaria the larger cells divide shortly *before* the smaller. On account of the size differences between the cells of the upper and lower quartets, we may describe the upper quartet of smaller cells (micromeres) as budded off from the lower quartet of macromeres. The quartets of micromeres are designated by small letters, the macromeres by capitals. Thus in the eight-cell stage we have the first quartet of micromeres, 1a, 1b, 1c, 1d, and the first quartet of macromeres, 1A, 1B, 1C, 1D. Successive quartets are designated by numerical coefficients, products of the division of the quartet cells by exponents. Thus in passing from eight cells to sixteen the macromeres bud off, this time in a læotropic direction, a second quartet of micromeres, 2a, 2b, 2c, 2d, the macromeres themselves remaining known now as 2A, 2B, 2C, 2D. Shortly thereafter the first quartet of micromeres divide, also læotropically, forming two groups of four cells each known as $1a^1$, $1b^1$, $1c^1$, $1d^1$, and $1a^2$, $1b^2$, $1c^2$, $1d^2$; the cells lying toward the animal pole are designated by the lower exponent.

The fifth cleavage, dividing the sixteen cells into thirty-two, in general resembles the preceding but is dexiotropic throughout. As the result we have the macromeres, 3A, 3B, 3C, 3D; a third quartet of micromeres, 3a, 3b, 3c, 3d; the second quartet of micromeres divides into $2a^1$, $2b^1$, $2c^1$, $2d^1$, and $2a^2$, $2b^2$, $2c^2$, $2d^2$, while the eight cells previously derived from the first quartet of micromeres now form $1a^{11}$, $1a^{12}$, $1a^{21}$, $1a^{22}$, $1b^{11}$, $1b^{12}$, $1b^{21}$, $1b^{22}$, $1c^{11}$, $1c^{12}$, $1c^{21}$, $1c^{22}$, $1d^{11}$, $1d^{12}$, $1d^{21}$, $1d^{22}$.

After thirty-two cells have been formed in this fairly regular fashion, the rhythm of cleavage becomes modified so that there follow stages of 40, 44, 45, 53, 61 cells, *etc.*

It is unnecessary for us to go farther with the details of these cleavages save in one particular. After the thirty-two-cell stage the macromeres divide again unequally giving off a group of *large* cells which contain most of the deutoplasm of the original ovum. In spite of their size relations these large cells are known as the fourth quartet of micromeres, 4a, 4b, 4c, 4d, and the remaining smaller cells as the fourth quartet of macromeres, 4A, 4B, 4C, 4D. This contradiction in terminology is justified by the later history of these cells.

We may now consider the fates of these various groups of blastomeres. Quoting from Surface, "From the first quartet [of micromeres] arises the ectoderm, covering the anterior and dorsal portions of the body. From cells of this quartet four strings of cells bud into the interior of the embryo and form the ganglion. The eyes arise in ectodermal cells of this quartet. The second quartet gives rise to the larger portion of the ectoderm on the ventral and posterior regions of the body. From cells of this quartet is formed most of the ectodermal pharynx. A portion of the second quartet is budded into the embryo and forms mesoderm. From this source arises probably only that mesoderm formed around the blastopore and which is later concerned in the structures of the pharynx.

"The third quartet consists of small cells from which apparently only ectoderm is derived. The individual divisions of these cells have not been traced very far, but there is every reason to believe that they form ectoderm only.

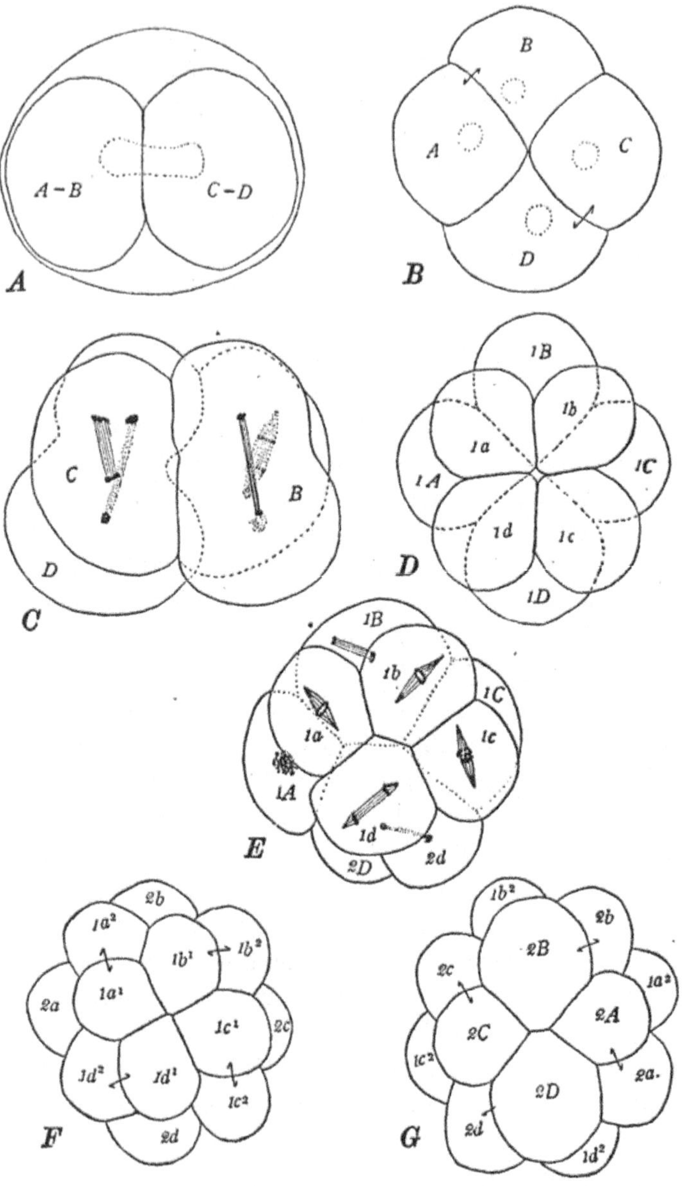

Fig. 119.—Cleavage and cell lineage in the Polyclad Turbellarian, *Planocera inquilina*. From Surface. *A*. Egg during the first cleavage; side view. The cell *C–D* is slightly larger than *A–B*. *B*. Four-cell stage, from animal pole. *C*. Formation of first quartet, from right side, showing spiral cleavage (dexiotropic). *D*. Eight-cell stage, from animal pole. *E*. Eight-cells dividing into sixteen, showing læotropic division. The division of the cells of the *D* quadrant is in advance of the others. *F*. Sixteen-cells, from animal pole. *G*. Sixteen-cells, from vegetal pole.

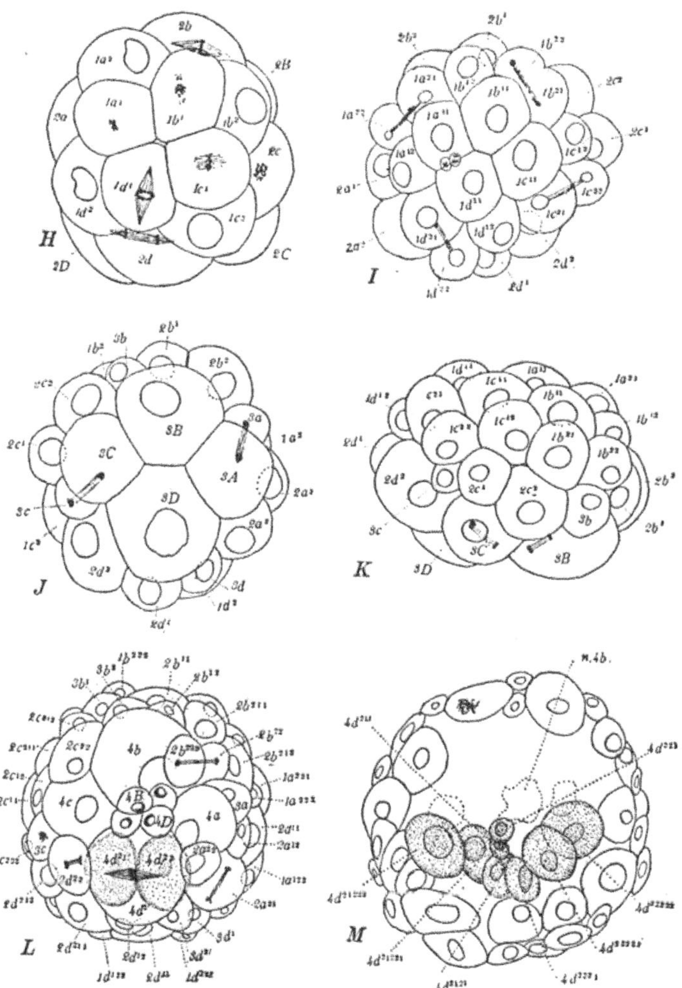

Fig. 120.—Continuation of Fig. 119. *H*. Dexiotropic division of 1a¹–1d¹ and of 2a–2d. From animal pole. *I*. Thirty-two-cells, from animal pole. *J*. Thirty-two-cells from vegetal pole. *K*. Thirty-two-cells from right side. *L*. Late cleavage showing the history of cells 4a–4d and 4A–4D. *M*. Optical section of a much later stage, viewed from near the vegetal pole. The mesoderm bands are stippled.

"The history of the fourth quartet is peculiar . . . . .
The posterior cell 4d is the mesentoblast, from which the alimentary canal and a portion of the mesoderm arise. The other three cells of the fourth quartet, 4a, 4b, 4c, do not divide as long as their history can be traced. They, however, break up into a large number of homogeneous yolk spheres which are absorbed by the endoderm cells. The large nuclei of these three cells can be traced until the alimentary canal is partly formed.

TABLE OF THE CELL LINEAGE OF PLANOCERA

Condensed from Surface.

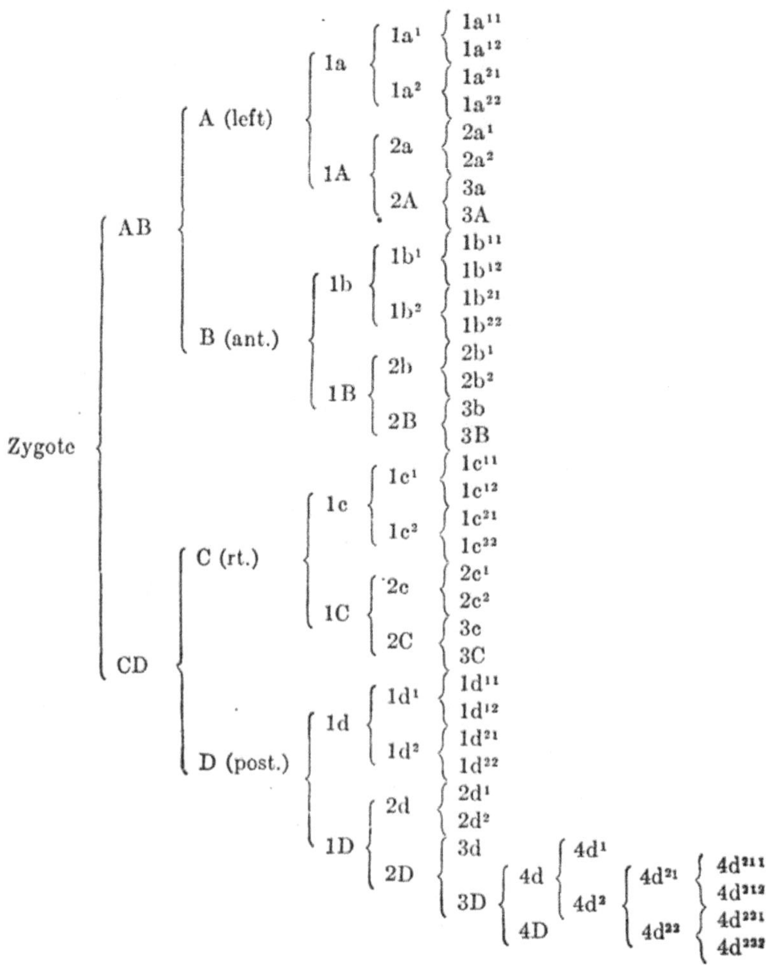

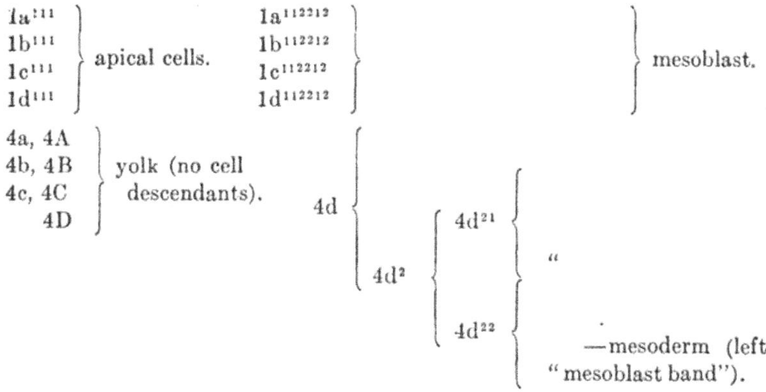

The remaining cells form covering ectoderm.
Ectoderm of first quartet—anterior and dorsal, including eyes.
Ectoderm of second quartet—posterior and ventral, including pharyngeal.
Mesoderm of second quartet—blastoporal (pharyngeal).

"The nuclei of the small macromeres [4A, 4B, 4C, 4D] show evidences of degeneration. These do not divide as long as they can be followed . . . . . and it seems probable that they degenerate without giving rise to any morphological structure."

This cell lineage of *Planocera* is summarized incompletely in the accompanying table.

It is interesting to compare with this lineage of *Planocera* that of *Ascaris*, described by Zur Strassen, which is somewhat less regular. This is particularly interesting as it shows clearly the history of the germ cells, which become wholly separate from somatic cells in the sixteen-cell stage. Cleavage of *Ascaris* (Fig. 121) is bilateral but more or less irregular, particularly in its rhythms, so that without attempting to apply the ordinary terminology completely we may summarize the early cell history in the table accompanying (p. 255).

The cell lineage of a considerable number of organisms has been definitely traced, often in much greater detail than we have indicated. The histories best known are found among the Platyhelminthes, Nemathelminthes, Nemertinea, Annulata, Trochelminthes, Mollusca, and Tunicata. Of course the eggs of many classes and phyla show no such regularity, for as we have pointed out, cleavage may be irregular as well as indeterminate. And in many of the groups named above, normal development

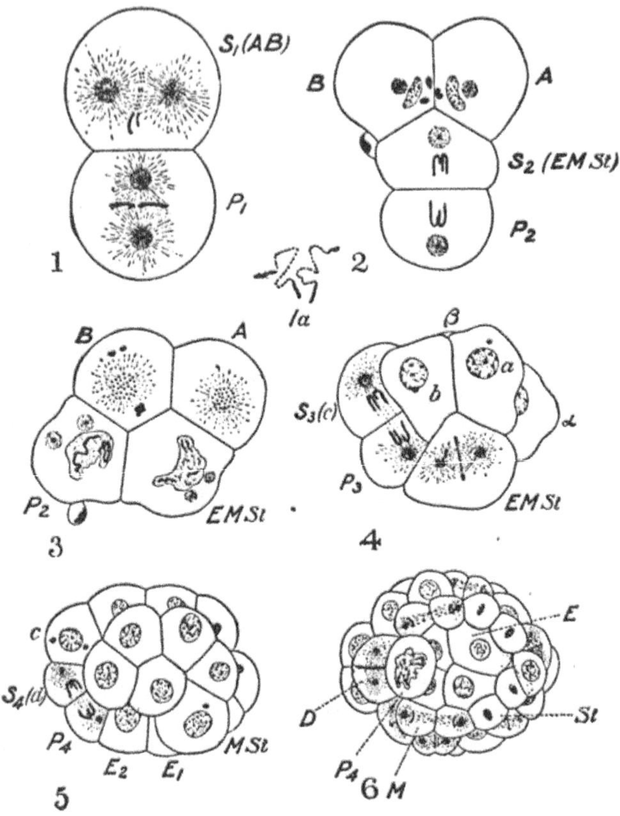

FIG. 121.—Cleavage in *Ascaris megalocephala bivalens*. From Jenkinson, after Boveri. 1. Division of the two-cell stage. Elimination of chromatin in the somatic cell $S_1(AB)$. 1a. Chromosomes of the cell $S_1(AB)$. 2. Four-cell stage (T-form). In A and B can be seen the eliminated chromatin. The cell P_1 has divided into a somatic cell $S_2(EMSt)$, in the descendants of which chromatin elimination occurs, and the cell P_2. 3. Four-cell stage (lozenge-form). A is anterior, A and B, dorsal. 4. Continued chromatin elimination in somatic cells. P_2 has divided into P_3, and $S_3(C)$—secondary ectoderm. a, b, primary ectoderm of right side, a, $β$, of left side. 5. The endoderm cell has been formed and has divided (E_1, E_2). P_3 has divided into P_4, the primordial germ cell, and $S_4(d)$, tertiary ectoderm. 6. Ventral view at the beginning of invagination. Elimination of chromatin in $S_4(D)$. The four endoderm cells (E) beginning to invaginate. On each side two mesoderm cells (M) in which granular chromosomes may be seen, and two stomodæal cells (St).

may occur even though interrupted by the removal of parts, by pressure, *etc.*

TABLE OF THE CELL LINEAGE OF ASCARIS

Modified from Zur Strassen

(Letters in parenthesis are the notation of Zur Strassen, Boveri and others)

$$
\text{AB } (S_1)
\begin{cases}
\text{A } (A) \begin{cases}
\text{a } (a) \begin{cases} a^1\ (aI) \\ a^2\ (aII) \end{cases} \\
a\ (\alpha) \begin{cases} a^1\ (\alpha I) \\ a^2 \end{cases}
\end{cases} \\[2em]
\text{B } (B) \begin{cases}
\text{b } (b) \\
b\ (\beta) \quad b^2
\end{cases}
\end{cases}
$$

$$
\text{CD } (P_1)
\begin{cases}
\text{C } (EMSt) \begin{cases}
\text{c } (MSt) \begin{cases} c^{11} \\ c^{12} \\ c^{21} \\ c^{22} \end{cases} \\
c\ (E) \left.\begin{cases} c^{11} \\ c^{12} \\ c^{21} \\ c^{22} \end{cases}\right\} \text{endoderm I.}
\end{cases} \\[3em]
\text{D } (P_2) \begin{cases}
\text{d } (S_3)\,((c)) \begin{cases}
d^1\ (c) \left.\begin{cases} d^{11} \\ d^{12} \end{cases}\right. \\
d^2\ (\nu) \left.\begin{cases} d^{21} \\ d^{22} \end{cases}\right\} \text{mesoderm III.}
\end{cases} \\
d\ (P_3) \begin{cases}
d^1\ (S_4)(D) \left.\begin{cases} d^{11} \\ d^{12} \end{cases}\right\} \text{mesoderm II.} \\
d^{2*}\ (P_4) \begin{cases}
d^{21}\ (G)\ (\text{ant.}) \begin{cases} d^{211}\ (\text{left}) \\ d^{212}\ (\text{rt.}) \end{cases} \\
d^{22}\ (G_1)\ (\text{post.}) \begin{cases} d^{221}\ (\text{left}) \\ d^{222}\ (\text{rt.}) \end{cases}
\end{cases}
\end{cases}
\end{cases}
\end{cases}
$$

primordial germ cells. (bracketing d^{211}–d^{222})

* Primordial germ cell.

The study of cleavage from this point of view discloses the fact, of the utmost importance in development, that blastomeres may be individually and specifically recognizable as morphological and morphogenetic units. They bear much the same relation to the whole cell group that the organs and tissues bear to the embryo or later organism. As distinct morphological and physiological units they represent real differentiations at a very early stage of development, and may truly be said to form embryonic rudiments of structures appearing later in the form of germ layers or derivatives of these.

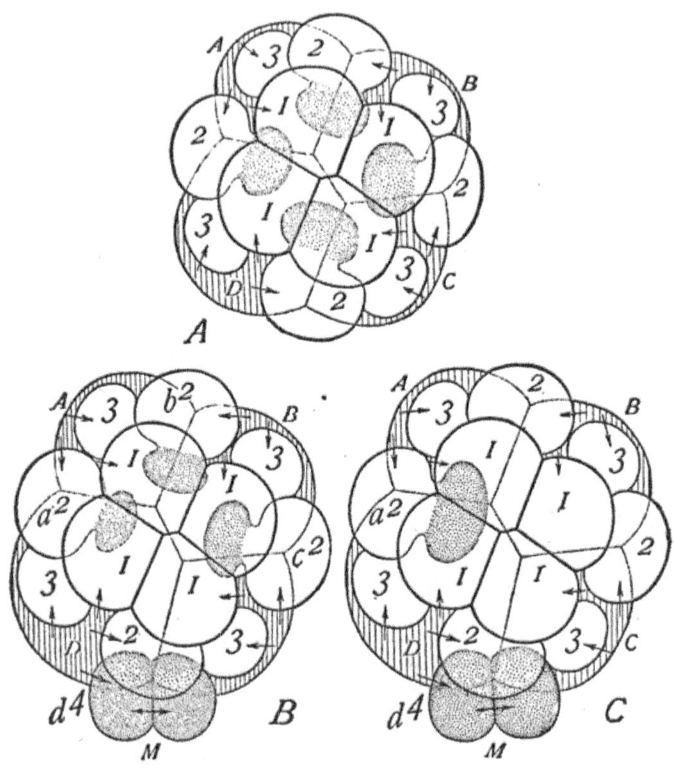

Fig. 122.—Diagrams illustrating the value of the quartets in three animals. From Wilson, "Cell." Ectoplasm is unshaded; mesoplasm is dotted; endoplasm is vertically ruled. *A*. The Polyclad, *Leptoplana*, showing mesoplasm formation in second quartet. (Compare *Planocera*, Fig. 120, where Surface finds mesoplasm in cell 4*d* descendants.) *B*. The Gasteropod, *Crepidula*. *C*. The Pelecypod, *Unio*.

Not only this but comparison of the cell lineages of different classes and phyla often brings out the fact that particular cells can be identified and compared in diverse groups of animals, making it possible to apply the idea of homology to blastomeres and groups of blastomeres in the early embryo, as well as to the organs and parts of the fully formed organism. *Cells* may be vestigial, rudimentary, and the like, in the same way that organs may be. The three or four successively formed quartets of micromeres or even an individual cell, for example that known as 4d, can be identified and homologized both in origin and in fate, in the phyla Platyhelminthes, Annulata, and Mollusca (Fig. 122). Wilson has written ("The Cell," *etc.*, page 416): "Thus we find that the cleavage of polyclades, annelids and gasteropods shows a really wonderful argeement in form, yet the individual cells differ markedly in prospective value. In all of these forms three quartets of micromeres are successively formed according to exactly the same remarkable law of alternation of the spirals; and, in all, the posterior cell of a fourth quartet lies at the hinder end of the embryo in precisely the same geometrical relation to the remainder of the embryo; yet in the gasteropods and annelids this cell gives rise to the mesoblast-bands and their products, in the polyclade to a part of the archenteron, while important differences also exist in the value of the other quartets." (It should be added that in the Polyclad, *Planocera*, the particular cell mentioned gives rise to mesoblast also.)

Such conditions also illustrate how the facts of embryology may have a certain value, often very great, as evidence upon phylogenetic problems.

Often these similarities of structure can be carried back into the pre-cleavage stage, and in the uncleaved zygote or ovum before fertilization, substances can be identified which later become contained within restricted groups of similar cells. So that cleavage is in part to be regarded as a process by which specific substances or regions of the egg become segregated in different regions of the embryo, where each continues its normal differentiation during later developmental stages and gives rise to specific tissues or organs. In other words the process of differentiation is not limited to the later stages of development following cleavage; it occurs during and even preceding cleavage. In *Ascaris*, for example, each of the two cells resulting from the first cleavage is specific; in other forms each of the four or eight cells has already become differentiated. In forms like the Ascidians (Conklin) true differentiation has commenced in the uncleaved ova and has been carried to a very pronounced degree. And it is not at all unlikely that these differentiations of the undivided egg may represent the most essential and most fundamental differentiations of the organsim. This aspect of cleavage cannot be discussed satisfactorily here without encroaching widely upon the subject of the next chapter and it is therefore left at this point.

It is to be noted, however, in conclusion that while cleavage may have an important chemical significance of general character, and a general physiological significance, yet the process is primarily a specific process of development and not mere cell multiplication. The process is closely related in all its details to the structure of the egg and also to the structure of the adult, since this too is similarly related to egg structure. Many of the details of cleavage do not occur according to physical laws based upon space and time relations of the parts, but departures from what we should expect on the basis of such laws may result, (a) from the historical factor of the relationships of organisms and the process of descent, (b) from the teleological factor, for cleavage has a prospective significance, looking forward, as well as a retrospective significance— cleavage has its promorphology as well as its morphology.

<h2 style="text-align:center">REFERENCES TO LITERATURE</h2>

BALFOUR, F. M., (See ref. Ch. III.)

BEARD, J., (See ref. Ch. III.)

BOVERI, T., Die Polarität von Ovocyte, Ei und Larve des *Strongylocentrotus lividus*. Zool. Jahrb. **14.** 1901.

BRESSLAU, E., Beiträge zur Entwicklungsgeschichte der Turbellarien. I. Die Entwicklung der Rhabdocölen und Alloicölen. Zeit. wiss. Zool. **76.** 1901.

CONKLIN, E. G., (See ref. Ch. II, III.)

DRIESCH, H., Entwicklungsmechanische Studien. VIII. Ueber Variation der Mikromerenbildung. (Wirkung von Verdünnung des Meerwasser). Mitt. Stat. Neapel. **11.** 1893.

Godlewski, E., Plasma und Kernsubstanz in der normalen und der durch äussere Factoren veränderten Entwicklung der Echiniden. Arch. Entw.-Mech. **26.** 1908.

Hargitt, C. W., Some Problems of Cœlenterate Ontogeny. Jour. Morph. **22.** 1911.

Hertwig, O., Das Problem der Befruchtung und der Isotropie des Eies, eine Theorie der Vererbung. Jena. Zeit. **18 (11).** 1885.

Korschelt und Heider, Lehrbuch, *etc.* III Abschnitt. Furchung und Keimblätterbildung. Jena. 1909.

Lillie, F. R., Adaptation in Cleavage. Woods Holl Biol. Lect. 1899.

Masing, E., (See ref. Ch. V.)

Moenkhaus, W. J., (See ref. Ch. II.)

Robert, A., Recherches sur le developpement des Troques. Arch. Zool. Exp. (III). **10.** 1903.

Rückert, J., Ueber das Selbständigbleiben der väterlichen und mütterlichen Kernsubstanz während der ersten Entwicklung des befruchteten *Cyclops*-Eies. Arch. mikr. Anat. **45.** 1895.

Sachs, J., Ueber die Anordnung der Zellen in jüngsten Pflanzentheilen. Arbeiten Bot. Inst. Würzburg. **2.** 1882.

Selenka, E., Die Keimblätter der Echinodermen. Stud. über Entw. II. Wiesbaden. 1883.

Surface, F. M., The Early Development of a Polyclad, *Planocera inquilina*, Wh. Proc. Acad. Nat. Sci. Philadelphia. 1907.

Watasé, S., Studies on Cephalopods. I. Cleavage of the Ovum. Jour. Morph. **4.** 1891.

Wilson, E. B., The Cell-Lineage of *Nereis.* Jour. Morph. **6.** 1892. Cell Lineage and Ancestral Reminiscence. Woods Holl Biol. Lect. 1899. (See also ref. Ch. II.)

Wilson, H. V., The Embryology of the Sea-Bass. (*Serranus atrarius.*) Bull. U. S. Fish Com. **9.** 1889. (1891).

Zeleny, C., Experiments on the Localization of Developmental Factors in the Nemertine Egg. Jour. Exp. Zool. **1.** 1904.

Ziegler, H. E., Experimentelle Studien über Zelltheilung. III. Die Furchungszellen von *Beroë ovata.* Arch. Entw.-Mech. **7.** 1898.

Zur Strassen, O., Embryonalentwickelung der *Ascaris megalocephala.* Arch. Entw.-Mech. **3.** 1896. Ueber die Lage der Centrosomen in ruhenden Zellen. Arch. Entw.-Mech. **12.** 1901.

CHAPTER VII

THE GERM CELLS AND THE PROCESSES OF DIFFERENTIATION, HEREDITY, AND SEX DETERMINATION

THE problem of heredity is the problem of development. The student of heredity is concerned primarily with the comparison of the traits of adult organisms as they appear in successive generations, and with the methods of the distribution of distinctively individual parental characteristics among successive generations of offspring. The student of development is concerned primarily with the genesis of the traits of the individual, with that continuous and orderly sequence of changes that gives to the single-celled zygote the final form of the fully matured animal.

It might seem, therefore, that any consideration of the problems of heredity is somewhat out of place in an account of the processes of development. At one time this might justly have been urged. But to-day the students of heredity and of embryology have in common much that is fundamental. Their interests meet in the ideas that the organism is a specific creature at every stage of its existence, from zygote to adult; that the qualities of each later stage are conditioned by those of an earlier; and so ultimately the structural and functional differentiations of the adult must be traced back to corresponding differentiations of the zygote, or even to pre-conjugation phases of the gametes. It is the common endeavor of the students of embryology and of genetics to answer the question why the egg of a star-fish develops into a star-fish, with the characteristics of its parents, rather than into a sea-urchin, although it may develop in the same dish of water with other eggs that do develop into sea-urchins.

Heredity is the fact of resemblance between offspring and parents, not the resemblance of adult stages alone, but the likeness at all corresponding ages. That the individual ova of *Asterias vulgaris* are alike, that the cleavage processes, blastulas, gastrulas, larvæ, and adolescent stages of all the individuals of this species are essentially alike, in structure and in behavior—all this is similarly the fact of heredity. A specific kind of protoplasm is never, whatever its form, anything other than that specific kind.

In other words, the interest of the embryologist in the problems as to why the ovum develops as it does, passes from one condition to another as it does, and finally produces the kind of adult that it does, is essentially an interest in the problem of heredity—the problem of organismal specificity. This is the central point of embryological study. Consequently we are fully justified in considering in this place, these general problems of the relation of the facts of cytology and embryology to the facts of parental and specific likeness of organisms. Failure to do so would mean the omission of the most vital topic around which much, perhaps it would not be going too far to say most, of recent embryological investigation has centered, and upon which it is to-day focussed.

The answers which the facts of embryology have to offer to these fundamental questions are still rather vague and uncertain. Most of them are stated in the potential mood and must still be framed as hypotheses. But although the facts here may be much clearer than their significance, we must attempt a statement of both; and it goes almost without saying that our treatment of both must be as brief and as elementary as possible; this is not the place for extended consideration of hypothetical views, however great the importance of the central ideas.

We may address ourselves, therefore, to a survey and brief analysis of the answers which have been given to the question why the organism develops in the way it does. First, let us recall the idea, stated briefly in the introductory chapter, that development is a form of behavior—a series of reactions. In any organic reaction the two factors of external and internal conditions are involved. The reactions of the ovum in cleaving, of the blastula in gastrulating, and the like, are of a general nature, *i.e.*, the external conditions involved may be considerably varied, within limits, and yet produce the same response on the part of the organism. Emphasis thus is placed upon the internal factors in the reactions of development, for on account of the definite character of habits of life and of spawning, the normal external conditions of development are sufficiently uniform to produce a series of reactions, a development, which is also uniform, *i.e.*, normal. Of course it is easily possible to alter, artificially, the external as well as the internal conditions of development and the results of such alteration often lead, as we shall see, to important ideas regarding normal or characteristic embryonic behavior.

Development is, then, a series of reactions, one condition leading to the next; and the primary factor in determining the quality of each reaction is the internal condition or structure, both morphological and physiological, of the organism, whether it be ovum, zygote, blastula, larva, or adolescent individual.

Throughout the efforts to solve the problem of individual development, the attempt has always been to explain existing differentiations as being dependent upon some preëxisting differentiation, related but of a different kind. Thus at one time the earliest differentiations that became visible in the developing organism were the germ layers, and these were consequently regarded as the fundamental differentiations of the embryo, determining its subsequent history. Next, differentiations among the cleavage cells were noted and emphasized as primary. Then as technique improved, and the subject of cytology developed, attention became focussed upon the nucleus and its organs not only as the centers of cell life, but as the structures primarily concerned in the differentiations of the developing egg cell. And lately, chiefly as a result of experimental analysis of the processes of development, rather than as the consequence of observation alone, structural differentiations of the cytoplasmic portion of the ovum have occupied the center of interest as determining factors in development. This succession of views represents rather the order of their general acceptance as working bases, than of their discovery and individual promulgation.

In accordance with this historical succession of ideas regarding the nature of the underlying differentiations in development, we may outline briefly three general hypotheses of the causes of differentiation. We may omit, as being now of historical interest chiefly, any further reference to the germ layers as the primary determiners of development, and may begin with the idea of Pflüger, that the egg and the blastomere group are homogeneous or *isotropic* throughout, and that the early developmental processes of cleavage are nothing more than indifferent multiplications of similar units, resulting in the formation of blocks out of which later differentiating structures may be built. During the early '90's this view was quite prevalent, and especially favored by such embryologists as Oscar Hertwig and Driesch, who developed it somewhat farther into what has been termed the "cell interaction" hypothesis. According to this hypothesis, while the cells of the blastomere group are essentially similar and equivalent in their potentialities ("prospective potency," Driesch), differentiation exists among them by virtue of their relative position in the cell group—not through any actual, individual, intracellular differentiation. Stated in the words of Wilson (The Cell, *etc.*, page 415), two sentences of Driesch summarize this view as follows: "The blastomeres of the sea-urchin are to be regarded as forming a uniform material, and they may be thrown about, like balls in a pile, without in the least degree impairing thereby the normal power of development." "The relative position of a blastomere in the whole determines in general what develops from it; if its position be changed, it gives rise to something different; in other words, its prospective value ["prospective significance," Driesch] is a function of its position."

In itself, then, the cell interaction hypothesis offers no explanation of differentiation or development, for it throws back upon some unknown factor the real cause of differentiation *through* position. Later investigation, chiefly experimental, has shown clearly that cell interactions alone play but a small part indeed in the process of differentiation, and has led to the search for that underlying factor or group of factors. And while Driesch himself concludes that no explanation is possible in known terms of matter or energy, and relies upon an unknown, and therefore metaphysical, factor, the great majority of embryologists believe that the question is still susceptible of further scientific analysis. We find two general hypotheses regarding the nature of the causes or conditions of differentiation, and since differentiations are always specific we may speak of these as also hypotheses as to the causes of those resemblances among generations of organisms which we call in a word, heredity.

The first of these is the hypothesis of "germinal localization" or "germinal, organ-forming regions" associated primarily with the names of His, Lankester, and Whitman. The essentials of this hypothesis in its present form may be stated as follows. The cytoplasm of the ovum before development (*i.e.*, cleavage) begins, has a definite structure or morphology of its own, such that particular regions or substances, by effecting specific developmental reactions, are seen to correspond with, or to lead to the formation of, particular tissues·or structures of later stages and of the fully developed organism. The cytoplasm is conceived as a mosaic-work of physiological units which have not only a definite morphology but a definite *pro*-morphology, looking toward the structure of the mature individual. This immediately suggests the old idea of "preformation" but it omits, of course, the naïve crudities of this conception and rests upon the idea that certain structures and regions of the egg cytoplasm are in direct genetic relation with corresponding structures and regions differentiating later by a true process of development or epigenesis. These germinal structures have specific reference, but not resemblance, to the parts of the mature organism. This predetermination may be only general to begin with, but it becomes more complete and specific as one condition succeeds another, the cytoplasmic structure of the ovum representing in the beginning all there is of definite, specific, organismal structure. This cytoplasmic structure of the germ is regarded as continuous from one generation of ova to the next, through the germ within the organism, and it thus serves as the physical basis of heredity.

As more or less opposed to this conception we have a group of hypotheses which may collectively be termed the hypothesis of "nuclear analysis." Here the nucleus alone is regarded as that part of the germ or zygote which bears a specific relation to later differentiations, and within the nucleus, the chromosomes are the elements chiefly concerned. Chromosomes are supposed to possess a structural predetermination that is *pro*morphological; they are unlike and individually behave specifically in determining the characteristics of developmental reactions. This hypothesis, in its various forms, is associated chiefly with the names of Nägeli, Roux, Weismann, DeVries, and Oscar Hertwig. It will be recognized as also preformational in its essentials; specific configurations of chromatin represent potentially, corresponding embryonic and adult traits, which become actual by a truly epigenetic series of developmental reactions.

It has been pointed out frequently that this hypothesis really transfers the idea of germinal localization from the cytoplasm to the nucleus; the structure of the cytoplasm is regarded as real, but as secondary and dependent upon the primary structure of the nuclear elements. These nuclear organs, the chromosomes, are thought to maintain a specific physiological continuity from one generation of ova to the next and thus to constitute a real physical basis of heredity.

These two general hypotheses have in common the idea of a fixed promorphological structure within the germ which becomes expressed epigenetically. They differ as regards the particular part of the germ whose promorphology is to be regarded as primary, and yet it is quite possible that both hypotheses contain elements of truth. It remains now for us to review some of the more significant facts of development bearing directly upon these ideas in order to determine, perhaps which, perhaps how much of each, is justified. In doing this we shall not attempt to relate the evidence directly to either hypothesis, leaving that for the reader; and in conclusion we shall attempt a summary which may serve to bring the two hypotheses together.

We may first describe certain facts associated chiefly with the hypothesis of germinal cytoplasmic localization.

In the first place it is perfectly clear that the ovum does possess a marked structure and organization, indicated in several ways. Occasionally the ovum may show external differentiations of form; such an egg as that of the squid (*Loligo*, Fig. 113) or the fly (*Musca*, Fig. 47), is obviously not only bilaterally symmetrical but it exhibits definite antero-posterior and dorso-ventral differentiation. In a few instances the eggs of a species are dimorphic, and while apparently the nuclei of 'both kinds are identical in structure, the total volume of one form may be three times that of the other. One of the very important and highly significant factors in the organization of the ovum is that of polarity, already described (*e.g.*, Figs. 42, 93). In many cases the polarity of the ovum can be traced back into oögonial stages, where it is seen to correspond with the polarity of the cells of the germinal epithelium (Mark). Polarity pervades the whole structure of the mature ovum and is expressed in a variety of ways, by the eccentric position of the nucleus, by the point at which the polar bodies are formed, by the disposition of the deutoplasm, and by the arrangement and distribution of a variety of formed substances, such as pigments, granules, and vacuoles of many kinds.

Living protoplasm, as we have seen, consists of a fundamental matrix or ground substance of rather uncertain form and composition, and suspended within this, particles and granules of many sizes, forms, and materials, the nature of which gives character to a particular region of protoplasm. These different kinds of substance are not distributed at random through the ovum, but they are localized in certain regions, as zones or layers, either horizontal or concentric.

There are at present two views as to the relation of these

substances to the fundamental polarity and general organization of the egg. In the opinion of some these substances are truly to be regarded as the initial developmental differentiations of the ovum. Each of these substances has a specific function in development and leads to the formation of a certain tissue or organ alone. Consequently these materials are known as "organ-forming substances." Before cleavage they usually assume a very definite symmetry of distribution, closely related to that of the later developing organism, and during cleavage they become distributed among certain specific groups of cells whose lineage can be traced directly to the rudiments of certain organs.

Thus in the egg of the Ascidian, *Cynthia* (*Styela*), fully described by Conklin and one of the best examples of this type of structure, at the close of the first cleavage there are five regions of protoplasm, present in amounts roughly proportional to the size of the parts to which they later give rise, and distinguishable by the character of their granular contents (Fig. 92). At the animal pole is a superficial region of comparatively clear protoplasm, the *ectoplasm*, from which the ectoderm develops; in the vegetal pole there is a dark gray region, the *endoplasm*, rich in yolk and later forming the endoderm; the mesoderm is formed from a crescentic region, the *mesoplasm*, located just below the equator, on the posterior side; this is characterized by its content of yellow pigment, and is divided into lighter and darker areas, forming respectively the mesenchyme and the tail muscles (myoplasm); a light gray crescent around the anterior border forms later the neural plate and notochord (neuroplasm, chordaplasm).

These substances are arranged symmetrically with reference to the first cleavage plane and this corresponds also to the median plane of the larva and adult, and thus from the first separates right and left sides of the body. The second cleavage plane, at right angles to the first, separates the yellow mesoplasm from the light gray neuroplasm and chordaplasm, and the third cleavage, horizontal, separates the clear ectoplasm from the other substances. The later cell lineage of this form

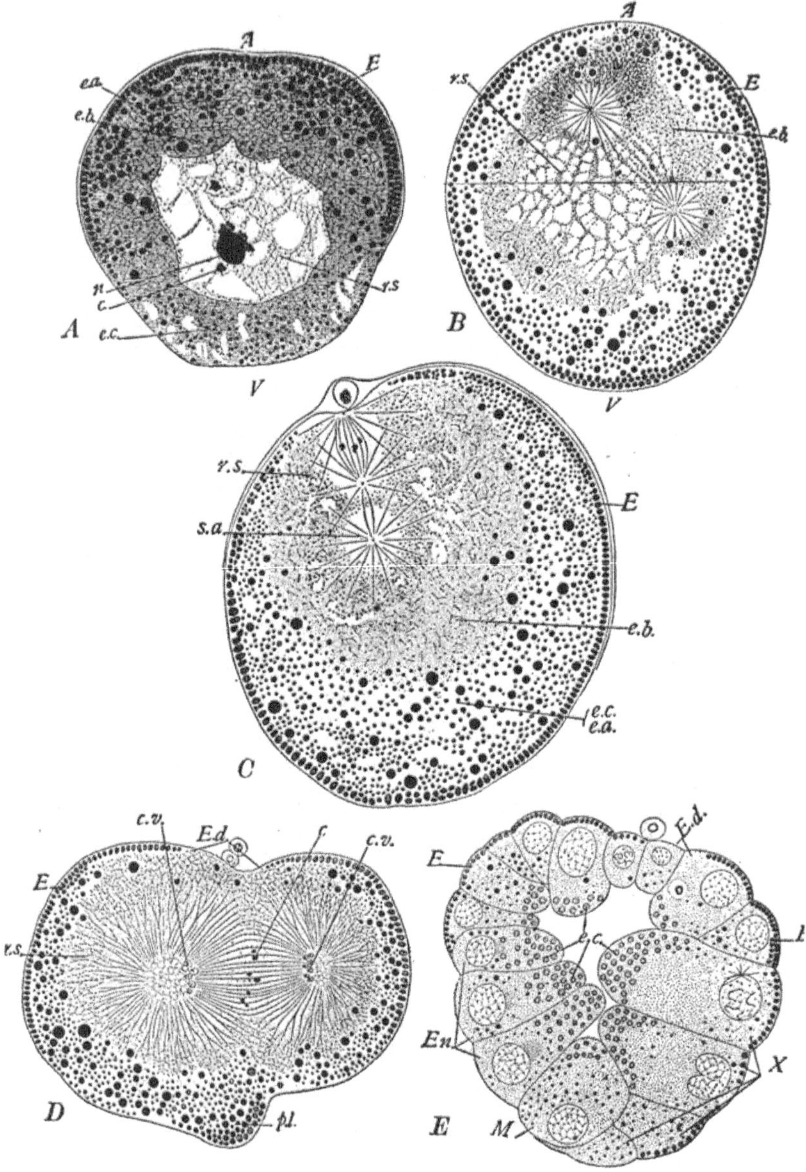

Fig. 123.—Changes in the structure of the egg of the Annulate, *Chætopterus*, during and after fertilization. From Lillie. *A*. Axial section through fully grown oöcyte, still within ovarian epithelium. Ectoplasm in upper two-thirds of egg only. *B*. Axial section through primary oöcyte, ten minutes after extrusion, three minutes after fertilization. Ectoplasm has already flowed to the vegetal pole, leaving an exposed area of endoplasm at the animal pole. Part of the *a*

is fully described by the same author and it is clear that each of these substances becomes contained within the cells forming the rudiment of the particular organ or tissue mentioned.

There are few other instances known where it is quite so easily possible to distinguish the specific formed substances in the egg. But in many ova, before cleavage it is possible to distinguish several kinds of substance; thus there are known in the eggs of certain Echinoderms four differentiated materials (Lyon), three in *Hydatina* (Whitney), *Dentalium* (Wilson), *Physa*, *Lymnæa* (Conklin), four in *Cumingia* (Morgan), three in the frog (McClendon), *etc.* (Figs. 42, 45, 86, 91, 123, 126, 129).

The term "organ forming" as applied to these materials does not rest alone upon the observation of normal development, but upon experimental grounds as well. For an example of this kind of evidence we may return to the work of Conklin on the egg of *Cynthia*. If one, say the right, of the blastomeres of the two-cell stage is injured so as to prevent its further development, the remaining blastomere develops as it would normally, *i.e.*, into the left half of an embryo and larva, containing approximately one-half the number of cells found in the normal organism at the corresponding stage (Fig. 124). Embryos derived from the two anterior cells of the four-cell stage never possess a tail or any muscle cells, while chorda cells and neural plate cells differentiate normally, and the endoderm and covering ectoderm are also typically formed (Fig. 125). Correspondingly, embryos derived from the two posterior cells have no chorda, nerve, or sensory cells, or gastral endoderm, but

endoplasm has also passed to the vegetal pole. The germinal vesicle has broken down, and the maturation spindle is in the process of formation, between the two asters. The residual substance of the germinal vesicle is clearly seen. *C.* Axial section through secondary oöcyte, thirty-two minutes after fertilization. *D.* Longitudinal section, first cleavage; late anaphase. Posterior end toward the left, anterior toward right. The ectoplasm of the polar lobe has been separated from the remainder. *E.* Sagittal section through stage of about sixty-four cells. The small upper cells are the apical cells. The ectoplasmic defect will be noted in the posterior apical cell to the observer's right. *A*, animal pole; *c*, chromatin; *c.v.*, chromosomal vesicles of daughter nuclei; *E*, ectoplasm; *e.a.*, *e.b.*, *e.c.*, endoplasm *a*, *b*, and *c*. *E.d.*, ectoplasmic defect; *En*, endoderm cells; *m*, mesoblast cell; *n*, nucleolus; *p.l.*, polar lobe; *r.s.*, residual substance of germinal vesicle; *s.a.*, sperm aster; *s.n.*, sperm nucleus; *V*, vegetal pole; *X*, derivatives of first somatoblast.

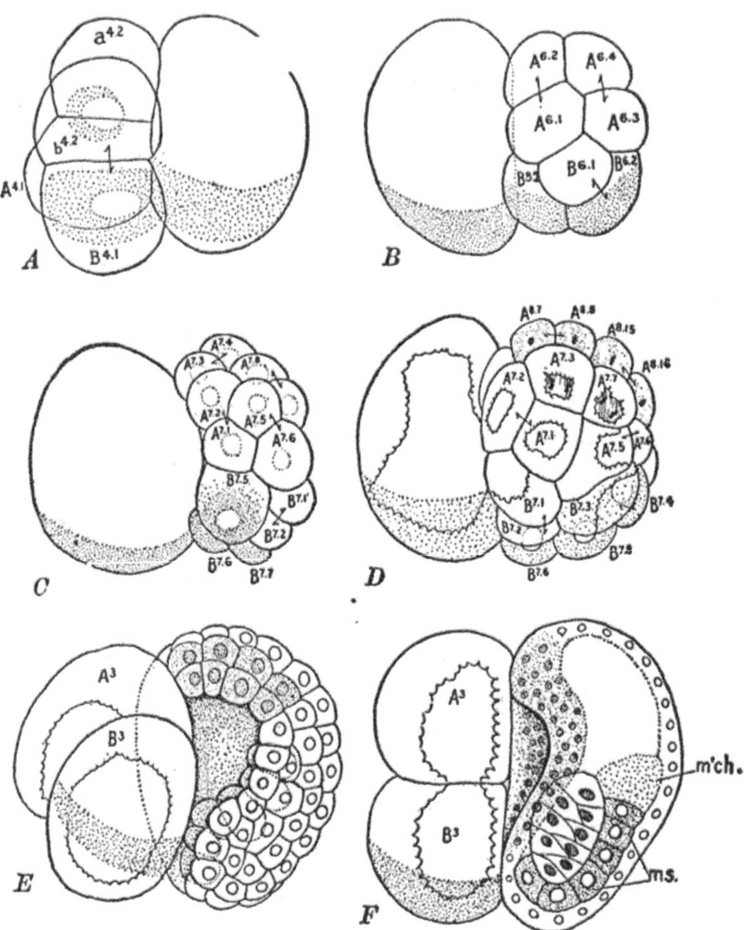

Fig. 124.—The development of one of the blastomeres of the two-cell stage of the Tunicate, *Cynthia*. From Conklin. The yellow material (see Figs. 91, 92) is stippled; the boundary between clear protoplasm and yolk is indicated by a crenated line.

A. Right half of eight-cell stage; posterior view. B. Right half of thirty-cell stage; dorsal view. C. Right half of forty-eight-cell stage; dorsal view. D. Right half of sixty-four- to seventy-six-cell stage; dorsal view. E. Right half of gastrula of about 220 cells derived from four-cell stage. The neural plate, chorda, and mesoderm cells, are present only on the right side, and in their normal positions and numbers. F. Right half of young tadpole; dorsal view. Derived from four-cell stage. The notochord consists of a small number of cells which are interdigitating; muscle-cells and mesenchyme lie on the right side of the chorda, but not on the left side, though the muscle cells have begun to grow around to the left side. The neural plate is normal in position but not in form. *m'ch.*, mesenchyme; *ms*, muscle cells.

The cell nomenclature in this and the following figure, differs from that described in Chapter VI. The right and left halves of the embryo are designated consist of a mass of muscle and mesenchyme cells with a double row of caudal endoderm cells, as in the corresponding region of a normal larva.   Equivalent results may be obtained by injuring one or three cells of the four-cell stage (Fig. 125).

The work of Roux, Fischel, Wilson, and many others has demonstrated similar localizations in the eggs of many forms— the frog, other Ascidians, several Molluscs, Annulates, and the Ctenophores, but we must limit ourselves to the mention of only a few interesting details of the experiments on these forms.

The Mollusca afford several very striking illustrations of the effects of the removal of parts of the egg or of blastomeres. The egg of *Dentalium*, as described by Wilson, has an upper clear area which normally forms the ectoderm, a middle reddish or brownish pigmented zone forming endoderm, and a lower clear area which during cleavage forms a peculiar "yolk lobe" or "polar lobe" (Fig. 126).   When this yolk lobe is entirely removed from the segmenting egg the development of the remainder proceeds as if it were present, and a larva is formed which lacks the apical organ and the entire post-trochal region (for explanation of terms see Fig. 126), and which develops later into an organism lacking those structures which would normally have been formed from this part of the egg and larva, namely, the foot, mantle, shell glands and shell, pedal ganglion, and apparently also cœlomic mesoblast.   Other Mollusca give essentially similar results although of course not all possess a yolk lobe; but removal of blastomeres is always followed by absence of specific parts in later development (*e.g.*, Wilson, Crampton, Conklin) (Fig. 126).

The blastomeres of several species of animals fall apart, or may be shaken apart easily, after a brief treatment with calcium-free sea water, a fact discovered by Herbst and applied by him

---

by the same letters, those referring to the right side being underscored. *A* and *B* refer, respectively, to the anterior and posterior hemispheres. After the third cleavage, all cells lying on the polar body side of that cleavage plane are desig-nated by lower case letters, while those on the opposite side of that plane continue to be designated by capitals. The first exponent following a letter indicates the generation to which the cell belongs. The second exponent refers to the position of the cell relative to the vegetal pole.

and many others to an analysis of this problem of localization. The blastomeres of many Echinoderms, Molluscs, *etc.*, can be thus separated, and it is a remarkable fact that one of two, four,

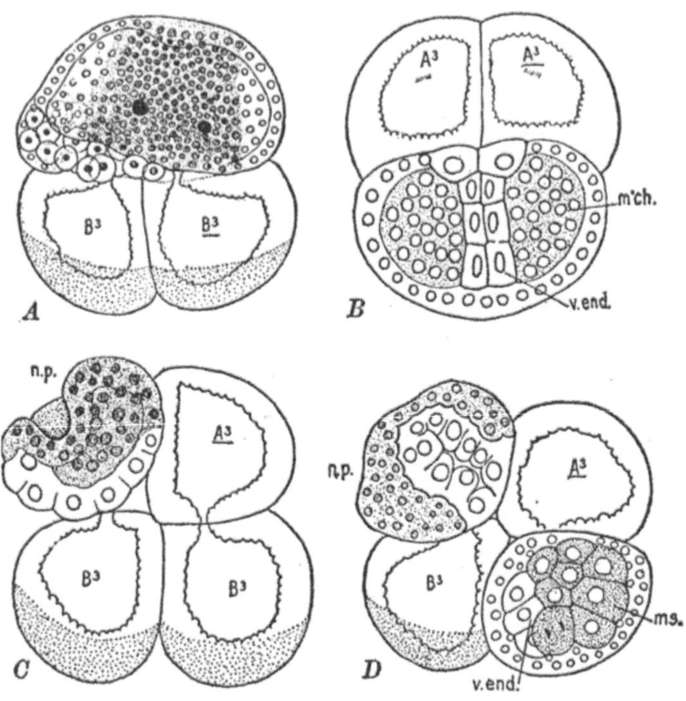

Fig. 125.—The development of blastomeres of the four-cell stage of *Cynthia.* From Conklin. *A.* Anterior half-embryo derived from two anterior blastomeres. The yellow crescent remains visible in the posterior, uninjured cells (B^3). Sense spots are present but the neural plate never forms a tube. The chorda cells lie in a heap at the left side. There is no trace of muscle substance or of a tail. *B.* Posterior half-embryo from the two posterior blastomeres. Dorsal view, focussed deeply upon the double row of ventral endoderm cells in the mid-line, a mass of mesenchyme cells on each side. No neural or chorda cells. *C.* Left anterior quarter embryo from cell *A*; dorsal view. An invagination of the ectoderm cells has the appearance of a gastrula, but is probably the invagination of the neural plate. *D.* Left anterior, and right posterior quarter-embryos, from cells *A* and *B*; dorsal view. The former shows thickened ectoderm cells, probably neural plate, around the endoderm cells; in the latter are eight muscle cells and three caudal endoderm cells. *m'ch*, mesenchyme; *ms*, muscle cells; *n.p.*, neural plate; *v.end.*, ventral endoderm.

eight, or even one of sixteen cells, continues to develop for some time and forms those parts, and only those, which it would have formed, had development of the entire cell group

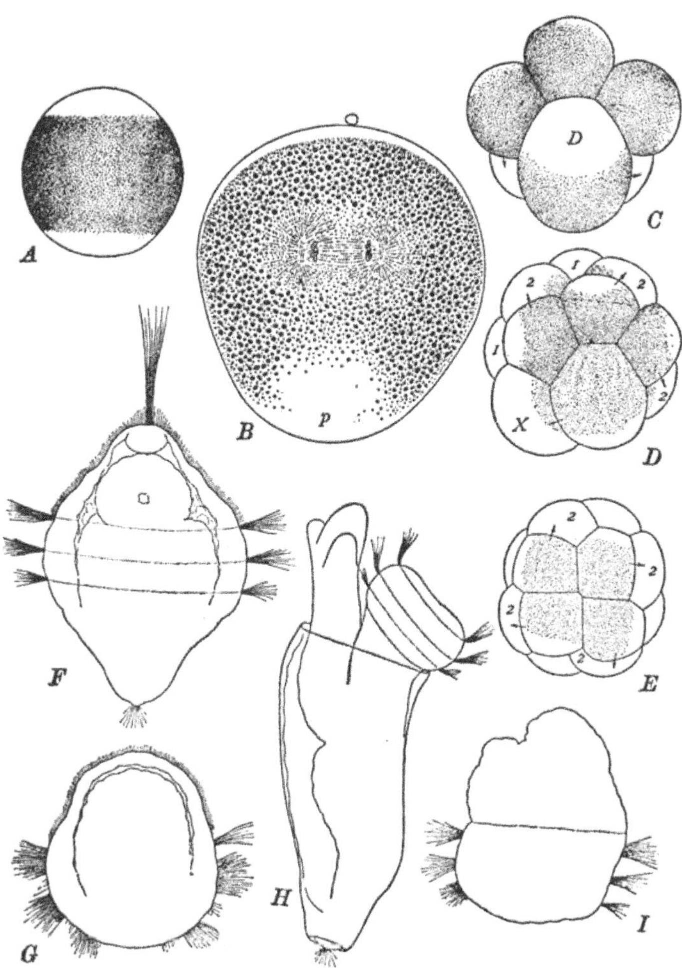

Fig. 126.—Development of the Mollusc, *Dentalium*, after removal of the "polar lobe." From Wilson. *A*. Egg twenty minutes after extrusion, and before maturation is completed, showing regional differentiation. *B*. Section through egg one hour after fertilization, showing the beginning of the formation of the polar lobe. *C*. Normal eight-cell stage, viewed from lower pole. The polar lobe is the light part of cell *D*. *D*. Normal sixteen-cell stage viewed from lower pole. The materials of the polar lobe are now contained in the cell marked *X*. *E*. Sixteen-cell stage of egg from which the polar lobe was removed during the first cleavage period. *F*. Normal trochophore of twenty-four hours. *G*. Trochophore of twenty-four hours, developed from "lobeless" egg. *H*. Normal larva of seventy-two hours, showing foot and shell. *I*. Seventy-two-hour larva from "lobeless" egg. *p*, polar lobe.

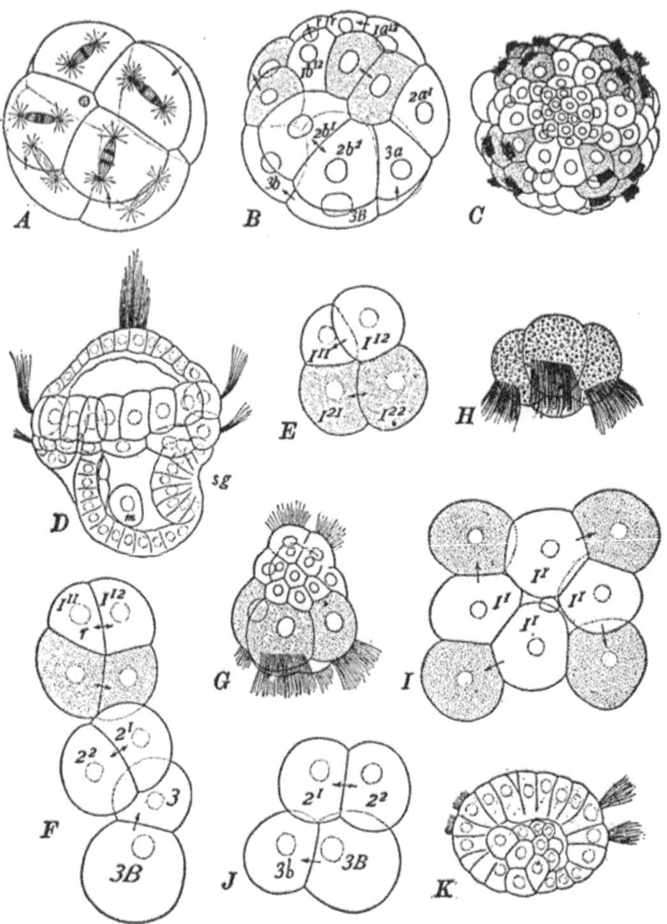

Fig. 127.—Cleavage of isolated blastomeres in the egg of the Mollusc, *Patella.*
From Wilson. *A–D,* × 167; *G, H,* × 242; others × 208. *A.* Normal eight-cell
stage, viewed from upper pole. Fourth cleavage in progress. *B.* Normal thirty-
two cell stage, from side. *C.* The so-called "ctenophore stage" (normal) viewed
from upper pole. The primary trochoblasts are ciliated. *D.* Normal trochophore of thirty hours, from left side. Body wall in section, prototrochal cells in
surface view. *E.* Second cleavage of an isolated micromere of the first quartet
(one of eight cells). *F.* Entire quadrant—products of first and second quartet
cells, being formed much as in the normal egg. *G.* Larva of twenty-four hours
from one of eight cells (micromere). From side, showing trochoblasts below,
apical cells above. *H.* Product of primary trochoblast isolated from sixteen-
cell stage. *I.* First division of isolated first quartet. *J.* Division of isolated
basal cell of eight-cell stage, showing typical arrangement of these four cells as
in the normal group of thirty-two. *K.* Larva of twenty-four hours, developed
from group like *I,* showing two secondary trochoblasts and two feebly ciliated
cells (? pre-anal cells). *m,* primary mesoblast cell; *s.g,* shell gland.

been occurring normally, although ultimately a normal larva may be formed (Fig. 127).

The idea that these differentiated materials of the cytoplasm really play the rôle of organ-forming substances in development, is opposed by some (Lillie, Morgan, and others). The opposed idea rests upon the experimental evidence that, briefly stated, the really primary and fundamental organization of the egg cytoplasm concerns the *ground substance* of the protoplasm; the arrangement of the various formed stuffs coincides with a similar and primary polarity and organization of this fundamental protoplasmic matrix. This structure is less manifest, but is really the factor which determines the arrangement of the suspended cytoplasmic and deutoplasmic granules and vacuoles. The correspondence between the arrangement of these stuffs and the organ-forming substances proper, is thus unessential, for the localization of the germ is primarily a localization of the ground substance. In other words, the varieties of material described by Conklin in *Cynthia*, for example, are only secondarily related to the later differentiation of particular organs or tissues, and their arrangement is dependent upon the same primary factor that determines the arrangement of the organs and tissues.

The evidence for this view is found chiefly in the results of certain experiments upon the eggs of *Chætopterus* (Lillie) and *Arbacia* (Morgan and Spooner). The granules which give character to the various regions of the cytoplasm differ in specific density, and consequently can be thrown, by centrifugal force, into abnormal regions of the ovum. When this is done normal cleavage and development may proceed, normal with respect to the *original polarity* of the ovum and not with respect to the new polarity as indicated by the altered arrangement of the plasmas.

To illustrate, the egg of the sea-urchin *Arbacia*, contains four different kinds of substance; one of these is distinguished by the presence of bright orange or reddish pigment. In normal development this substance lies toward the lower pole and becomes localized in the lower quartet, so that when the micromeres form here, they are composed of this material. The micromeres, which later form the mesenchyme, always appear at the pole opposite the micropyle (Fig. 109), which marks the point of attachment of the ovum in the ovarian germinal epithelium; this is also the point at which gastrulation commences. The centrifuge brings about a stratification of these substances which is independent of the polarity of the ovum, since the ovum may assume any position with reference to the

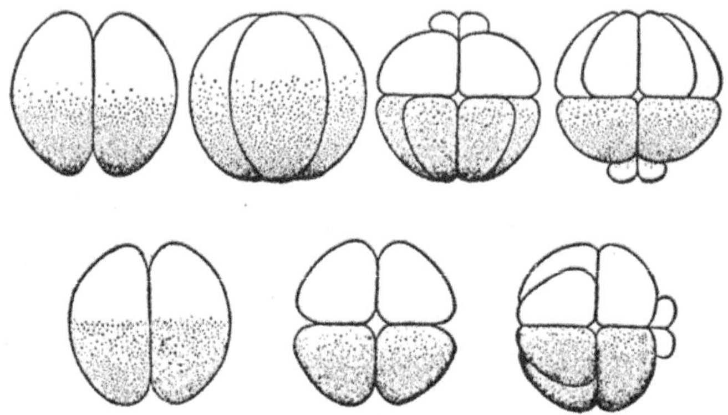

Fig. 128.—Normal cleavage in the sea-urchin, *Arbacia*, following abnormal distribution of egg substances by centrifuging. From Morgan and Spooner. The figures are turned so that the pigment (dotted area) is downward. The location of the cleavage planes, and the position of the micromeres, which always mark the invaginating pole also, are independent of the induced stratification of the egg substances.

axis of rotation of the machine. The pigmented protoplasm may be thrown to any part of the cell. But Morgan has found that the cleavage of eggs with abnormally distributed substances proceeds normally with reference to the original polarity of the ovum and not according to the induced arrangement. The micromeres, for example, continue to form opposite the micropyle, and gastrulation occurs here, as usual, although the pigmented protoplasm may occupy some remote and unusual position in the cell (Fig. 128). Perfectly typical larvæ develop from such eggs, normal save in the distribution of pigment. Development and differentiation thus seem to be quite independent of the so-called "formative stuffs," which are, in such instances, evidently not "organ forming."

Several other forms are known to give similar results. One of the clearest instances is to be seen in the Lamellibranch, *Cumingia*, also described by Morgan. The egg of *Cumingia* contains, besides the clear protoplasm, three kinds of formed substance, yolk, pigment, and oil. With the centrifuge these can be thrown to any part of the cell whatever, and yet cleavage and development proceed normally with reference to the

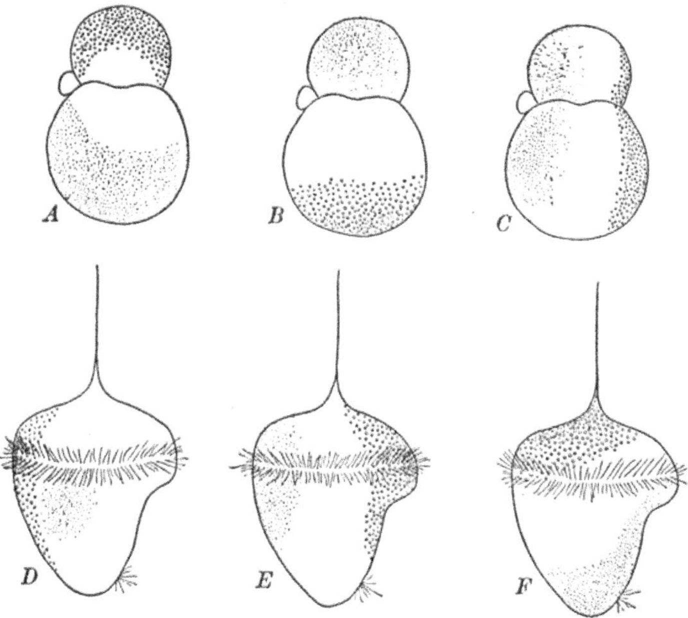

Fig. 129.—Normal development of the Pelecypod, *Cumingia*, following abnormal arrangement of the egg substances by centrifuging. After Morgan. The pigment is indicated by stipples, the oil by small circles. *A*. Two-cell stage with oil in small cell. *B*. Same with oil in large cell. *C*. Same with oil in both cells. *D*. Normal trochophore, showing usual distribution of pigment and oil. *E*. Trochophore with oil on oral side, and yet normal. *F*. Normal trochophore with oil aboral and interior.

original polarity and not at all to the actual distribution of these substances (Fig. 129). It should be noted that, although this has not been definitely determined for *Cumingia*, the Mollusca in general are excellent examples of determinately cleaving eggs and the removal of parts of ova is followed by definitely corresponding defects in embryo and larva.

Such experiments as these seem to indicate clearly that the

determining structure of the ovum is really that of the underlying protoplasmic ground substance, and that the arrangement of the various formed substances coincides with this, is determined by it in the first place, but is not always or necessarily concerned directly in the later differentiations of the ovum. The defects following removal of parts of the unsegmented ovum, or of blastomeres result therefore from the loss of parts of this underlying structure, and not from the loss of the formed materials or "formative stuffs," which in such cases at least turn out not to be "formative."

Lillie has shown that carefully graduated centrifuging reveals the existence in the egg of *Chœtopterus*, of certain regional differentiations of the ground substance, indicated by the differences in the ease with which the granules of various sizes, and other structures, such as parts of the mitotic figure, pass through it. These regions are not otherwise visible but Lillie suggests that since they. are undoubtedly real they may represent or mark in some way the primary organization of the cytoplasm.

According to this view of organization the term "organ-forming substances" for the visibly differentiated substances is a misnomer. For if these are removed to abnormal positions within the cell they are then not related to the formation of the same structures that they are in normal development.

At present it seems difficult, though not impossible as we shall see later, to reconcile these two views as to the real seat of the primary organization of the cytoplasm of the ovum; the balance of evidence appears to favor the conception of organization as a condition of the fundamental ground substance of protoplasm. But in any event it is perfectly clear that the cytoplasm is organized definitely.

We should call attention in passing to the fact that many of the results described above indicate that cleavage is not to be regarded always as a developmental process of primary importance. Conklin has called attention to the fact that in *Cynthia* the early cell boundaries do not always coincide with the limits of the various kinds of cytoplasm. The determination of the structure of the egg and the localization of these materials, precede cleavage and are independent of it. Certain pressure experiments to be mentioned shortly, illustrate the independence of cleavage and determination, and the centrifuging experiments similarly demonstrate the independence of cleavage and the distribution of the cytoplasmic stuffs. Indeed Lillie describes the formation of a trochophore-like larva from the egg of *Chætopterus* in which cleavage had been artificially prevented; this embryo formed external cilia and certain other

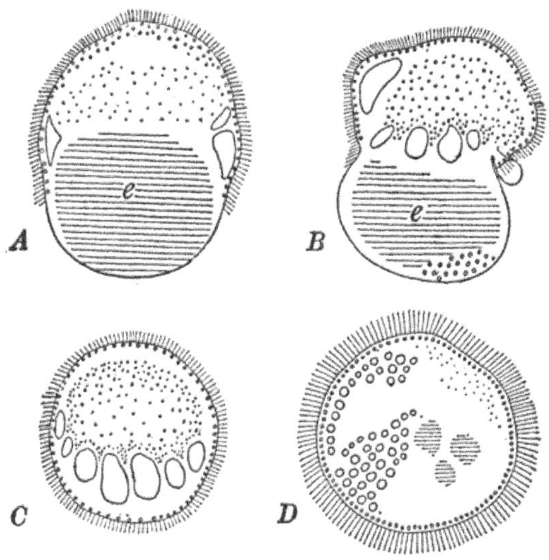

Fig. 130.—Development and differentiation in the absence of cell division, in *Chætopterus*. From Lillie. *A, B, C.* Ciliated, uninucleated unsegmented eggs, about twenty-three hours old. The vacuoles are about in the position of the prototroch of the larva. *D.* Ciliated unsegmented egg about twenty-eight hours old; most of the endoplasm has been consumed. *e*, endoplasm.

differentiated structures in the complete absence of cell divisions (Fig. 130). The normal processes of development are varied and more or less independent of each other, while having common reference to some general underlying condition.

We must now consider the facts of development in a considerable group of eggs which do not show any such results as those described in the foregoing pages. Although these eggs possess more or less differentiated regions of cytoplasm, yet removal of parts or of blastomeres is not followed by any structural defect. For example, the blastomeres of many Cœlenterates (Haeckel, Zoja, Maas, Wilson) may be separated when in the two-, four-, eight-, or even, in some cases, in the sixteen-cell stage, and from such isolated blastomeres typically formed embryos and even free-swimming larvæ develop, normal in every respect save that of size, being respectively approximately one-half, one

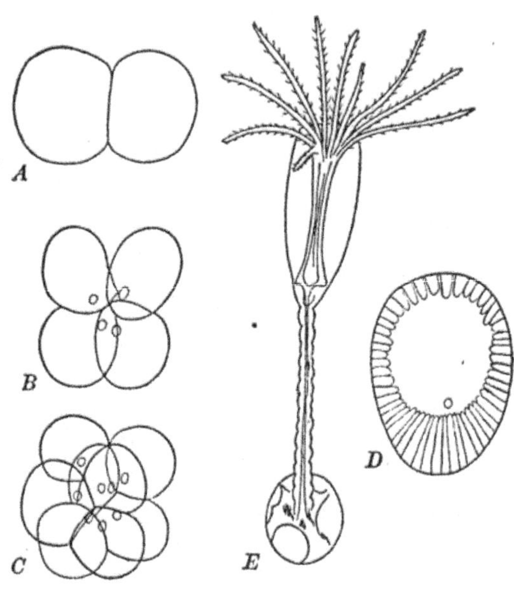

Fig. 131.—Normal development of one of the blastomeres of the two-cell stage of the Hydroid, *Clytia flavidula*. After Zoja. *A*. Two-cells. *B*. Four-cells. *C*. Eight-cells. *D*. Blastula. *E*. Young polype.

fourth, one-eighth, or one-sixteenth the normal size (Fig. 131). This is true to a certain extent also of some of the Teleosts and of Amphibians, the Nemerteans, and Echinoderms (Figs. 132, 133); in the last named forms even portions of the blastula or gastrula (Driesch) may give rise to normal but diminutive larvæ (Fig. 134). It seems very apparent that if cytoplasmic localization occurs at all in such cases, it must be of a very different kind from that described above.

This is an appropriate place to mention certain experiments of a different kind bearing upon this same problem. Eggs may be subjected, during their early cleavages, to deforming pressure so that the planes of cell division appear in abnormal relations to one another and to the egg as a whole (Hertwig, Born). Thus in the sea-urchin (Driesch) the blastomeres of the eight-cell stage instead of forming a spheroidal group may be forced into the form of a flat plate (Fig. 135). When released from the pressure such eggs form perfectly typical

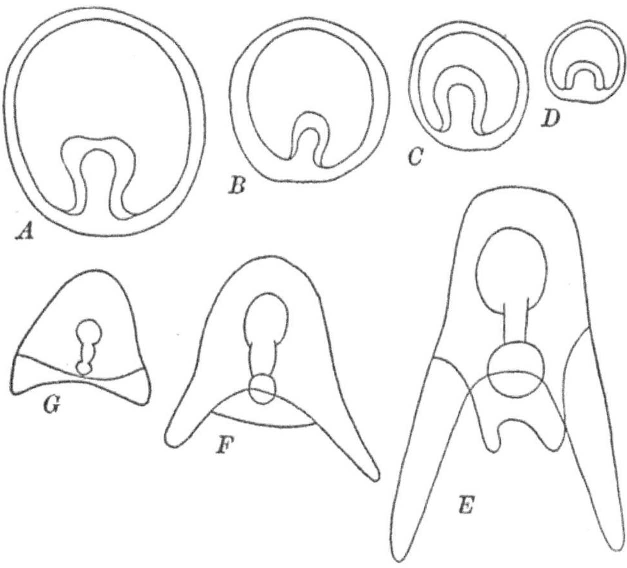

Fig. 132.—Gastrulæ and plutei from isolated blastomeres of the sea-urchins, *Echinus* (*A–D*), and *Sphærechinus* (*E–G*). After Driesch. *A*. Gastrula from entire egg. *B*. Gastrula from one blastomere of the two-cell stage. *C*. Gastrula from one blastomere of the four-cell stage. *D*. Gastrula from one blastomere of the eight-cell stage. *E*. Normal pluteus. *F*. Pluteus from one blastomere of the two-cell stage. *G*. Pluteus from one blastomere of the four-cell stage.

larvæ. And even in a form like the Annulate, *Nereis*, whose cleavage is determinate and whose blastomeres are highly differentiated, Wilson has found that when the egg, subjected to pressure, became divided by vertical planes into a flat plate of eight cells, each one contained substance normally found only in the macromeres of the lower pole; when released these eight cells divided into sixteen, eight micromeres and eight macromeres, instead of into the normal twelve and four respectively. And from these, normal larvæ developed; the eight macromeres developed as the normal four would have done, although under normal conditions four of the eight cells and nuclei would have formed the first quartet, giving rise to the apical nerve cells and anterior band of ciliated cells.

Furthermore, the experiment of bringing about the coalescence of parts of two eggs, or even of two complete eggs, has

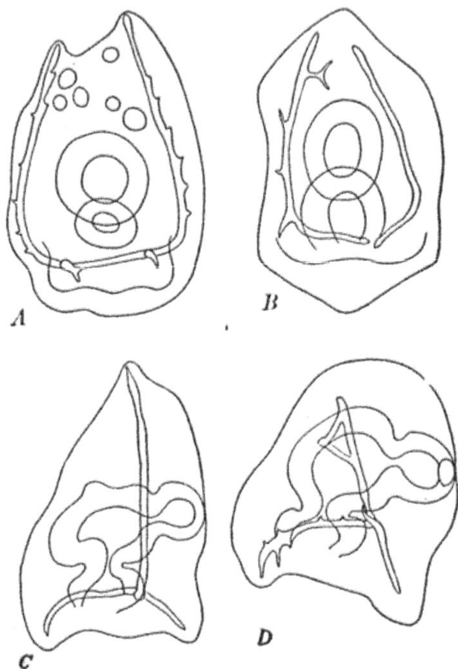

Fig. 133.—Four normal but diminutive plutei from the isolated blastomeres of the four-cell stage of the sea-urchin, *Strongylocentrotus*. After Boveri. *A, B.* in oral view. *C, D,* lateral view.

been accomplished with *Ascaris* (Sala, Zur Strassen) and with the sea-urchin (Driesch). The result is again the development of a normal larva, of very large size when two entire eggs are fused (Fig. 136). Even when two blastulas coalesce the final result may be a single larva, though with some doubling of parts. One especially interesting point is that normal development may result even though the parts of the two blastulas may be of different species, in somewhat different stages of development, and no micromeres included in the mass (Garbowsky).

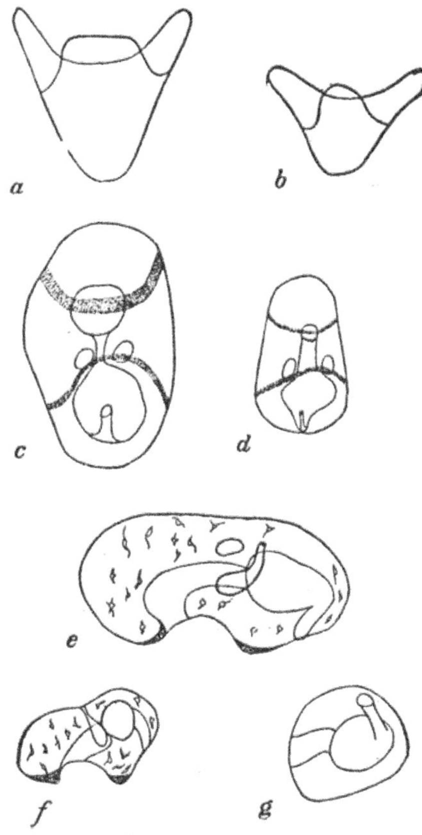

Fig. 134.—Normal but diminutive larvæ of Echinoderms, derived from portions of gastrulæ. From Jenkinson, after Driesch. *a.* Normal pluteus of *Sphærechinus.* *b.* Pluteus of same from portion of gastrula. *c,e.* Normal bipennaria of *Asterias glacialis.* *d,f.* Bipennaria of same, from vegetative half of gastrula. *g.* Larva of *Asterias* with typical three-parted gut, but no cœlom, from vegetative half of gastrula, removed *after* development of the cœlomic sacs.

While such results as these are at first sight opposed to the hypothesis of germinal localization, yet it is quite possible to reconcile the differences between such extreme forms as the Echinoderms, where one of eight or sixteen cells finally forms a typical larva one-eighth or one-sixteenth normal size, and the

Ascidian, where one of four, eight, or sixteen cells gives rise, not to a complete diminutive larva, but to a group of differentiated tissue cells of the same kind that would normally have been formed from the particular cell, had it remained *in situ* in the normal group.

The discordance of these results may have one of two meanings. First, it may mean that in such eggs as those of the Echinoderms and Amphioxus a process of regeneration or

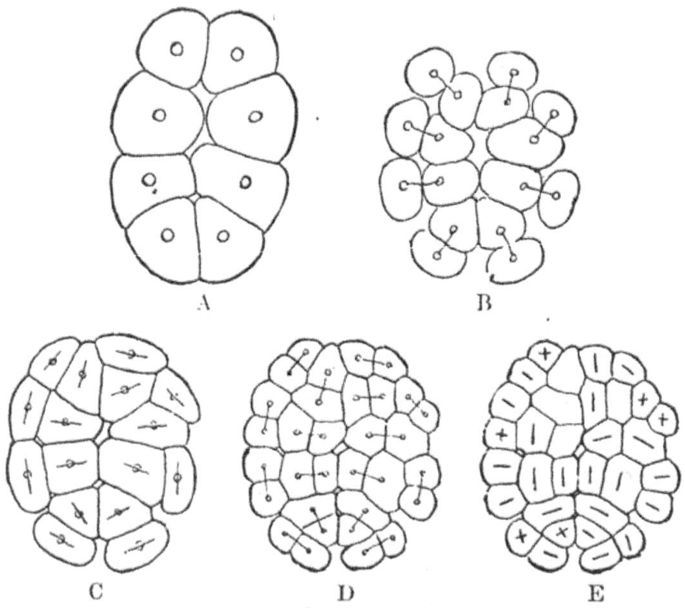

Fig. 135.—Cleavage in the egg of the sea-urchin, *Echinus micro-tuberculatus*, under pressure. From O. Hertwig, after Ziegler. *A, B.* Eight- and sixteen-cell stages. *C.* Sixteen-cell stage preparing for division. *D.* Thirty-two-cell stage, in the form of a flat plate. *E.* Thirty-two-cells preparing for next division. Crosses mark cells in which the spindle is vertical or oblique, to the plane of the cell group.

regulation goes on. And, just as many adult organisms are easily capable of restoring or regenerating lost parts, so the embryo or even the ovum may have the property of reforming parts artificially removed. This process has received the special term of *post-generation* (Roux). Such a possibility is indicated by the classic experiment of Roux upon the egg of the frog. Here, if one of the two blastomeres is destroyed the remaining one, if undisturbed, develops into a half-embryo; but if the egg is inverted after the injury of one blastomere, then during the consequent rearrangement of the substances of the uninjured cell, through the action of gravity, the organization is restored to the normal and a small normal embryo subsequently develops.   This shows that the uninjured half of

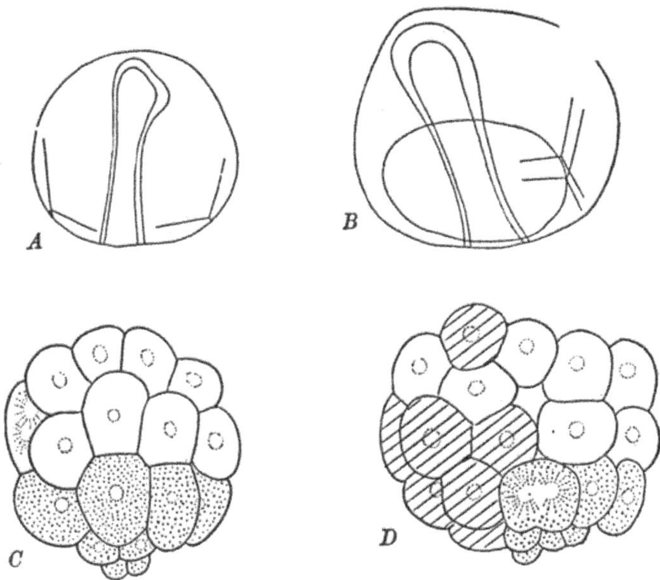

Fig. 136.—Fusion of Echinoderm larvæ. *A, B. Sphærechinus.* After Driesch. *C, D. Psammechinus miliaris.* After Garbowski. *A.* Normal gastrula. *B.* Single gastrula derived from the fusion of two normal blastulæ, showing single, large gut and doubled spicule. *C.* Normal stage of thirty-two-cells. *D.* Organism formed by the coalescence of parts of two organisms in different stages. The cells ruled obliquely were part of an eight-cell stage, stained intravitally with neutral red. The remaining cells were part of a normal thirty-two-cell stage. In both *C* and *D* the stippling marks the cells derived from the vegetative half of the egg.

the egg does possess the potentiality of developing as a complete egg.

Or second, the contrast between the two extreme cases mentioned may mean that localization results from a progressive process of true development. Of course, in all organisms, sooner or later, groups of cells become specifically differentiated as particular tissues and organs or parts of organs. And similarly there comes a time in the history of any cell group when, once started on its course of differentiation, return or redifferentiation in another direction is impossible.

The formation of localized germinal areas of cytoplasm is to be regarded as a process of development, and in the eggs of different species this process may be carried forward at relatively different times with respect to fertilization, cleavage, and other early developmental phases. The most important steps in cytoplasmic localization of the germ may be completed while maturation and fertilization are going on, prior to the first cleavage (Ascidians); or localization may be accomplished during cleavage (*Cerebratulus*), or not until the gastrula or post-gastrula stages (Echinoderms).

This idea is not essentially different from that of post-generation in certain respects, for regeneration and regulation are after all essentially processes of development, deferred development. The two differ however in that, according to the former view localization is really present throughout the early stages and disturbances are followed by an active process of regulation; according to the latter, localization is not determined during the earlier stages and when it does appear, the parts of the egg remaining after the removal or injury of parts, behave as a complete and normal unit, no regulation being necessary.

There is evidence for both of these views, and both may be true at the same time. The second appears to be the more widely applicable. Regulation seems more likely to occur during comparatively late phases of localization. Evidence of regulation following the separation of blastomeres is afforded by such eggs as those of the Echinoderms, where the isolated blastomere continues to segment for a time as if it were part of a normal cell group, but gradually its cell products assume the characters of a typical whole group and finally give rise to a normal embryo and larva. In other cases (*Amphioxus*) separated blastomeres develop from the beginning like whole eggs, and no regulation is necessary. The results of deformation by pressure also indicate that localization is subject to a regulatory process which may occur even in a comparatively late stage in cleavage.

That the "organization" of the cytoplasm results from a progressive developmental process is clearly evidenced by the experiments of Wilson, Yatsu, and Zeleny· on the egg of *Cerebratulus* before cleavage. If portions of this egg are removed before maturation has begun, while the egg nucleus is still in the form of an intact germinal vesicle, no defects are seen in the resulting larva. Entrance of the spermatozoön is followed by maturation and a general rearrangement of the substances of the cytoplasm, one result of which is the formation of a cap of clear protoplasm at the animal pole. Removal of this substance prior to or during the first cleavage, often produces no later abnormality. The separated blastomeres of the two-cell stage, however, while for a time cleaving like halves, soon assume the character of wholes. Those of the four-cell stage continue longer to behave like parts, even through the blastula stage, although ultimately they may form typical free-swimming larvæ. The degree of defect corresponds in a general way with the stage to which cleavage has progressed at the time of separation. Larvæ developed from eggs without the upper quartet, which contains the clear protoplasm mentioned, have typically formed enteron, but lack the apical organ. Larvæ from this upper quartet have the apical organ but are without enteron. And the same is true when, in the sixteen-cell stage, the upper and lower octets develop separately. Parts of the blastula continue to develop for a time and form only the restricted cell groups to which they give rise in normal development.

.Such facts seem clearly to mean that cytoplasmic germinal localization may be complete in later stages, but incomplete or absent in the earlier, that it is truly a process or result of development and not a primary determiner of the course of development, not a fixed thing persisting from generation to generation, which might be regarded as the physical basis of heredity.

The conception of cytoplasmic localization as a progressive process, *i.e.*, as one factor or link in the chain of developmental events, immediately raises the question as to what condition then lies back of this, and determines the character of the progressive steps or reactions. This leads us directly to the second chief view as to the fundamental character of the specific organization of the ovum, that is, to the hypothesis of "*nuclear analysis*" or *nuclear determination*, and to this we may now give our attention. To state them again, the essentials of this hypothesis are, that the real germinal localization of the ovum is to be sought in the nucleus, that the organization of the cytoplasm is preceded and its character determined primarily, by the organization of the nucleus, that this organization is continuous from one generation to the next and is so to be regarded as representing the physical basis of heredity. Polarity and other cytoplasmic differentiations, certainly exist in the ovum, even before fertilization or cleavage, but the only structural differentiation of the ovum which is invariably marked out at all stages of the organism's existence, is the differentiation between nucleus and cytoplasm. And while not alone development, but all the normal life processes of the cell are the results of interaction between nucleus and cytoplasm, both being essential, yet the action of the nucleus is primary and seems to determine the particularity of the cell actions.

This general subject of nuclear determination is enormously complex and has been the occasion for whole volumes; our account of it must perforce be brief and therefore more or less fragmentary and dogmatic.

The search for the underlying causes of development is in part a search for elements or conditions that are comparatively fixed and that remain continuous from generation to generation through the individual waves of species life. Specificity is continuous; are there structural elements or conditions correspondingly fixed and constant, not having to develop anew in each individual ontogeny? Are there structures in the germ cells which determine the direction of development and thus represent (using this word in a very broad sense) the organs and parts of the developing embryo?

In the endeavor to answer these questions the nuclei of the germ cells at once compel attention as containing organs whose morphology appears to be constant and specific at all stages of the individual life history, and through successive generations. Thus the chromosomes at once become the foci of observation and discussion, and the hypothesis of nuclear determination becomes, to a considerable extent, the hypothesis of the specificity of the chromosomes.

This conception has already been outlined in Chapter II; the chromosomes are believed to be differentiated functionally, in a specific manner so that each chromosome of the nucleus represents a center of activity of a particular character.   That is to say each chromosome, either individually or as a component of a unified group, determines a specific form of reaction with the cytoplasm, or rather influences in a particular way certain of the reactions constantly occurring between nucleus and cytoplasm.   And the final result of these reactions is the production of certain structural and physiological characteristics of the embryo or mature organism.   Thus, leaving out all the intermediate chain of processes or reactions, there is an actual correspondence between certain traits of the mature organism and certain chromosomal characters of the gametic nuclei.   Of course the chromosomal characters determine only the first step in the development of the corresponding trait; but this in turn determines the next, and so on.   And since the quality of one step or reaction in development is determined by the preceding, we are correct in relating directly the character of the final steps in development with the factor that first determined the trend of reaction.

Emphasis is thus placed upon the physiological character of the relation between chromosome and later structure, and care must be exercised constantly, in the discussion of this subject, to guard against a conception of this relation which is too strictly morphological, and which might suggest too strongly the conception of development as preformational.   A wrong interpretation of the modern view of the chromosome relation leads to a rather strict preformational view; but such an idea does not to-day represent the hypothesis fairly.   What is formed, or preformed, in the germ is a certain arrangement or configuration of the chromatic substance, which in its reactions with the cytoplasm produces new and specific conditions, these lead to others, and so on through development.

The conception of the determinative character of the chromosomes must now be modified to include the idea that each chromosome is not a simple unit, homogeneous either morphologically or physiologically. Each chromosome is to be regarded as made up of a series or group of elements which singly are simple and homogeneous, and behave as physiological units or "determiners." These may or may not correspond with the chromioles, or granules of chromatin, of which the chromosome is composed; and while it is true that they have never been positively identified as units or determiners, some such bodies must be present in the chromosome according to this hypothesis. Such determiners, although apparently necessary hypothetical units cannot be described; they may prove not to be definite particles at all, but rather dynamic relations, or configurations of substance. So for practical and descriptive purposes we are nearly limited to the chromosomes.

It is quite likely that the chromosomes may not be the only factors in the determination of development, there may be a whole series of factors back of these, and we know that a whole series of factors follows after. But if they are proved to be necessary links in a chain of determining factors, then they are causes of differentiation, and if they are found to be the earliest visible differentiations with which later differentiations somehow correspond, then we may refer to them as *the* causes of specific differentiation. At some future time it may indeed be possible to push the analysis of the factors of differentiation still farther back; such a possibility is in no wise excluded by the chromosome hypothesis as it stands to-day.

One of the obvious requirements of any hypothesis of differentiation and heredity is that it must readily allow interpretation, in cytological terms, of the enormously complex phenomena of alternative or Mendelian heredity. Most of the traits of an organism are the property of the species, common to all the individuals of a specific group. But there are other charac-

ters that are family possessions and may or may not be inherited
by individuals.   These individual characteristics are, in many
cases, comparatively late developments.   The early characters
are those of the larger group; those of the species appear later,
and finally the family and individual traits.   The whole subject of Mendelism has developed into an extremely complicated
system, in directions largely unforeseen.   And yet it is hardly
too much to say that the cytology of the germ cells and their
nuclei has on the whole fairly kept pace, and it is in most instances quite possible to parallel the facts of Mendelism with the
facts of chromosome behavior.   We shall return briefly to this
subject in a more appropriate connection.

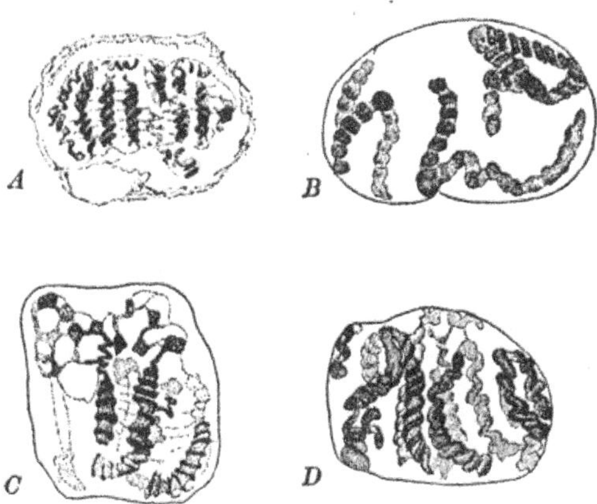

Fig. 137.—The structure of chromosomes.   A, after K. C. Schneider, others
after Bonnevie.   A. Nucleus from epidermis of Salamander larva, in telophase.
B. Prophase of first cleavage of *Ascaris megalocephala bivalens*.   C. Nucleus
from cleavage stage of same.   D. Interkinesis in *Amphiuma*.

Let us now repeat, from this particular point of view the
general ideas regarding the chromosomes mentioned in Chapter
II, emphasizing certain topics and adding a few details which
bear directly upon the relation of the chromosomes to the processes of development and heredity.

The chromosomes are not structurally homogeneous masses,
but are built up of certain granules which often have a definite
arrangement giving the chromosome as a whole a general
structure.   This structure has been variously described (Fig.
137); in some cases it seems to be a cylinder of chromatin

granules with a core of differentiated substance, probably linin; in other cases the granules are arranged in a linear series wound in a definite spiral; but in still other cases the granules seem to have little if any regular disposition. These granules are very close to the limit of vision, indeed often they are invisible, and in most cases it is difficult to make confident assertions regarding their arrangement. Of course their invisibility on account of minuteness does not prove that they are not present. Indeed many, following Weismann, postulate an elaborate series of representative particles within the nucleus: the chromosomes or "idants," are divided into "ids" or chromatin granules; the ids are then assumed to be formed of groups of "determinants," and these in turn are thought to be composed of the really elementary, self-propagating protoplasmic units, the "biophores." Of this series, the idants and ids are visibly known; the determinants and biophores are invisible hypothetical bodies, postulated to aid in relating many of the complicated facts of heredity to certain cytological facts. The assumption of a final determining unit that is, and seemingly must remain, invisible has proved fortunate as affording a convenient shelter against criticism, for such an assumption partly removes the question from scientific treatment. We shall lose little and gain much by considering here merely those elements of the nucleus that can be identified and whose behavior can be traced to some extent, *i.e.*, the chromosomes and chromatin granules.

Our ideas in this field are, to a remarkable degree, the outgrowth of the pioneer work of Weismann. Although based upon the properties of hypothetical units whose behavior was outlined upon purely hypothetical grounds, his conceptions of the relation between chromosome behavior and the facts of development and heredity, formulated more than thirty years ago, before the science of cytology was established, have a distinctly modern aspect. The remarkable convergence of the facts of heredity, of development, and of cytology, which have become known subsequently to the formulation of Weismann's hypotheses, constitutes splendid evidence of the keenness of this great embryologist.

We may here suggest again the character of the evidence for regarding the chromosomes of the germ nuclei of the very greatest importance as factors controlling or directing the process of development (*i.e.*, the process of heredity), facts of such constancy and universality that they must have some meaning.

First of all comes the fact of the very high degree of morphological constancy of these organs throughout the tissues of the species, not merely of the individual. This constancy, always considering corresponding ages of course, concerns their number, size, and form, and proves them to be specific organs in a real sense. They are, with a few easily explained exceptions, present in pairs of similar elements, whose history can be traced back to their derivation in groups of unpaired elements in the male and female gametes. During mitosis the distribution of the chromosomes to the daughter cells is never a haphazard process, but the whole process of mitosis appears to be an adaptation toward securing their equal division, and the distribution to the daughter nuclei of groups of similar morphological composition.

It is unnecessary to repeat here any of the evidence outlined in Chapter II for the idea of the genetic continuity of the chromosomes from cell to cell. We saw there that it is the chromatin granules which may be regarded as actually morphologically continuous, and that these may seem to become similarly associated every time that the chromosomes visibly appear. What determines their constant association in the reforming chromosomes of each cell generation is to be answered only hypothetically if at all. But in spite of the contrary belief held by some, the chromosomes may be regarded as genetically continuous individual elements, although the details of their composition may vary slightly from generation to generation. And after all it may be that the most important continuity is that of the chromatin granules, supposing them to be qualitatively unlike and to play the rôle of specific determiners.

The validity of the chromosome hypothesis has always been strongly indicated by the essential facts of syngamy, namely, the formation of the zygotic nucleus from equal contributions of chromosomes from the male and female parents. This is the only portion of the new organism which is derived equally from both parents, and while it is true that a small amount of cytoplasm accompanies the sperm nucleus in its entrance into the ovum, this varies considerably in amount in different forms. This latter fact together with the general fact of the primary importance of the nucleus in all aspects of cell activity, combine to enhance the significance of the equal derivation of the chromosomes (Fig. 106), especially in view of the further fact that *on the whole* the family and individual traits of organisms are inherited with equal likelihood from either parent.

The behavior of the chromosomes throughout the maturation process affords many highly interesting and significant parallels between chromosome behavior and the facts of heredity. Interest centers here in the phenomena of synapsis and the "reducing" divisions.

Any precise interpretation of these two phenomena seems impossible until more is known with certainty regarding the behavior here of the chromatin *granules*, but the phenomena themselves are readily interpretable in the light of the facts of alternative, or Mendelian heredity.

In synapsis we see the final union of pairs of chromosomes introduced into a single nucleus at the time of fertilization, but remaining distinct throughout the life of the hybrid generation, until the time when the hybrid organism forms its gametes. Synapsis is not a haphazard junction of chromosomes, but an orderly union of elements of paternal and maternal origin, similar in size, in details of form, and probably also in function. The bivalent chromosomes thus formed are, in consequence of their derivation from two individuals, not quite homogeneous throughout. Following synapsis come two divisions of each chromosome, and in most organisms one of these apparently divides the chromosome equally, into two similar parts (equation division), while the other divides each of the daughter chromosomes dissimilarly (reducing division), the dissimilarity resulting from the relation of the plane of division to the plan of arrangement of its dissimilar component granules. The relation of the reducing division to the chromosome depends upon the character of the synapsis, whether telosynapsis or parasynapsis, and also upon the behavior of the chromatin granules in all these events, and it is difficult to be certain of this. It is safe to say, however, that in most cases each bivalent chromosome, composed in equal parts of substance from each parent, clearly separates into four elements, two having one composition, two another. These elements are then distributed to separate gametes, so that with respect to the composition of each separate chromosome, the gametes produced by an organism are of two kinds, approximately equal numerically. This accords perfectly with the facts of Mendelian heredity, upon the supposition that there is a correspondence between chromatic elements and organismal traits. This may be made somewhat clearer with the aid of a diagram: see Fig. 80.

In the process of maturation, therefore, it is easily possible to find a mechanism which permits the segregation of characteristics in the germ cells and their distribution to separate organisms in regular Mendelian ratios. One important correspondence should not be overlooked. In Mendelian heredity the individual qualities of the parents may not appear separately until the first generation *after* the hybrids. This is possibly related to the fact that the parental chromosomes undergo synapsis and subsequent redistribution first in the germ cells formed by the hybrid, and the segregated elements are, therefore, distributed separately first in the organisms formed from these hybrids, *i.e.*, in the F_1 generation.

The conclusion resulting from the study of Mendelian heredity, that the organism is a sum of "unit characters" which in the organism interact with one another, so as to produce a physiological whole, but which in heredity are more or less clearly separable units, affords strong evidence for the general hypothesis of the representative particle composition of the germ nuclei. Chromosomes might thus represent groups of such "units" or in occasional instances perhaps, single units, although this must be the case only rarely, for the total number of unit characters is far in excess of the number of chromosomes.

That the chromosome of the Metazoan is really made up of a group of unit determiners, is also indicated by the behavior of the Protozoan nucleus in maturation. In most of the simpler Protozoa where the maturation phenomena appear, there is no indication of definite elements like chromosomes in the nucleus. But in many of the Ciliates, in which vegetative chromatin and reproductive chromatin become sharply separated, the latter, or idiochromatin, is seen to be formed into definite bodies. Thus in *Paramœcium*, as observed by Calkins and Cull (Fig. 82), the micronucleus (idiochromatin) becomes resolved into a large number—more than 200—chromatin granules (idiochromidia) whose definite behavior can be traced. Their behavior is complex, but the result is that each idiochromidium is divided longitudinally and transversely, and the resulting daughter-bodies may, therefore, be dissimilar. After fertilization the division of the zygotic nucleus brings about the division of each chromidium and the distribution of the halves to the two daughter cells.

The very large number of separate elements in these "gametic" nuclei may indicate that each corresponds with a single character, or with a smaller group of characters than in the Metazoan, and that therefore the chromosome of the Metazoan must be an enormously complex affair. All of this lends weight to the idea that "chromosomes, the characteristic structures of the nucleus in mitosis, have had an evolution no less surely than has the nervous system, digestive system, or supporting system of the higher animals, and that the chromosomes of the protozoa have the same relation to the chromosomes of the metazoa that the organization of the protozoan body has to that of the metazoan, *i.e.*, a unit structure." (Calkins, "Protozoology," page 171). Admitting the representative particle composition of the chromosomes, it must of course follow that their evolution in the Metazoa, parallels the evolution of adult form and structure.

If this is true, then the chromosomes of the Metazoan germ cells must each represent a congeries of determiners, the form

of association of which might differ in different species, as widely
as the groups of characteristics of the adults differ.

The question as to just how the chromatic determiners (assuming their existence) really do affect the quality of the
reactions of the developing organism, is still practically un-
touched. To some it seems necessary to postulate the asymmetrical distribution of the chromatin granules through successive mitoses, so that certain kinds of granules or "determiners" become distributed to certain cells and regions, directly
effecting there specific reactions. No such form of distribution
has been observed, though indeed it has not been sought in a
thorough fashion. In tissues whose differentiation is fairly
advanced there are certainly characteristic and specific nuclear
appearances which indicate that the nuclei as well as the
cytoplasms have undergone a real differentiation, but whether
this is related to chromosome or granule structure remains
undemonstrated.

If any such sorting out of determiners occurs it must be at
widely divergent stages of development in the various groups,
on account of the variety of the results in the way of specific
embryonic defect following the removal and pressure experiments described above. Indeed the results of the pressure
experiments referred to, become highly significant from this
point of view, for it will be remembered that the presence of a
completely "foreign" nucleus may in some cases not influence
the particular form of differentiation of the cytoplasm. To
say, in such cases as these and in the removal experiments, that
regeneration may occur and the proper determiners be reformed,
does not offer much that is helpful in the way of a solution of
this particular problem, for it would necessitate the assumption
of some mechanism back of the "determining" particles, by
which they themselves are formed and determined.

The fact that parts, even small bits, of a fully developed and
differentiated organism may finally, through a process of regulation, give rise to a complete organism again, or that in many
plants, buds, bits of stem or leaf, may similarly give rise to a
completely formed organism capable of developing typical

germ cells, renders extremely unlikely any strictly morphological conception of the relation between strictly differentiated germinal determiners and the formation of certain tissues or organs. The ideas cannot be overemphasized or repeated too often that, while the thing or the relation that we call a determiner may sometimes have a morphological expression in the germ, essentially the relation is physiological—functional—probably chemical or energetic (dynamic), and that the reactions or interactions of whole groups and masses of particles or systems are involved in determining the intermediate and final results of development.

One further group of observations must be considered in connection with the possibility of the primary character of the nuclear control of development and heredity. During the process of development there occurs a constant giving off of substance from the nucleus to the cytoplasm (Figs. 138, 32). At every mitosis only a part of the chromatic substance is formed into chromosomes, while the remainder passes into the cytoplasm and dissolves, and of course the whole fluid content of the nucleus, the nuclear sap, is discharged into the cytoplasm. Herbst has emphasized the importance of the nuclear sap as an important determining factor in development and heredity.

In many instances, substances discharged from the nucleus into the cytoplasm of the oögonia, especially during their growth period, are directly concerned in the formation of specific materials and bodies of the mature ovum. And later in development Conklin has observed that the cilia of the superficial cells of *Crepidula* develop only when certain chromatic granules reach that region, a single cilium then differentiating opposite each granule. We have already mentioned the fact, described by Wilson, that in *Cerebratulus* the effects of the removal of parts of the egg cytoplasm before the germinal vesicle has broken down, are very different from the effects of the removal of similar portions after the contents of the vesicle have been discharged into the surrounding cytoplasm during the initial stages of maturation.

But the evidence for the hypothesis of nuclear determination

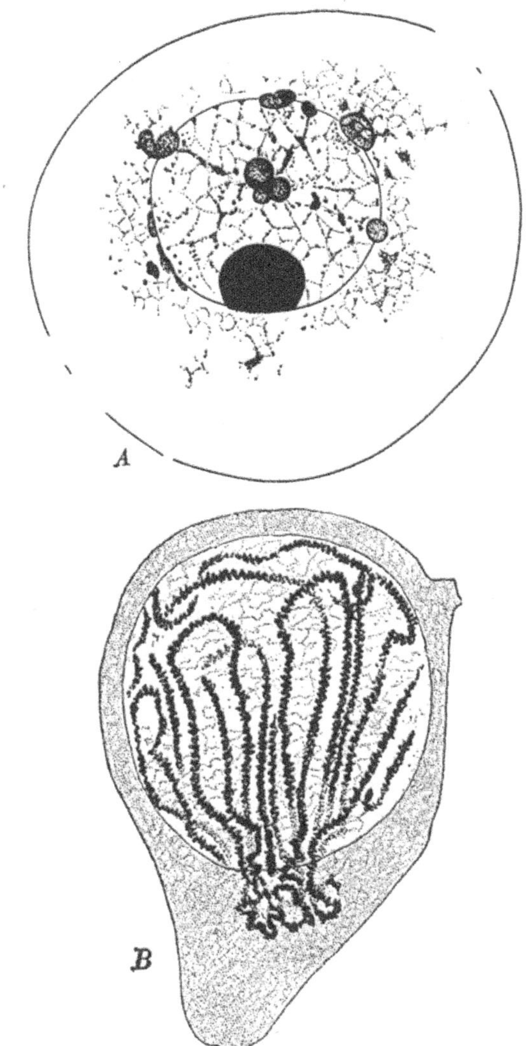

Fig. 138.   *A*. Chromatin extrusion from the nucleus into the cytoplasm in the oöcyte of the Medusa, *Pelagia noctiluca*. After Schaxel.   *B*. Extrusion of chromatin into the cytoplasm during the maturation of the oöcyte of *Proteus anguineus*. After Jörgensen.   × 1080.

is not altogether purely observational; there is some experimental evidence as well, although largely indirect and possibly
none is positively conclusive.

For example, Boveri has described the results of dispermy in
the sea-urchin egg. When two spermatozoa enter the ovum the
result is frequently the formation of three or four centrosomes
and asters connected with one another by spindles, upon which
the chromosome groups are usually drawn in abnormal com-

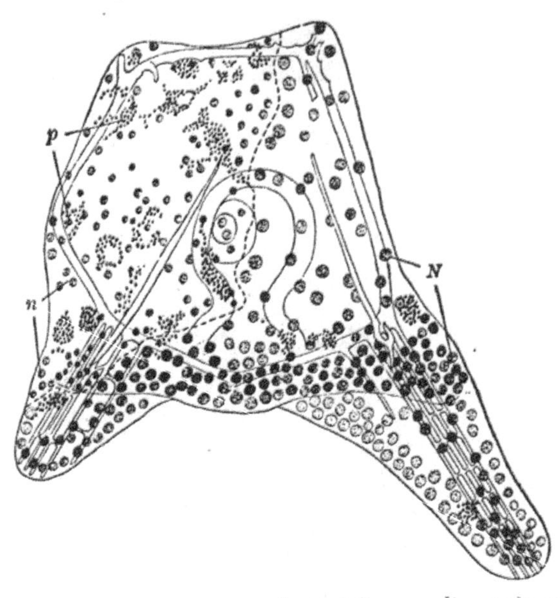

Fig. 139.—Larva of *Sphærechinus*, derived from a dispermic egg, showing
differences in nuclear size, distribution of pigment, *etc.* The dashed line marks
the separation of the two portions of the larva. After Boveri (reconstructed
from two figures). *n*, small nuclei; *N*, large nuclei; *p*, pigment.

binations, so that when such an ovum cleaves it separates
into three or four cells containing nuclei whose composition is
therefore abnormal. In many such cases Boveri finds that
each of the three or four cells forms a group of cell descendants
which can be identified by the presence or absence of certain
characters or by unusual combinations of characters (Fig. 139),
so that the entire embryo may be said to consist of three or
four regions, each with certain distinctive characteristics.
Furthermore, in some instances, the cells of the various fractions

may be characterized by unusually large or small nuclei, indicating the presence of larger or smaller amounts of chromatin (numbers of chromosomes) than usual; the microscopic examination of these multipolar spindles shows that the chromosomes may be distributed with great irregularity in the first division.

More striking are the results following the separation of the blastomeres of such dispermic eggs. The isolated cells of a four-cell stage resulting from normal fertilization, develop normally, producing four similar but small normal larvæ (Fig. 133). But the isolated cells of one of these three-cell stages develop dissimilarly, each with certain defects; and just as any possible combination of chromosomes may have occurred in each of the three original cells, so all possible combinations of characters are found in the larvæ developing from such cells when isolated. Boveri believes that this warrants the conclusion that, while the presence or absence of certain chromosomes may not result in the presence or absence of specific traits, yet a certain *combination* of chromosomes is essential for normal development, a fact which would mean only the physiological specificity of the individual chromosomes.

Perhaps the most striking experimental results are those obtained by fertilizing the eggs of one species, with the sperm of another species, genus, or even phylum. In the first place, Boveri in 1889 reported that non-nucleated egg fragments of *Sphærechinus* (one of the sea-urchins), fertilized with the sperm of *Echinus*, developed into larvæ exhibiting only paternal characters. This appeared to afford strong evidence that the characteristics of the nucleus rather than of the cytoplasm determine the course of development. Later attempts (Seeliger, Morgan, Boveri) to confirm these facts led to inconclusive results. Indeed exactly opposed results were obtained by several investigators (Driesch, Loeb, Godlewski, Hagedoorn). Eggs of one species of Echinoderm fertilized with the sperm of another species, genus or class, of Echinoderm, or even with Molluscan sperm, resulted in the development of larvæ possessing wholly or largely the *maternal* characters. These results indicated just as strongly that the nuclear com-

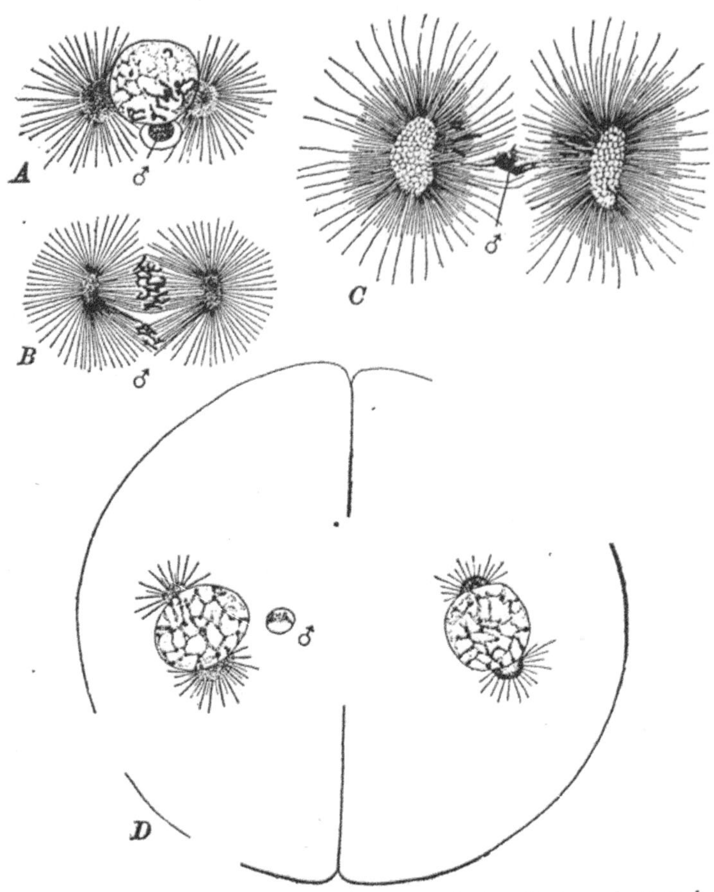

Fig. 140.—History of the paternal chromatin during the first cleavage in the pseudohybrid sea-urchin, *Sphærechinus* ♀ × *Strongylocentrotus* ♂. After Herbst. *A*. First cleavage figure. Sperm and egg pronuclei associated, but not fused. Chromosomes beginning to form in the egg pronucleus. *B*. Chromosomes in both pronuclei; in separate groups with separate spindles. *C*. Anaphase of first cleavage. Maternal chromosomes reaching the poles. Paternal chromatin (chromosomes no longer) forming an irregular mass, spun out on the spindle between the maternal chromosome groups. *D*. Division completed. Daughter nuclei reconstructed and consisting entirely of maternal chromatin. One of the cells contains a small vesicle consisting of the paternal chromatin, which takes no further share in cleavage. ♂, chromatin of the sperm pronucleus.

position is of the lesser importance in determining the development of specific traits, and of course seriously affected the validity of the hypothesis of nuclear determination.

These experiments, however, curiously turned out to afford very strong evidence in support of this hypothesis. For other workers (Herbst, Kupelwieser, Bataillon) showed that in many of these, and in other instances, the nucleus of the spermatozoön did not actually fuse with the egg nucleus, but remained either partly or wholly inactive, taking little or no share in the formation of the mitotic figures of the first and subsequent cleavages (Fig. 140). The resulting larvæ therefore were not truly hybrids; the spermatozoön had merely stimulated the egg to develop, as in artificial parthenogenesis, but itself took no part in the formation of the nuclear structures of the larva. In the absence of microscopic examination of the embryo, therefore, it is impossible to place any emphasis upon the development of purely maternal or paternal characters under such conditions.

Fortunately such cytological evidence is now provided extensively through the work of Baltzer, who has traced the nuclear history of many forms of Echinoderm hybrids. It appears that part or all of the paternal chromatin, never the maternal, may be thrown out of the nuclei of such "hybrids" (pseudohybrids). Such an elimination of paternal chromatin may occur during the very first cleavage, or it may be delayed until the blastula or even early gastrula stage (Fig. 141). The examination of a long series of hybrids, showing all degrees of purity of the maternal characters, leads Baltzer to the conclusion that the degree to which paternal characters appear in the resulting hybrids, is closely parallel to the relative amount of paternal chromatin which is retained within the nuclei of the organism. Where the fusion of the sperm and egg nuclei remains complete, the hybrids have intermediate characters; where little or no chromatin from the spermatozoön is retained in the nuclei, there appear, chiefly or alone, maternal characters. Only in the case of the fertilization of the eggs of a sea-urchin (*Strongylocentrotus*) with the sperm of a Crinoid (*Antedon*) has it been shown that the fusion of the germ nuclei really occurred

(Godlewski, Baltzer) while the larvæ resulting from this cross exhibited certain characters which were purely maternal; but this result is wholly inconclusive as evidence opposing the hypothesis of nuclear control.

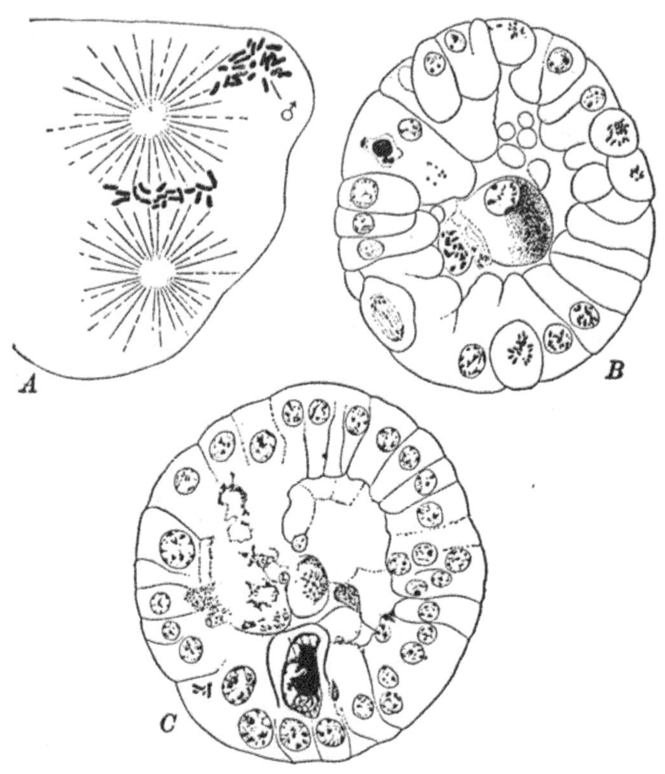

Fig. 141.—History of the paternal chromatin in the pseudohybrids of the sea-urchins, *Strongylocentrotus* ♀ × *Sphærechinus* ♂. After Baltzer. *A.* Cleavage cell showing paternal chromatin (♂) outside the division figure. *B.* Early blastula. *C.* Late blastula, showing the elimination of the paternal chromatin in the irregular cells and spaces within the blastocœl (for normal blastula see Fig. 109).

For it has been found, in the first place, that external conditions often determine whether certain characters shall be paternal or maternal in their qualities (Vernon, Tennent). Under certain conditions of temperature, alkalinity, *etc.*, the larva may exhibit paternal resemblances, while under other conditions maternal resemblances may appear in the same cross. And in the second place, the phenomenon of "dominance" appears even in these early stages of development, and a hybrid may show certain clearly maternal characters and yet in other respects closely resemble the paternal type (Steinbruck, Driesch, Boveri, Loeb and Moore). Great variability is often the rule and frequently it is impossible to say whether either parental trait really appears purely. It should be pointed out, first, that it frequently happens in Mendelian inheritance that true hybrids are either purely maternal or purely paternal with respect to single traits, and second, that only after synapsis, which occurs in the germ cells of the mature hybrid organism, are the paternal and maternal chromosomes really brought into complete relation.

On the whole, then, while there are some few results difficult of favorable interpretation, we may say that the evidence from hybridization, though at first distinctly opposed to the hypothesis of nuclear determination, at present affords the strongest support of this hypothesis, and indicates that normally the characters of a hybrid are determined by both of the germ nuclei, and that when nuclear material from only one parent is functional the characters of the so-called hybrid are determined thereby.

We might mention one further possible interpretation of some of the results opposed to this conclusion, and emphasized by Conklin and others, namely, that in some cases, at least, the fundamental or general traits of an organism may be determined immediately by the cytoplasmic structure of the ovum alone or chiefly, while the nuclei are equally concerned in the determination of the more particular specific or individual traits, often appearing relatively late in development. Such a possibility seems to be indicated by many of the facts of germinal localization already described, and it may be that some of the results indicated above, non-conformable with the hypothesis of nuclear determination, point in the same direction. Conklin writes (Science, XXVII, 89–99): "At the time of fertilization the hereditary potencies of the two germ cells are not equal, all the early development, including the polarity, symmetry, type of cleavage, and the relative positions and properties of future organs being predetermined in the cytoplasm of the egg cell, while only the differentiations of later development are influenced by the sperm. In short, the egg cytoplasm fixes the type of development and the sperm and egg nuclei supply only the details." And yet we should not overlook this fact, of basic importance, that these fundamental cytoplasmic differentiations have resulted from interactions between the cytoplasm and the nucleus of the oögonial cell, and that the nucleus of the oögonial cell, and egg cell, is itself originally derived in equal parts, from paternal and maternal ancestry. And further many of the conditions of polarity, symmetry, and the like, may in some cases be determined or altered by the entrance and subsequent activity of the spermatozoön within the ovum.

Probably altogether the most striking evidence in support of the hypothesis under consideration is to be found in some of the recent work upon the association of sex with chromosome characters. The nature of this association is so particular and significant that certain chromosomes are actually regarded by many as representing sex "determiners." This relation, besides affording striking evidence in this connection, is of very great importance in itself, and we may therefore consider it at somewhat greater length than this connection alone would justify.

During recent years many instances have come to light, of a variation in the number of chromosomes in different individuals of a single species. With but very few exceptions these numerical differences are associated with difference in sex, and when any such difference exists it is usually found that the cells of the female contain one or more chromosomes in excess of the number found in the male. In some species then, the chromosome number may be uneven in one sex, and therefore not all the chromosomes are paired structures. In other cases the equivalent diversity of the chromosome groups is indicated by size differences between the members of a certain pair.

Differences of these kinds are now known in many scores of species of many groups, from the lower worms to man. It is clearly impossible to include here any extended history or survey of this fascinating subject and we can do little more than describe a typical instance or two, and then mention some comparisons which may throw some light, from this point of view, upon the general interpretation of the chromosome problem.

Since the larger number of the known instances of this relation are found among the Arthropoda, particularly the Insecta, we may select our first illustration from this group. The number of chromosomes in the somatic cells of the common squash bug, *Anasa tristis* (Henking, Paulmier), is twenty-two in the female, and twenty-one in the male. How does this difference come about?

For an answer to this question we must observe the behavior of the chromosomes during the process of maturation of the germ cells. In the process of oögenesis, preparatory to the first oöcyte division, synapsis occurs normally, and eleven bivalent chromosomes are formed. The succeeding steps in oögenesis are not unusual and the result is the formation, in each ovum, of a group of eleven univalent chromosomes representing every pair of the original somatic or oögonial group.

The events of spermatogenesis do not run quite parallel, however. In the somatic and spermatogonial cells, twenty-one chromosomes are present, *i.e.*, ten pairs plus one, and in the division of these cells every chromosome divides in the usual way (Fig. 142). In synapsis the paired elements fuse, forming ten bivalent chromosomes, but the odd chromosome, or X-chromosome remains free, and usually quite apart from the other chromosomes. This X-element, or *idiochromosome*, may be distinct, even throughout the growth period of the spermatogonia, and during the two spermatocyte divisions it can be identified in many species as a nucleolus-like body, indeed formerly it was described as a chromatin nucleolus. The behavior of this body during the maturation divisions is entirely unusual. During the first spermatocyte division the idiochromosome, although univalent, divides just as the ten bivalent elements do, and eleven chromosomes consequently pass into the nuclei of the secondary spermatocytes. But during the

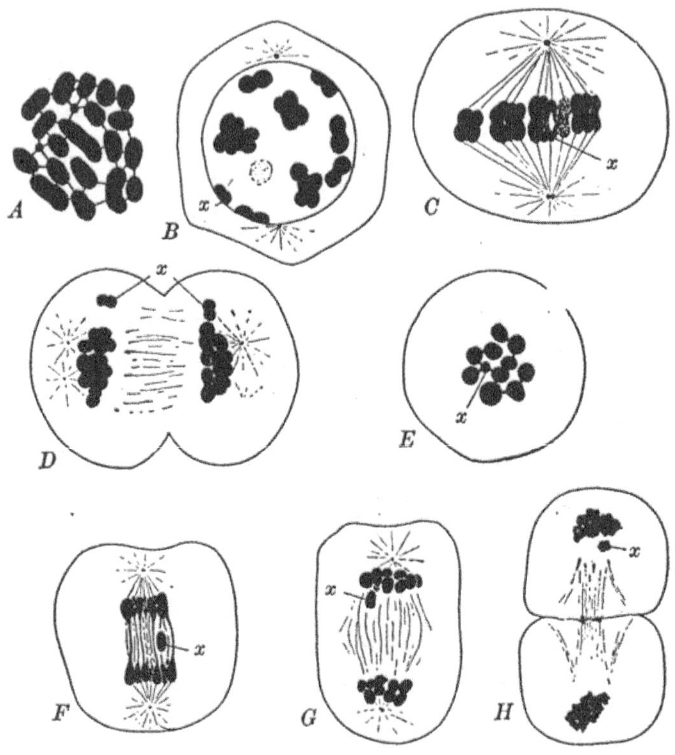

Fig. 142.—Maturation during the spermatogenesis of the squash-bug, *Anasa tristis*, showing the behavior of the X-chromosome or idiochromosome. *A*, after Wilson, others after Paulmier. *A*. Spermatogonium. Polar view of equatorial plate showing twenty-one chromosomes (ten pairs, plus one). The X-chromosome is not distinguishable at this time. *B*. Primary spermatocyte. Tetrads formed. *C*. Equatorial plate of first spermatocyte division. X-chromosome divided. *D*. Anaphase of same division. The daughter X-chromosomes have also diverged. *E*. Equatorial plate of second spermatocyte division. *F*. Metaphase of same division. The X-chromosome lies, undivided, between the two groups of daughter chromosomes. *G*. Anaphase of same division. The undivided X-chromosome has passed to the upper pole, lagging behind the others. *H*. Telophase of same division. X-chromosome still distinct.

mitosis of these secondary spermatocytes the idiochromosome fails to divide and passes as a whole to one pole of the spindle (Fig. 142, *F, G, H*). The result is that the nuclei of one-half of the spermatids, and therefore of one-half of the spermatozoa,

contain eleven chromosomes, while the other half contain but
ten, lacking the idiochromosome.

Since the nuclei of all the ova contain eleven chromosomes

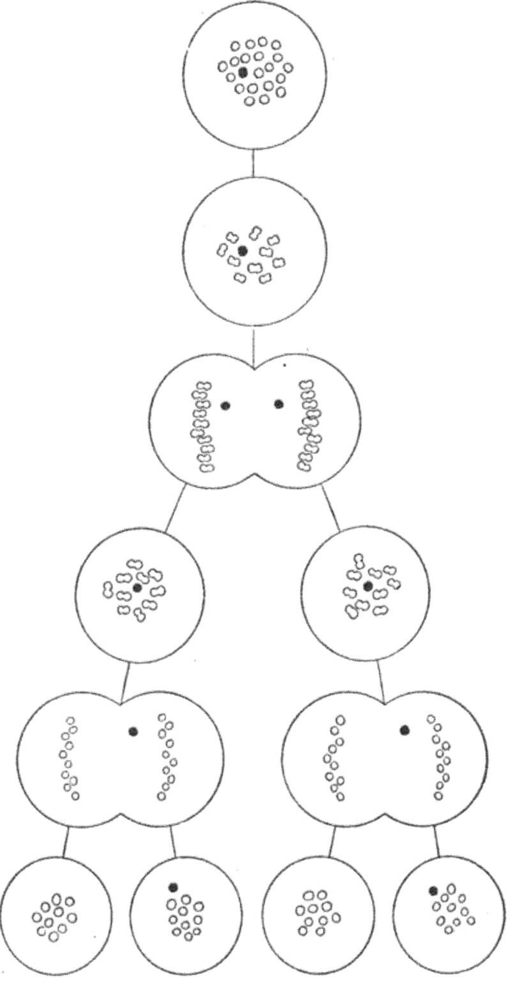

Fig. 143.—Diagram illustrating the behavior of the X-chromosome during
maturation. The X-element is shown in solid black. The essentials are the
same in cases where X is a multiple element, or where it is paired with a Y-element (see text).

there are but two possibilities in fertilization. The egg with its
eleven chromosomes may be fertilized by a sperm with ten,

giving a somatic group of twenty-one; or the egg with its eleven chromosomes may be fertilized by a sperm with eleven, giving a somatic group of twenty-two. And since there are equal numbers of ten- and eleven-chromosome spermatozoa, there will be approximately equal numbers of zygotes with twenty-one and twenty-two chromosomes. These relations are shown in diagrammatic form in Figs. 143, 144.

Since this numerical difference between the somatic chromosome groups is constantly associated with sex-difference, males possessing twenty-one, females twenty-two chromosomes, it may be said that the presence of the idiochromosome is in

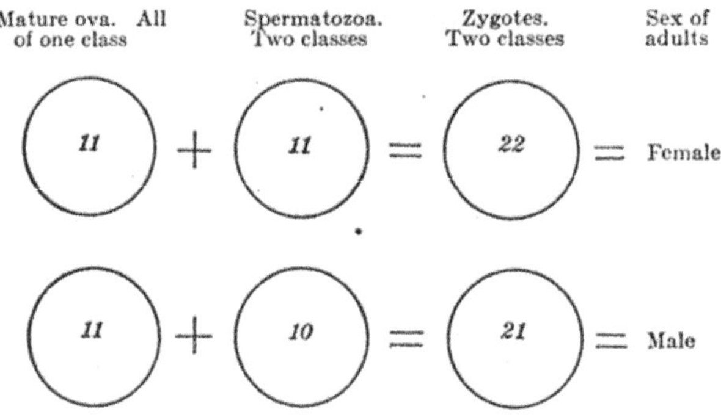

Fig. 144.—Diagram of the relations of chromosome number and sex during fertilization in *Anasa*. The essentials are the same in cases where X is a multiple element, or where it is paired with a Y-element.

some way connected with the determination of the female, or the lack of it the male, sex.

Since Paulmier's description of these events in *Anasa*, in 1899, a great many similar instances have come to light, and more recently quite a variety of conditions related to this but more or less dissimilar in details, have become known. The X-chromosome or idiochromosome has been described under many different names such as *accessory chromosome, allosome, heterotropic chromosome, heterochromosome, monosome, etc.* Such an unequal distribution of the chromosomes was first observed by Henking in 1890, and in 1902 McClung described similar processes in several of the locusts and grasshoppers (Orthoptera), and first suggested the possible relation between this "accessory chromosome" and the determination of sex. The work of Wilson, Stevens, Montgomery, Payne, Guyer, Morgan, and many others, has made known the presence of these elements in a whole host of Insects, including most of the orders, in Myriopods, Arachnids, and Copepods. Among the lower forms, Nematodes (Boveri, Edwards, Boring, Gulick), *Sagitta* (Stevens), and Echinoderms (Baltzer) are now known to possess idiochromosomes. And more recently some of the Chordates have been added to the ever increasing list, for idiochromosomes have been described in the common fowl, guinea fowl, and rat (Guyer), the guinea pig (Stevens) and even in man, where Guyer has reported two idiochromosomes, half the sperm containing twelve, and half ten, chromosomes; the number of chromosomes in the human somatic cells is, therefore, twenty-two in the male, and twenty-four in the female.

Not all of the forms included in the above list exhibit this phenomenon as simply as it occurs in the case of *Anasa;* we may mention two general modifications of this typical condition. In many Coleoptera, Diptera, and Hemiptera, the idiochromosome is not strictly an unpaired element for during the spermatocyte divisions, and in the spermatids, it is paired with a very small chromosome called the Y-element; together with X this makes up an XY bivalent chromosome which behaves like any bivalent chromosome in the preliminaries to the first spermatocyte division (Fig. 145). In the spermatozoa, therefore, half the nuclei contain the large idiochromosome (X), and half the small one (Y). The relation to sex is what might be expected, namely, the females contain the large X, the males the small Y. In *Metapodius,* one of the Orthoptera, this small Y-element may be either present or absent apparently, although it is possible that it may be present and fused with another chromosome when it is said to be absent.

In *Ascaris megalocephala* it seems clear that a small X-element, no Y-element being present, may thus appear either as a separate body, or fused with one of the other chromosomes (Boring, Boveri, Edwards). Such an attachment of the idiochromosome to a certain one of the ordinary chromosomes is

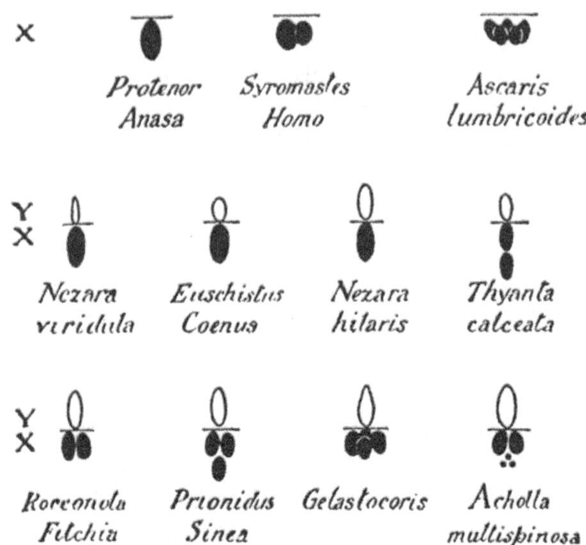

Fig. 145.—Diagrams illustrating some of the variations in the X- and Y-chromosomes. From Wilson. X, either as a simple or as a multiple element, may or may not be paired with a Y-chromosome. ·

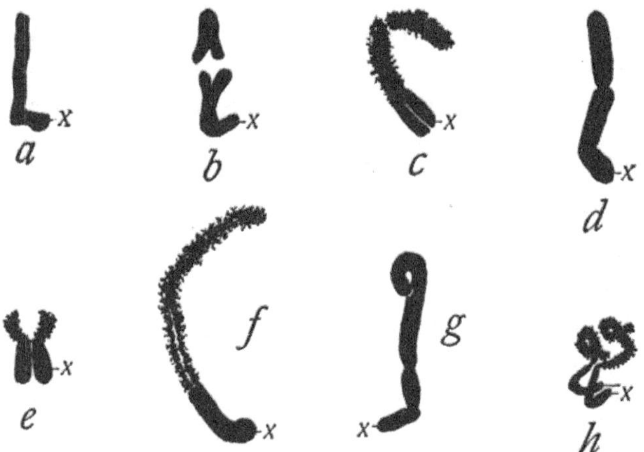

Fig. 146.—Compound chromosome-groups, formed by the union of the X-chromosome with other chromosomes, in the Orthoptera. From Wilson, a, b, after de Sinéty, others after McClung. a. Triad group, first spermatocyte division of *Leptynia*, metaphase. b. Division of similar triad in *Dixippus*. c. Triad group formed by union of the X-chromosome with one of the bivalent chromosomes, first spermatocyte prophase, *Hesperotettix*. d. The same element from a metaphase group. e. The same element in the ensuing interkinesis. f. The compound element of *Mermiria*, from a first spermatocyte prophase. g. The same element in the metaphase (now, according to McClung, united to a second bivalent chromosome, to form a pentad element). h. The same element after its division, in the ensuing telophase. ·

well known as the constant relation in some Insects (Sinéty, McClung), and among these forms various degrees of the intimacy of the association occur (Fig. 146).

Several stages can be found in the gradual increase in the relative size of the Y-element, until in such forms as *Nezara hilaris*, one of the Hemiptera-Heteroptera described by Wilson, X and Y are nearly equal (Fig. 145), and finally in some of the Lepidoptera and other forms, X and Y are quite equal and indistinguishable from one another, although the XY pair may be distinguished from the other chromosomes by staining properties and behavior.

We thus reach through gradual transitions a condition where the spermatozoa are no longer dimorphic with respect to chromosome content. The conditions of such a series suggest, however, the possibility that spermatozoa that visibly appear morphologically alike, may after all be physiologically dimorphic as regards chromosome characters; such an assumption must of course be made with respect to traits other than sex, which are inherited in an alternative fashion.

Another series of modifications of the *Anasa*-type is illustrated by various genera of Hemiptera, where the X-chromosome is represented by more than a single element. Such a series has been described by Payne, and is readily derived from the condition in such a form as *Euschistus* (Fig. 145), with unequal X- and Y-elements. Thus in *Fitchia* and several others, Y is a fairly large chromosome while X is represented by two somewhat smaller chromosomes; in *Prionidus, Sinea, etc.* there are three X-elements to one Y; in *Gelastocoris* there are four X and one Y, and in *Acholla* five X and one Y. In still another series (Fig. 145) Y is entirely absent and X is represented by several chromosomes—two in *Phylloxera* (Morgan), *Syromastes* (Wilson), *Agalena* (Wallace), and man (Guyer), five in *Ascaris lumbricoides* according to Edwards.

In all of these cases where X is a multiple element, different species show greatly varying relations among the members of the X-group; they may be approximately equal in size or very unequal, but it is important that the size relations, whatever they are, are constant within a given species.

One further general condition of the idiochromosomes must be noted. In all of the instances mentioned above the spermatozoa are the dimorphic gametes, *i.e.*, the males are "digametic" (Wilson's term). In a few species, however, it is the female that is digametic, and while the spermatozoa are all alike, the ova are of two classes with respect to certain modified chromosomes which may properly be regarded as homologous with the X- and Y-elements of the spermatozoa. Thus in two of the sea-urchins, *Strongylocentrotus*, and *Echinus*, Baltzer

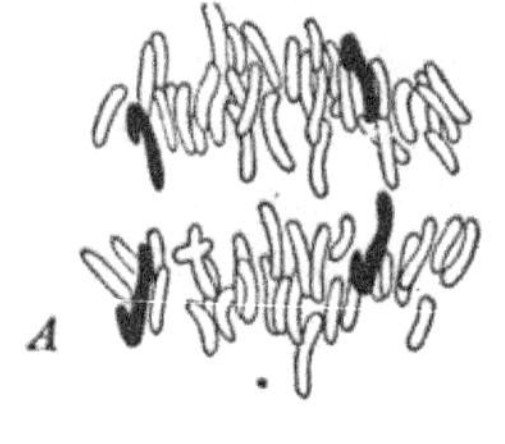

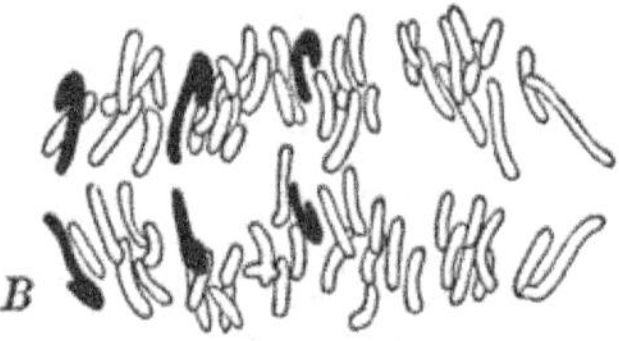

Fig. 147.—Chromosomes in the sea-urchin, *Strongylocentrotus lividus*. After Baltzer. × 2610. *A*. First cleavage spindle (reconstructed from two drawings). No small hooks present. *B*. First cleavage spindle with small hooks. The modified chromosomes, both long and short hooks, are shown in solid black.

finds the chromosome groups of the spermatids all alike, with eighteen components uniformly differentiated; the nuclei of some of the ova, however, are characterized by the presence of a single modified component which is absent from the remainder (Fig. 147). This obviously corresponds with the X-element of the dimorphic sperm. There is also very good reason for believing the female digametic in some of the Lepidoptera; this is of considerable interest in view of the fact that repeated investigation has thus far failed to disclose idiochromosomes in the spermatocytes of these Insects.

The idea that the relation between the chromosomes and sex characters parallels that between the chromosomes and any other traits involves the conclusion that sex is a character, or group of characters, inherited in the same way that other bodily traits are. And this conclusion may now be accepted. Indeed there are in the field several hypotheses as to the precise statement. of a Mendelian formula according to which sex is inherited, and while no one of them has a preponderance of evidence in its favor, the fundamental fact of sex heredity is clear.

There are extant scores of hypotheses regarding the factors and processes involved in sex determination, depending upon the action of conditions outside of the germ itself. These must be abandoned when the facts now known to be true of the germinal structure of a comparatively limited number of species, gain a wider applicability. For the sex of an organism, as well as other fundamental characters, appears to be already determined in the zygote, and all that external conditions can do toward determining sex is to alter sex ratios by affecting differentially (selectively) the gametes or immature organisms of a certain sex.

There is some evidence of other kinds that sex is determined in the gamete and not by external conditions. In certain cases a single egg indirectly gives rise to a number of embryos or larvæ (multiple embryo formation) which are all of one sex, either male or female. Silvestri describes such a case in the development of a Hymenopter, *Litomastix*, parasitic in the larva of a Lepidopter, *Plusia*, where as many as one thousand embryos, all of one sex, are thus formed. And there is good reason for believing that the embryos of one of the armadillos, described by Newman and Patterson, are all derived from a single ovum, and these are always of one.sex only, either male or female. There is also the familiar example of the bees (Dzierzon), where unfertilized eggs develop parthenogenetically into males (drones), while the fertilized ova produce females (queen and workers); the same thing is apparently true of most ants. And in one of the rotifers, *Hydatina*, a certain kind of female produces eggs which if fertilized produce females, if unfertilized produce males.

One further point is to be mentioned in connection with the general hypothesis of nuclear determination. This is in connection with that curious form of Mendelian heredity known as "sex-limited" heredity, where certain characters are exhibited by the individuals of one sex only, although transmitted by the individuals of the other sex without being exhibited by them. Such a form of heredity can readily be explained upon the chromosome hypothesis, upon the simple assumption of a close relation between the "determiner" for the sex-limited character and the sex-determining element.

Any further consideration of the problems of sex determination and heredity would lead us too far afield; more extended treatment must be had from other sources. In conclusion then, it is hardly necessary to point out that the constancy of the form and of the complicated behavior of these idiochromosomes affords very striking confirmation of the hypothesis of the specificity and genetic continuity of the chromosomes. While it is possible that their form and behavior are determined by underlying conditions, such conditions cannot be directly observed and can only be postulated. Taken in connection with the facts mentioned in Chapter II, and with the results drawn from the development of dispermic eggs, and from hybridization, they amount to practical demonstration of some form of chromosomal specificity in development.

As to the question whether the idiochromosomes are in particular the sex determinants, several views may be held in the absence of conclusive experimental demonstration of the precise relation. It has been held in some quarters that sex is determined by the relative *amount* of chromatin received into the nucleus of the zygote, irrespective of its content in certain chromosomal elements. This is hardly tenable however in view of many contradictory conditions. Others have suggested that the dimorphism of the gametes is merely associated with other more fundamental diversities, and that sex differentiation and gamete differentiation are related only because both are related to some primary differentiation. Still others hold to the idea that the idiochromosomes actually determine by their presence or absence, the nature of the reactions of development, so that finally organisms with female or male characteristics are formed. The most adequately justified and most conservative view seems to be that the nature of the interrelations of the components of the whole chromosome group, among themselves and to the cytoplasm, is modified by the presence or absence of certain elements so that in one case the primary and secondary female characters develop, in the other case the male characters.

Returning now to the general subject of this chapter, namely, the factors determining the course of development and the process of heredity, we come to another extremely important subject. We have thus far emphasized the importance of the internal factors of development. But we have defined development as a series of reactions between internal and external factors. The omission hitherto of specific reference to the external conditions of development is not because these are of lesser general importance. Alterations in the conditions of gravity, pressure, temperature, light, moisture, and chemical composition of the surrounding medium may, each or all affect the course of development, either in a general or in a specific way. A great deal is known of the results of modifying such conditions and a rather full discussion of these effects would be in order, were the results susceptible of more definite and more uniformly applicable statement. For normal development, normal environing conditions are necessary. However, slight variations in external conditions rarely produce effects comparable with those following slight variations in the internal conditions. That is to say, slight variations in external conditions are "normal." When the modification of external conditions is sufficiently marked to produce visible effects upon development, these are frequently so marked as to be regarded as distinct abnormalities, and the organism so affected is rarely able to complete its development to maturity.

The natural environment ordinarily varies within rather narrow limits, frequently on account of the ovipositing habits of the adult, and changes within these limits rarely affect the course of development, so that for the subjects of heredity and differentiation we should inquire here only into the question to what extent external conditions are necessary factors in carrying on the life of the organism as it exists in the form of an egg or embryo. The particulars of development and heredity are referable to internal characteristics which determine the specific or individual quality of the reactions between organism and environment.

We may proceed, therefore, to mention a few illustrations of the effects of alterations in external conditions of development, not attempting to do more than to suggest the nature of the work accomplished in this field; an adequate survey falls outside the scope of such a text as this. (For a convenient summary, see, *e.g.*, Jenkinson, "Experimental Embryology," Oxford, 1909.)

More is known regarding the effects upon development, of chemical substances than of other conditions. While a few forms, such as the minnow, *Fundulus*, are able to develop normally in media so widely unlike, physically and chemically, as sea water and distilled water, this and other forms show specific effects of the presence or absence of certain salts alone. Thus in *Fundulus* Stockard has shown that the presence of certain amounts of magnesium salts brings about the fusion of the optic vesicle regions, so that one-eyed monsters develop, apparently normal in other respects (Fig. 148). The eggs and embryos of the Echinoderms offer many striking facts in this connection. We have already noted that the alkalinity of the sea water may determine the appearance of paternal or maternal characters in hybrid Echinoderm larvæ. Herbst and others have shown that the absence of potassium salts is fatal or very harmful to Echinoderm larvæ apparently on account of the resulting diminution in the process of water absorption; the absence of calcium causes a tendency for the blastomeres to fall apart; magnesium and the sulphates are necessary for the normal differentiation of the alimentary tract; the production of ciliary movement depends upon the presence of magnesium, and an excess of calcium results in the hypertrophy of the cilia; sulphates are necessary also for the establishment of the fundamental structure of the embryo and for the formation of pigment; magnesium, sulphates, and calcium carbonate are

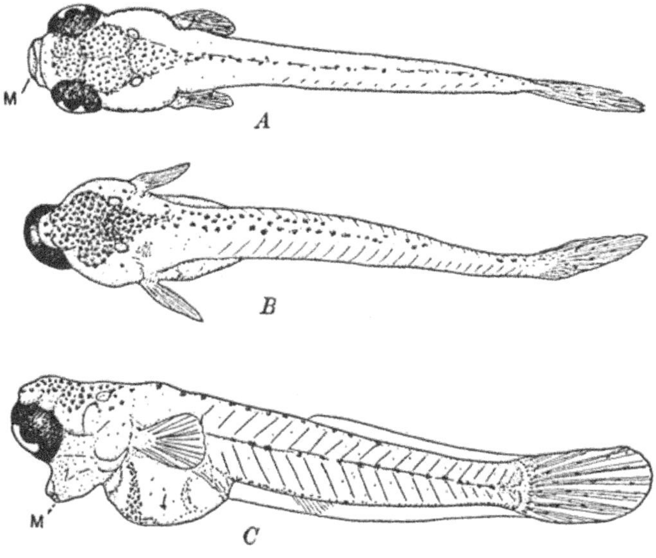

Fig. 148.—Effects of magnesium chloride upon the development of the Teleost, *Fundulus*. From Stockard. *A*. Normal fish, eight days after hatching. *M*, mouth. *B*, *C*. Two views of fish, showing the fusion of the optic vesicles as the result of treatment with $MgCl_2$.

necessary for the development of a normal skeleton (Fig. 149).

Certain optima exist for moisture, density, pressure, light and temperature; in development as in later life, deviations from the optimum condition, in either direction, affect the rate of development rather than its character. The direction of gravity takes an essential part in determining normal development in a few cases, but ordinarily development is independent of this factor.

In general all of these conditions are involved not so much in the regulation of development in specific and particular directions, as in determining whether it shall proceed at all or not. Modifications of development produced by effective variations in these conditions are often so extreme that the phenomena of heredity are scarcely apparent and usually the modified organism does not come to maturity.

We may now attempt to summarize a conservative conception of the relation of the structure of the germ cells to the processes of development and heredity.　The zygote is an organism, morphologically and physiologically specific.　It possesses

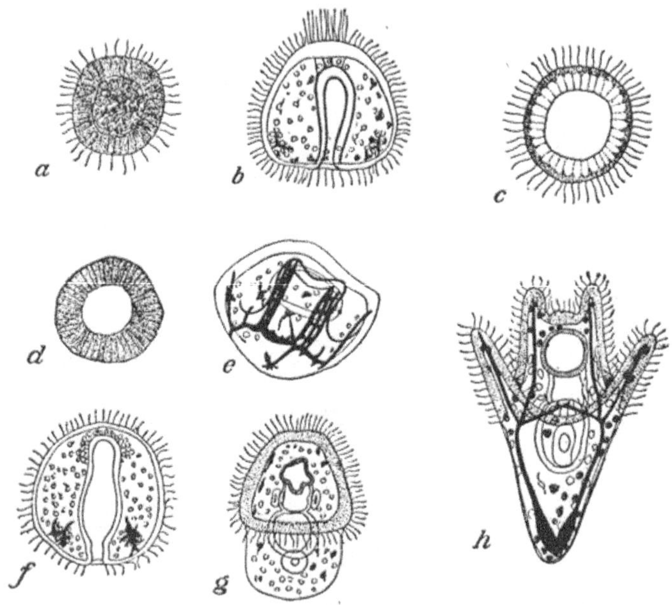

Fig. 149.—Effects of chemical alteration of the surrounding medium, upon the development of the sea-urchin. From Jenkinson. *a.* Without OH; ciliated solid blastula of *Sphærechinus.* *b.* KOH has been added. *c.* Normal blastula of *Sphærechinus.* *d.* Blastula in a K-free medium. *e.* Reared in K-free medium and replaced in normal sea-water (*Sphærechinus*). *f.* *Sphærechinus* larva from a medium devoid of Mg. *g.* *Echinus* pluteus with three-parted gut, mouth, and cœlomic sacs, but neither skeleton nor arms; reared without $CaCO_3$ or $CaSO_4$. *h.* Normal pluteus of *Echinus.*

polarity, symmetry, various forms of differentiated substance, even organs, composed of subsidiary elements and capable of performing definite and highly varied and specialized functions. This organism in its parts and as a whole does certain things, makes certain reactions, in a word, develops.　The quality of the developmental reactions is determined primarily by the conditions within the organism itself, and as it reacts, as the organism develops step by step, these internal conditions rapidly change. These reactions on the part of the organism fall into two groups. (1) Reactions between the organism (*i.e.*, cytoplasm *and* nucleus, whether the organism consists of one cell or many) and its environing stimuli. (2) Reactions between the nucleus and cytoplasm of each cell. The idea of reaction must involve two factors, but while equally necessary for reaction, they are not necessarily of equal value in determining or controlling the *quality* of the reaction. A great many organisms react to light; but the quality of the reaction is determined primarily by the organism.

The whole structure of the cytoplasm may play a large part in determining the quality of the reactions of the egg, but this cytoplasmic structure is itself the result of a series of interactions between cytoplasm and nucleus, and the action of the latter is of primary importance in affecting the quality of the result. Going one step further, what the nucleus does is determined by its structure, and this is also the result of interactions of its parts with one another, and with the cytoplasm, which is its environment; and here again certain elements of the nucleus, namely, the chromosomes, seem to be of primary importance in determining the quality of the interaction. The most important of the concrete, visible organs of the nucleus are the chromosomes. And when we attempt to analyze the behavior of these components we are met by the same problem—what determines the structure and behavior of these? Two answers have been offered. First, that here we reach the limit of analysis, that the chromosomes are autonomic, self-perpetuating, self-regulating bodies, whose morphology and behavior are *the* determining factors in all that happens in the life of the organism. Second, that the chromosomes are themselves made up of still more fundamental units, the chromatin granules; that these are the autonomic, self-perpetuating, finally determinative units in development, and chromosome structure is the result of the primary activity of these bodies.

Logically there is no reason why we must stop with the chromatin granules. And history, enumerating germ layers, cleavage cells, cytoplasmic organ-forming substances, chromosomes, and chromioles (chromatin granules), warns us against the idea that we must seek or hope to find ultimate particles, concrete, definable, and representatively determinative in function. In fact a few students of the problem frankly declare their belief that the idea of any sort of representative particle mechanism is futile, that the regulation of the processes of development and heredity depends upon interrelations which are not susceptible of interpretation in terms of any material basis.

Scientifically, however, we can to-day go no further back than the chromosomes, for here we find the most fundamental units whose actual behavior can be correlated with the facts of development. To say scientifically, that the chromosomes are (to-day known to be) the determining elements in development and heredity, is not to deny the existence of other bodies or conditions which may determine the existence and qualities of the chromosomes. Granules we know, but of their behavior we know little, and this little cannot at present be correlated with the facts of development. Of the real existence of elements underlying the granules we know nothing whatever. Assumption of the reality of such bodies or conditions may be a logical necessity, but to-day it carries us beyond the boundaries of observed fact.

To repeat a statement made on an earlier page, if the existence and activity of the chromosomes can be shown to be a necessary link in the processes of development and heredity, and if these can be shown to be the simplest and most nearly primary factors whose behavior can be correlated with these processes, then we shall be justified in saying that the chromosomes are to-day the determining factors in development and heredity.

<h2 style="text-align:center">REFERENCES TO LITERATURE</h2>

BALTZER, F., Die Chromosomen von *Strongylocentrotus lividus* und *Echinus microtuberculatus*. Arch. Zellf. **2.** 1909. Ueber die Beziehung zwischen dem Chromatin und der Entwicklung und

Vererbungsrichtung bei Echinodermenbastarden.   Arch. Zellf. **5.** 1910.

Bonnevie, K., (See ref. Ch. II.)

Born, G., Ueber Druckversuche an Froscheiern.   Anat. Anz. **8.** 1893.

Boveri, T., Ein geschlechtlich erzeugter Organismus ohne mütterliche Eigenschaften. Sitz.-Ber. Ges. Morph. Phys. München. **5.** 1889. English translation in Amer. Nat. 27. 1893. Ueber die Befruchtungs- und Entwickelungsfähigkeit kernloser Seeigel-Eier und über die Möglichkeit ihrer Bastardirung. Arch. Entw.-Mech. **2.** 1895. Die ˙Entwicklung von *Ascaris megalocephala*, mit besonderer Rücksicht auf die Kernverhältnisse. Festschr. f. v. Kupffer. Jena. 1899. Ueber mehrpolige Mitosen als Mittel zur Analyse des Zellkerns. Verh. phys.-med. Ges. Würzburg. **35.** 1902. Das Problem der Befruchtung. Jena. 1902. Protoplasmadifferenzierung als auslösender Faktor für Kernverschiedenheit. Sitz.-Ber. phys.-med. Ges. Würzburg. **37.** 1904. Ergebnisse über die Konstitution der chromatischen Substanz des Zellkerns. Jena. 1904. Zellenstudien V. Ueber die Abhängigkeit der Kerngrösse und Zellenzahl der Seeigel-Larven von der Chromosomenzahl der Ausgangszellen. Jena. Zeit. **39 (32).** 1905. Zellenstudien VI. Die Entwickelung dispermer Seeigel-Eier. Ein Beitrag zur Befruchtungslehre und zur Theorie des Kernes. Jena. Zeit. **43 (36).** 1908. Ueber "Geschlechtschromosomen" bei Nematoden. Arch. Zellf. **4.** 1909. Die Potenzen der Ascaris-Blastomeren bei abgeänderter Furchung. Zugleich ein Beitrag zur Frage qualitativ-ungleicher Chromosomen-Theilung. Festschr. f. R. Hertwig. Jena. 1910.

Buchner, P., Das accessorische Chromosom in Spermatogenese und Ovogenese der Orthopteren zugleich ein Beitrag zur Kenntnis der Reduktion. Arch. Zellf. **3.** 1909.

Castle, W. E., A Mendelian View of Sex-Heredity. Science. **29.** 1909.

Child, C. M., Some Considerations regarding so-called Formative Substances. Biol. Bull. **11.** 1906.

Conklin, E. G. Mosaic Development in Ascidian Eggs. Jour. Exp. Zool. **2.** 1905. Organ Forming Substances in the Eggs of Ascidians. Biol. Bull. **8.** 1905. Does Half of an Ascidian Egg give Rise to a whole Larva? Arch. Entw.-Mech. **21.** 1906. The Mechanism of Heredity. Science. **27.** 1908. The Effects of Centrifugal Force upon the Organization and Development of the Eggs of Fresh Water Pulmonates. Jour. Exp. Zool. **9.** 1910. The Organization of the Egg and the Development of Single Blastomeres of *Phallusia mamillata*. Jour. Exp. Zool. **10.** 1911. (See also Ref. Ch. II, III.)

CRAMPTON, H. E., Experimental Studies on Gasteropod Development. Arch. Entw.-Mech. **3.** 1896.

DRIESCH, H., Entwicklungsmechanische Studien. IV. Experimentelle Veränderungen des Typus der Furchung und ihre Folgen. (Wirkung von Wärmezufuhr und von Druck). Zeit. wiss. Zool. **55.** 1892. Zur Analysis der Potenzen embryonaler Organzellen. Arch. Entw.-Mech. **2.** 1895. Resultate und Probleme der Entwickelungsphysiologie der Tiere. Ergebnisse Anat. u. Entw. **8.** 1898 (1899). Die Localisation morphogenetischer Vorgänge. Ein Beweis vitalistischen Geschehens. Arch. Entw.-Mech. **8.** 1899. Die isolirten Blastomeren des Echinidenkeimes. Arch. Entw.-Mech. **10.** 1900. Studien über das Regulationsvermögen der Organismen. IV. Die Verschmelzung der Individualität bei Echinidenkeimen. Id. 1900. Zur Cytologie parthenogenetischer Larven von *Strongylocentrotus*. Id. **19.** 1905. Altes und Neues zur Entwicklungsphysiologie des jungen Asteridenkeimes. Id. **20.** 1905. Die Entwickelungsphysiologie. 1905-1908. Ergebnisse Anat. u. Entw. **17.** 1907 (1909). The Science and Philosophy of the Organism. London. 1908.

EDWARDS C. L., The Idiochromosomes in *Ascaris megalocephala* and *Ascaris lumbricoides*. Arch. Zellf. **5.** 1910.

FICK, R., (See ref. Ch. II.)

FISCHEL, A., Experimentelle Untersuchungen am Ctenophoreneı. I. Von der Entwickelung isolirter Eitheile. Arch. Entw.-Mech. **6.** 1897. Id. IV. Ueber den Entwickelungsgang und Organisationsstufe des Ctenophoreneıes. Id. **7.** 1898.

GARBOWSKI, T., Ueber Blastomerentransplantation bei Seeigeln. Bull. Internat. Acad. Sci. Cracovie. 1904 (1905).

GODLEWSKI, E., Das Vererbungsproblem im Lichte der Entwicklungsmechanik betrachtet. Leipzig. 1909. (See ref. Ch. V.)

GROSS, J., Heterochromosomen und Geschlechtsbestimmung bei Insecten. Zool. Jahrb. (Abth. f. Allgem. Zool. u. Physiol.). **32.** 1912.

GUYER, M. F., Deficiencies of the Chromosome Theory of Heredity. Univ. Cincinnati. (II). **5.** 1909. Accessory Chromosomes in Man. Biol. Bull. **19.** 1910. Nucleus and Cytoplasm in Heredity. Amer. Nat. **45.** 1911.

HÄCKER, V., Die Chromosomen als angenommene Vererbungsträger. Ergebnisse und Fortschritte der Zoologie. **1.** 1907. Ergebnisse und *Ausblicke in der Keimzellenforschung. Zeit. Indukt. Abstamm. Vererbungslehre. **3.** 1910. Allgemeine Vererbungslehre. 2 Auf. Braunnschweig. 1912.

HEFFNER, B., A Study of Chromosomes of *Toxopneustes variegatus* which show individual Peculiarities of Form. Biol. Bull. **19.** 1910.

HEGNER, R. W., Germ-Cell Determinants and their Significance. Amer. Nat. **45.** 1911.

HENKING, H., Untersuchungen über die ersten Entwicklungsvorgänge in den Eiern der Insekten. II. Ueber Spermatogenese und deren Beziehung zur Eientwicklung bei *Pyrrhocoris apterus.* Zeit. wiss. Zool. **51.** 1891.

HERBST, C., Ueber das Auseinandergehen von Furchungs- und Gewebezellen in kalkfreiem Medium. Arch. Entw.-Mech. **9.** 1900. Formative Reizen in der thierischen Ontogenese. Leipzig. 1901. Vererbungsstudien. VI. Die cytologischen Grundlagen der Verschiebung der Vererbungsrichtung nach der mütterlichen Seite. Arch. Entw.-Mech. **27.** 1909. Vererbungsstudien. VII. Same title, 2. Mitteilung. Id. **34.** 1912.

HERTWIG, O., Das Problem der Befruchtung und der Isotropie des Eies, eine Theorie der Vererbung. Jena. Zeit. **18 (11).** 1885. Ältere und neurere Entwicklungstheorien. Berlin. 1892. Ueber den Werth der ersten Furchungszellen für die Organbildung des Embryo. Experimentelle Studien am Frosch- und Tritonei. Arch. mikr. Anat. **42.** 1893. Der Kampf um Kernfragen der Entwicklungs- und Vererbungslehre. Jena. 1909.

HERTWIG, R., Ueber neue Probleme der Zellenlehre. Arch. Zellf. **1.** 1908.

HIS, W., Unsere Körperform und das physiologische Problem ihrer Entstehung. Leipzig. 1874.

JENKINSON, J. W., Experimental Embryology. Oxford. 1909.

KORSCHELT UND HEIDER, Lehrbuch, *etc.* I. Abschnitt. Experimentelle Entwicklungsgeschichte. Jena. 1902.

KUPELWIESER, H., (See ref. Ch. V.)

LANKESTER, E. R., Notes on Embryology and Classification. London. 1877.

LILLIE, F. R., Differentiation without Cleavage in the Egg of the Annelid *Chætopterus pergamentaceous.* Arch. Entw.-Mech. **14.** 1902. Observations and Experiments concerning the elementary Phenomena of Embryonic Development in *Chætopterus.* Jour. Exp. Zool. **3.** 1906. Polarity and Bilaterality of the Annelid Egg. Experiments with Centrifugal Force. Biol. Bull. **16.** 1909.

LOEB, J., KING, W. O. R., and MOORE, A. R., Ueber Dominanzerscheinungen bei den hybriden Pluteen des Seeigels. Arch. Entw.-Mech. **29.** 1910.

LYON, E. P., Some Results of centrifugalizing the Eggs of *Arbacia.* Proc. Amer. Physiol. Soc. in Amer. Jour. Physiol. **15.** 1905.

McClendon, J. F., Cytological and Chemical Studies of Centrifugalized Frog Eggs. Arch. Entw.-Mech. **27.** 1909.

McClung, C. E., A peculiar Nuclear Element in the Male Reproductive Cells of Insects. Zool. Bull. **2.** 1899. The Accessory Chromosome—Sex Determinant? Biol. Bull. **3.** 1902. The Chromosome Complex of Orthopteran Spermatocytes. Id. **9.** 1905.

Maas, O., Experimentelle Untersuchungen über die Eifurchung. Sitz.-Ber. Ges. Morph. Physiol. München. **17.** 1902. Einführung in die experimentelle Entwickelungsgeschichte (Entwickelungsmechanik). Wiesbaden. 1903.

Meves, F., Die Chondriosomen als Träger erblicher Anlagen. Cytologische Studien am Hühnerembryo. Arch. mikr. Anat. **72.** 1908.

Montgomery, T. H. Jr., Chromosomes in the Spermatogenesis of the Hemiptera-Heteroptera. Trans. Am. Phil. Soc. **21.** 1906. The Terminology of Aberrant Chromosomes and their Behavior in certain Hemiptera. Science. **23.** 1906. Are Particular Chromosomes Sex Determinants? Biol. Bull. **19.** 1910. (See also ref. Ch. II.)

Morgan, T. H., A Biological and Cytological Study of Sex Determination in Phylloxerans and Aphids. Jour. Exp. Zool. **7.** 1909. The Effects of Altering the Position of the Cleavage Planes in Eggs with Precocious Specification. Arch. Entw.-Mech. **29.** 1910. Cytological Studies of Centrifuged Eggs. Jour. Exp. Zool. **9.** 1910. Chromosomes and Heredity. Amer. Nat. **44.** 1910.

Morgan, T. H., and Spooner, G. B., The Polarity of the Centrifuged Egg. Arch. Entw.-Mech. **28.** 1909.

Nägeli, C., Mechanisch-physiologische Theorien der Abstammungslehre. München. 1884.

Paulmier, F. C., The Spermatogenesis of *Anasa tristis.* Jour. Morph. **15.** Suppl. 1899. (For important corrections and additions, see Wilson, E. B., Studies. II. 1905—Full title below.)

Payne, F., Some New Types of Chromosome Distribution and their Relation to Sex. Biol. Bull. **16.** 1909.

Pflüger, E., Ueber den Einfluss der Schwerkraft auf die Theilung der Zellen. Arch. Gesam. Physiol. **31, 32.** 1883.

Przibram, H., Experimentelle Biologie der Seeigel. (In Bronn's Klass. u. Ordn. II. **3**). Leipzig. 1902. Experimental Zoology. I. Embryogeny. Cambridge. 1908.

Roux, W., Ueber die Bedeutung der Kernteilungsfiguren. Leipzig. 1883, 1895: Ueber die verschiedene Entwickelung isolirter erster· Blastomeren. Arch. Entw.-Mech. **1.** 1895. Einleitung. Arch. Entw.-Mech. **1.** 1894. English Translation by W. M. Wheeler, "The Problems, Methods, and Scope of Developmental Mechanics." Woods Holl Biol. Lect. 1895.

Sala, L., Experimentelle Untersuchungen über die Reifung und Befruchtung der Eier bei *Ascaris megalocephala.* Arch. mikr. Anat. **44.** 1895.

Schaxel, J., Die Eibildung der Meduse *Pelagia noctiluca, etc.* Festsch. f. R. Hertwig. Jena. 1910. Versuch einer cytologischen Analysis der Entwicklungsvorgänge. I. Die Geschlechtszellenbildung und die normale Entwicklung von *Aricia foetida,* Clap. Zool. Jahrb. **34.** 1912.

Schleip, W., Geschlechtsbestimmende Ursachen im Tierreich. Ergeb. und Fortschr. Zool. **3.** 1912.

Schneider, K. C., (See ref. Ch. II.)

Silvestri, F., Un nuovo interessantissimo caso di germinogonia (poliembrionia specifica) in un Imenottero parassita endofago con particolare destino dei globuli polari e dimorphismo larvale. Atti Accad. Lincei Rend. (V.) **14.** 1905.

Sinéty, R. de, (Ref. in Ch. IV.)

Stevens, N. M., A Study of the Germ Cells of Certain Diptera, with Reference to the Heterochromosomes and the Phenomena of Synapsis. Jour. Exp. Zool. **5.** 1908. Further Studies on the Chromosomes of the Coleoptera. Id. **6.** 1909.

Stockard, C. R., The Development of artificially produced Cyclopean Fish—"The Magnesium Embryo." Jour. Exp. Zool. **6.** 1909.

Sutton, W. S., On the Morphology of the Chromosome Group in *Brachystola magna.* Biol. Bull. **4.** 1902. The Chromosomes in Heredity. Id. 1903.

Tennent, D. H., The Dominance of Maternal or of Paternal Characters in Echinoderm Hybrids. Arch. Entw.-Mech. **29.** 1910. Echinoderm Hybridization. Publ. Carnegie Inst. No. 132. Washington. 1911.

Vernon, H. M., The Effect of Environment on the Development of Echinoderm Larvæ. An Experimental Inquiry into the Causes of Variation. Phil. Trans. Roy. Soc. **186, B.** 1895. Cross Fertilisation among Echinoids. Arch. Entw.-Mech. **9.** 1900.

DeVries, H., Intracelluläre Pangenesis. Jena. 1889.

Weismann, A., (See ref. Ch. I.)

Whitman, C. O., The Embryology of *Clepsine.* Q.J.M.S. **18.** 1878. Evolution and Epigenesis. Woods Holl Biol. Lect. 1894.

Wilson, E. B., Experimental Studies on Germinal Localization. I. The Germ-Regions in the Eggs of *Dentalium.* II. Experiments on the Cleavage Mosaic in *Patella* and *Dentalium.* Jour. Exp. Zool. **1.** 1904. Experiments on Cleavage and Localization in the Nemertine Egg. Arch. Entw.-Mech. **16.** 1903. Studies on Chromosomes. I. The Behavior of the Idiochromosomes in Hemiptera. Jour. Exp. Zool. **2.** 1905. II. The paired Microchromosomes, Idiochromosomes and Heterotropic Chromosomes in the Hemiptera. Id. 1905. III. The sexual Differences of the Chromosome Groups in Hemiptera, with some Considerations on the Determination and Heredity of Sex. Id. **3.** 1906. IV. The "Accessory" Chromosome in *Syromastes* and *Pyrrhocoris*, with a Comparative Review of the Types of Sexual Differences of the Chromosome Groups. Id. **6.** 1909. V. The Chromosomes of *Metapodius*. A Contribution to the Hypothesis of the Genetic Continuity of Chromosomes. Id. **6.** 1909. VI. A new Type of Chromosome-Combination in *Metapodius*. Id. **9.** 1910. VII. A Review of the Chromosomes of *Nezara*. with some more general Considerations. Jour. Morph. **22.** 1911. The Chromosomes in Relation to the Determination of Sex. Science Progress. **4.** 1910. The Sex-Chromosomes. Arch mikr. Anat. **77.** 1911.

YATSU, N., Experiments on Germinal Localization in the Egg of *Cerebratulus*. Jour. Coll. Sci. Imp. Univ. Tokyo. **27.** 1910. Observations and Experiments on the Ctenophore Egg. II. Notes on Early Cleavage Stages and Experiments on Cleavage. Ann. Zool. Japonensis. **7.** 1911.

ZELENY, C., (Ref. Ch. VI.)

ZOJA, R., Sullo sviluppo dei blastomeri isolati dallé uova di alcune meduse (e di altri organismi). Arch. Entw.-Mech. **1** and **2.** 1895.

ZUR STRASSEN, O., Ueber die Riesenbildung bei Ascaris-Eiern. Arch. Entw.-Mech. **7.** 1898.

CHAPTER VIII

THE BLASTULA, GASTRULA, AND GERM LAYERS

MORPHOGENETIC PROCESSES

IN this chapter we shall endeavor to summarize the more general processes of early development which lead to the formation, out of the group of cleavage cells, of an embryo possessing the beginnings of the chief organs or systems. This will carry us from the formation of the blastula, through the important events of gastrulation and germ layer formation, and the varied processes by which tissue and organ rudiments are set apart and differentiated.

We shall give particular attention, indeed practically shall limit ourselves, to a descriptive morphological account of these events. This is done, not as a matter of choice, but because the experimental results of the functional analysis of these processes are so fragmentary and so scattered, that the attempt at their summary, in a text of this character, seems unwise. This is partly because the efforts to analyze these processes experimentally have been delayed until the similar problems of cell organization and cleavage should have been more satisfactorily solved. These topics we have already considered. We should say, however, that what has already been accomplished in the way of describing these later phenomena of the mechanics of development (*Entwicklungsmechanik*, · Roux) bears out the general conclusions indicated by the earlier processes, namely, that while the action of external conditions as stimuli is essential, their place normally is chiefly that of affording the general conditions of life and development. The actual quality and the really significant details of the later, as of the earlier phenomena of development and differentiation, depend primarily upon internal conditions and relations.

And further, there is little reason for supposing that here too the essential determinative conditions are other than nuclear, although from the nature of the case, the evidence must be less direct.

We shall limit ourselves in another direction also. It is obviously impossible to give a brief, and at the same time an adequate, account of the extremely diverse methods of gastrulation, germ layer formation, *etc.*, in the Metazoa as a whole. We shall, therefore, confine ourselves here largely to the consideration of these events among the Chordata. This will enable us to give a more adequate consideration to the topics selected.

We have seen that while no definite termination can be placed to the period of cleavage, there is rather general, though arbitrary, agreement that cleavage may be said to have terminated when the blastomeres become arranged as a more or less definite epithelium or layer, bounding an internal space of some sort; and further that this may, in some cases, also be marked by the attainment of a certain nuclear-cytoplasmic relation. The organism exhibiting these characteristics is known as the *blastula.* In this stage the organism is essentially a *monodermic* structure, that is, it is composed of a single tissue or layer of somewhat similar cells. In the simplest form of blastula this layer is but one cell thick (Amphioxus, Fig. 150, *A*), but in most of the Chordate blastulas the wall is many cells in thickness.

Like the cleavage pattern, though to a much greater extent, the general form of the blastula is largely determined by the amount of yolk or deutoplasm contained in the ovum; and the form of the blastula, in turn, largely determines the form of the gastrula, and the methods of gastrulation and germ layer formation. We may for convenience, therefore, distinguish three general forms of blastulas, although intergradations are not infrequent. When the ova are nearly homolecithal, and cleavage is total and adequal, as in Amphioxus (Fig. 150, *A*), the blastula is practically a hollow sphere *(cœloblastula)*. Its wall is a simple epithelium, one cell in thickness, and its cavity,

the *blastocœl* or *segmentation cavity*, is large and nearly central,
though not quite, for nearly always the cells at the vegetal pole
are larger than those at the animal pole, and the wall consequently thicker in the former region. This form of blastula is
commonly regarded as primitive, though hardly typical of the
Chordates in general, for among these it is found only in
Amphioxus.

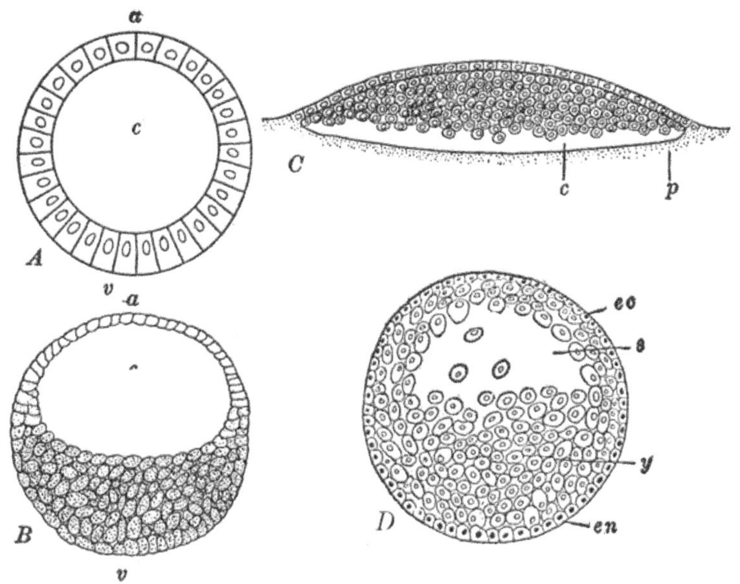

Fig. 150.—Types of Chordate blastulæ. *A*. Amphioxus (cœloblastula).
B. *Petromyzon*. After von Kupffer. *C*. *Noturus* (Teleost) (discoblastula).
D. *Triton* (Urodele). After Greil. *a*, animal pole; *c*, blastocœl; *p*, periblast;
v, vegetal pole.

More usually the Chordate ovum contains a considerable
amount of yolk, as in the Ganoids and Amphibia. And here
the cleavage, though nearly or quite complete in most cases, is
decidedly unequal and the form of the blastula is considerably
modified in consequence. Here the wall of the blastula is
several or many cells thick; the cells of the animal pole are
quite small, while the yolk-containing cells, of the vegetal
hemisphere, are very large. The blastocœl is therefore quite
eccentric, displaced toward the animal pole, and usually
much reduced in size (Fig. 150, *B, D*).

The most highly modified type of Chordate blastula is found in those forms with extremely meroblastic, telolecithal eggs, where cleavage is of the discoid type. This condition is common to the Elasmobranchs (Fig. 158), the true Teleosts (Fig. 150, *C*), and to the higher Craniates—the Reptiles, Birds. In reality this is, in a modified way, characteristic of the Mammals also, for although the Mammalian ovum is nearly alecithal, it is clearly derived from the Reptilian condition, and many features of its development show unmistakably the effects of a large yolk content previously present, but now lost in correlation with the newly acquired source of nourishment possessed by the Mammalian embryo. The result of discoid cleavage is the formation of a small mass of living active cells, the *blastoderm*, or *blastodisc*, or *germ disc*, lying upon the surface of the yolk mass. The blastula instead of being spherical, has therefore the form of a circular disc, the cellular elements of which can really be compared, at first, only with the cells of the animal pole of the spherical blastula, the unsegmented yolk representing, in this stage, the large cells of the vegetal pole of such a blastula. In comparing these two types of blastulas we may imagine that the ordinary spherical blastula has been cut in two horizontally, through or just above its equator, and the animal hemisphere flattened out, its circumference being thereby somewhat extended. This form of blastula (*discoblastula*) is several cells in thickness and is usually separated from the underlying yolk by a shallow space called the *sub-germinal cavity* which represents the blastocœl (Fig. 150, *C*). While the yolk mass is usually by no means wholly devoid of nuclei, these may or may not be associated with true cellular structures, and even when present at this stage these rarely give rise to any structural parts of the definitive embryo when this forms. The Mammalian blastula diverges widely from any of these conditions and on account of its very special character further reference to it may be omitted here.

The next important step in development consists essentially in the conversion of this monodermic blastula into a *didermic* organism, that is, one in which the cells are arranged in two, more or less distinct, tissues or layers. This process is known as *gastrulation*, and the didermic embryo itself is called a *gastrula.*

Among the lower forms the process of gastrulation often remains a simple one, involving little more than the mere rearrangement of the cells of the blastula into two nearly homogeneous layers. But in the higher forms, such as the Chordata, with which we are dealing, the process is greatly complicated by the precocious formation of the rudiments of the chief axial structures of the later developing embryo, as well as by the differentiation of a third tissue, or *intermediate layer* between the other two. The establishment of the rudiments of the axial notochord and central nervous system, characteristic structures of the Chordate embryo, is termed *notogenesis.* These rudiments are formed out of the substance of the two primary layers of the gastrula. But the formation of a tissue between these converts the didermic embryo into a *tridermic* organism.

It is possible to analyze the whole process of development during this period into these three subsidiary processes, gastrulation, notogenesis, and middle layer formation, and in some instances they may occur somewhat separately and successively. But usually there is much overlapping, and the attempt to describe the process of gastrulation by itself would give a very incomplete and incorrect view of the events of this period. Consequently we shall describe all three of these processes together.

Three general types of gastrulas and modes of gastrulation may be found, corresponding with the three types of blastulas and ova mentioned above. Again the simplest and probably the most primitive, though not the most typical, condition is found in Amphioxus. On the posterior side of the blastula, in the region just between animal and vegetal hemispheres, Cerfontaine has described a small group of cells marked by a tendency to rapid and continued multiplication (Fig. 151, *A*). This region of active proliferation gradually extends laterally around the blastula and ultimately forms a nearly complete ring, though this is not until the blastula has become converted into the gastrula. This specialized group of cells may be termed the *germ ring*, for it is evidently equivalent to the structure already known by that name in the Fishes, Amphibia, and other forms. At the time this rapid proliferation commences, the vegetal hemisphere becomes flattened. The large cells of this region then arch up slightly into the blastocœl and soon begin to fold, or swing, inward about their postero-ventral margin as a relatively fixed point (Fig. 151, *B*, *C*, *D*). This motion is made possible by the rapid extension of a sheet of cells which come off from the germ ring and which are thus drawn in, to line the inside of the animal hemisphere. Without going into details here, we may say that finally the inturning of the vegetal cells becomes complete (Fig. 151, *E*) and the blastula is converted into a cup-like structure, widely open toward one side (the posterior or postero-dorsal).

The wall of the organism is now composed of two layers or epithelia, the original blastocœlic cavity is nearly or quite obliterated, and a new cavity is formed, lined by the inturned cells, and widely open to the outside. This structure is the didermic *gastrula*. The two cell-layers composing its wall are the *primary germ layers*. The newly established gastrular cavity is the *archenteron* or *primitive gut cavity*. The superficial layer of cells, including the original animal hemisphere of the blastula and also some cells derived from the proliferating area, is known as the *outer germ layer*, or *ectoderm*, or *ectoblast*, or *epiblast;* the layer lining the archenteron, partly the cells of the vegetal hemisphere of the blastula and partly the cells derived from the proliferating region, is known as the *inner germ layer*, or *endoderm*, or *entoblast*, or *hypoblast*. The opening of the archenteron to the outside is the *blastopore;* the periphery of the blastopore is spoken of as its *margin* or *lip*, and we have seen that it is the region largely occupied by the germ ring: *it is here that the two primary germ layers are directly continuous with one another.*

The process of infolding, such as is carried out here by the vegetal cells which come to line the ventral region of the archen-

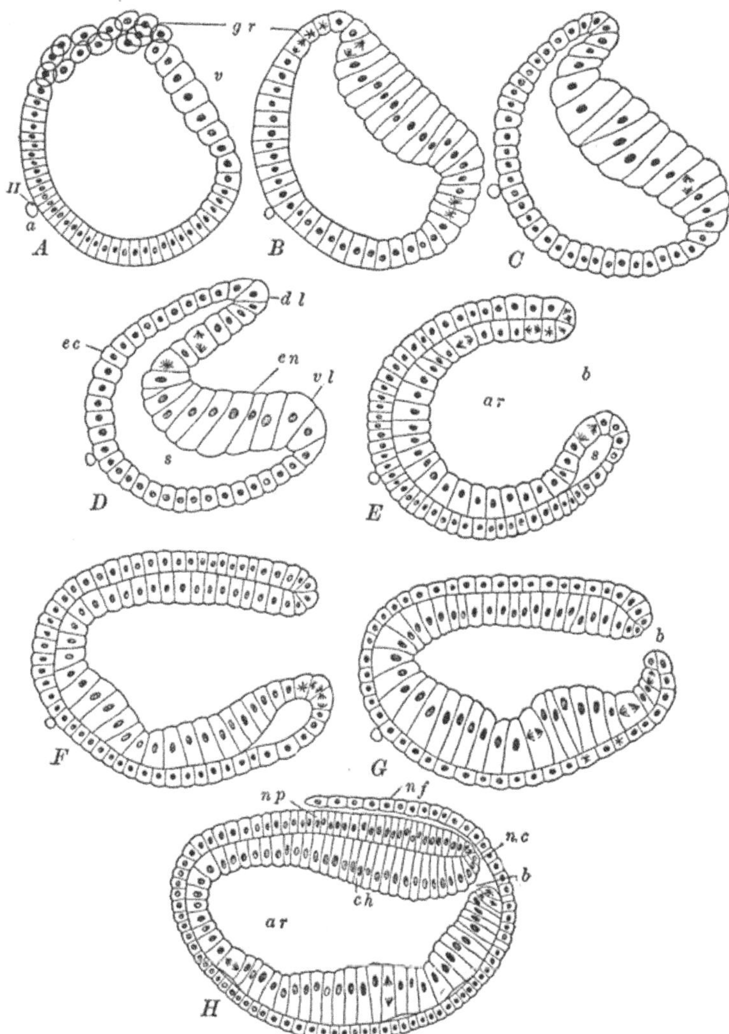

Fig. 151.—Gastrulation in Amphioxus. After Cerfontaine. *A*. Blastula showing flattening of the vegetal pole and the rapid proliferation of cells in the postero-dorsal region (germ ring). *B*. Flattening more pronounced; mitoses in cells of germ ring. *C*. Commencement of the infolding (invagination) of the cells of the vegetal pole. *D*. Continued infolding and inflection, or involution, of ectoderm cells in the dorsal lip of the blastopore. The blastocœl becoming obliterated and the archenteron being established. *E*. Invagination complete. Continued involution in the dorsal lip of blastopore. Mitoses in germ ring. *F*. Constriction of blastopore and commencement of elongation of the gastrula. Remnants of blastocœl in ventral lip of blastopore. *G*. Gastrulation completed. Continued elongation, and narrowing of blastopore. H. Neurenteric canal

teron, is known as *invagination*, and in some of the lower forms
the endoderm is wholly formed by this process. The process
of inturning, such as results in the lining of the dorsal region
of the archenteron by the cells derived from the margin of the
blastopore, is known as *involution*. In some forms the endoderm is largely formed by this process. Usually this is accompanied by the growth of the margin of the blastopore, or germ
ring, on over and past the involuted region, so that a layer of
cells is continually being overgrown, leading to more extended
gastrulation; this process of overgrowth may be termed *epiboly*.
The blastopore of the Amphioxus gastrula, at first widely
open, soon closes rapidly on account of the growth and epiboly
of its lip. This process leads also to the rapid elongation of the
gastrula (Fig. 151, *F-H*).

Although we have included involution and epiboly as gastrulation processes they are here more properly to be regarded
as processes leading to the formation of the notochord and the
intermediate layer. For as the gastrula continues to elongate
these structures begin to differentiate in the anterior part of the
roof of the archenteron, from that part of the inner layer formed
by involution and epiboly. Along the dorso-lateral regions of
the archenteron appear a pair of folds out of the archenteron,
each containing a narrow groove (Fig. 152, *B, C*). These folds
are the rudiments of the intermediate layer, or *mesoderm, or mesoblast;* and the grooves, known as the *enterocœlic grooves*, are the
rudiments of the *cœlom*, the cavity of the mesoderm. That
portion of the archenteric roof between the mesoderm folds,
later becomes folded outward and forms the rudiment of the
notochord. The remainder of the archenteric wall, namely the
ventral and ventro-lateral regions formed by the *invaginated*
portion of the inner layer, forms the *primitive gut* or *enteron.*
Whether we choose to call the involuted region really endoderm
or not, is immaterial; if we do, then we must say that the chorda

<hr>

established by overgrowth of neural folds. Continued mitosis in germ ring.
a, animal pole; *ar,* archenteron; *b,* blastoporal opening; *ch,* rudiment of notochord; *dl,* dorsal lip of blastopore; *ec,* ectoderm; *en,* endoderm; *gr,* germ ring;
nc, neuenteric canal; *nf,* neural fold; *np,* neural plate; *s,* blastocœl or segmentation cavity; *v,* vegetal pole; *vl,* ventral lip of blastopore; *II,* second polar body.

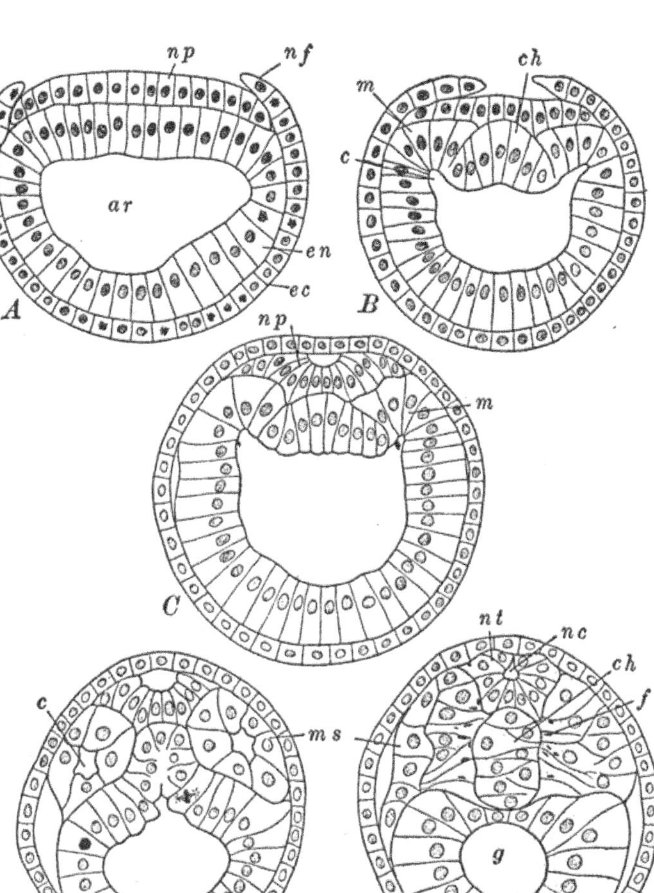

Fig. 152.—Transverse sections through young embryos of Amphioxus, showing formation of nerve cord, notochord, and mesoderm. ? After Cerfontaine. *A*. Commencement of the growth of the superficial ectoderm (neural folds) above the neural plate (medullary plate). *B*. Continued growth of the ectoderm over the neural plate. Differentiation of the notochord, and first indications of mesoderm and enterocœlic cavities. *C*. Section through middle of larva with two somites. Neural plate folding into a tube. *D*. Section through first pair of mesodermal somites, now completely constricted off. *E*. Section through middle of larva with nine pairs of somites. Neural plate folded into a tube. Notochord completely separated. In the inner cells of the somites, muscle fibrillæ are forming (compare Fig. 153). *ar*, archenteron; *c*, enterocœl; *ch*, notochord; *ec*, ectoderm; *en*, endoderm; *f*, muscle fibrillæ; *g*, gut cavity; *m*, mesoderm (gastral); *ms*, mesodermal somite; *nc*, neural canal; *nf*, neural fold; *np*, neural plate (medullary plate); *nt*, neural tube.

and mesoderm are both derived from endoderm and the process
of involution is to be regarded as a true gastrulation process.

But only the lesser part of the mesoderm is formed in the way
just described. This part of the mesoderm is known as the
gastral or *parachordal*, or *axial* mesoderm. If we trace the
mesoderm folds, just described, posteriorly, we can follow them
into the region of the germ ring or blastopore lip which has now
become considerably thickened on ac-
count of its contraction, and consists of
a more or less undifferentiated mass of
cells. This mass now passes almost en
tirely around the blastopore, laterally
and toward the ventral side. The
rapid proliferation of the cells of the
germ ring has early led to the disappearance of the original simple epithelial arrangement of its cells (Fig. 153),
but as it moves backward, with the
elongation of the gastrula, it leaves
behind it (*i.e.*, anteriorly from it) its cell
products, which rapidly become differentiated into certain layers. On the
surface of the embryo a layer is left
which is directly continuous with the
ectoderm of the original gastrula derived from the animal hemisphere of
the blastula. Another layer is differentiated lining the archenteron and

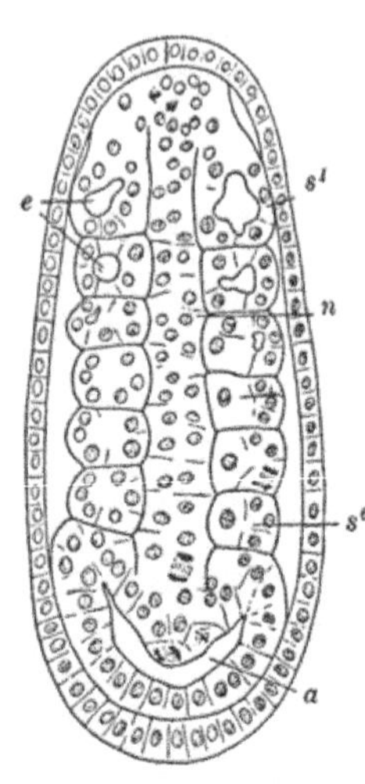

Fig. 153.—Frontal section
through larva of Amphioxus,
with six pairs of mesodermal
somites, at the level of the
notochord and somites.
After Cerfontaine. *a*,
archenteron ; *e*, enterocœl;
n, notochord; s^1, s^6, first and
sixth mesodermal somites.

continuous with the layer already there, and consisting of a ventral region continuous with the invaginated layer or true gut
endoderm, a dorsal region continuous with the involuted layer
forming the rudiment of the notochord, while dorso-laterally,
between these two regions, the inner sheet is continuous with
the mesodermal rudiments described above. In other words,
out of the germ ring there are gradually differentiated, true covering ectoderm, gut endoderm, chordal cells, and mesoderm. The
mesoderm formed in this way, directly from the germ ring, is

called *peristomial* mesoderm, to distinguish it from the gastral mesoderm formed in connection with the enterocœlic grooves. Very soon, as we have seen, the region which gives rise to the peristomial mesoderm comes to extend nearly to the ventral side of the blastopore region. The cœlomic spaces of this peristomial mesoderm are not formed as derivations of the archenteron; they result from rearrangements of the mesodermal cells, and are entirely independent of the enterocœlic portions of the cœlom in their origin, although the two become continuous later (Fig. 153). These two forms of mesoderm are directly continuous with one another and have indeed a common primary origin, the germ ring or margin of the blastopore. If we recognize the essential difference between them as that of time of formation, the altered circumstances surrounding the formation of each due to this time difference, become of secondary importance as regards our real conception of the mesoderm and its relations to the other germ layers. Thus the relation of both chorda and mesoderm proper to the cells of the monodermic blastula is the same as that of the endoderm proper.

Stated in a word then, the gastrulation of Amphioxus is a combination of invagination and involution, accompanied by epiboly, and the processes of notogenesis and mesoderm formation are intimately bound up with the formation of the inner layer.

Having become familiar now with the general ideas of gastrulation and the terminology of the process we may consider the remaining forms of this process in the Chordates much more briefly.

Our second type of gastrulation, as it occurs in the Amphibia and Ganoid Fishes, may be easily understood by comparison with the preceding. The chief differences result from the accumulation of yolk in the vegetal pole of the ovum and blastula, and the consequent comparative inertness of this region. That is, the chief modifications of the typical process of gastrulation appear in respect to the behavior of the lower pole, destined to form the inner layer of the gastrula.

In the Amphibian blastula, the form of which was described

above, a region of more actively dividing cells can be distinguished extending around the fully formed blastula just above its equator, that is, between animal and vegetal poles (Fig. 154); this is termed the *germ ring*, although it is frequently not a complete ring, being interrupted on the anterior side of the blastula. Gastrulation commences by the true invagination of cells just below the germ ring on the posterior side. The inertness of the large yolk-filled cells of the vegetal hemisphere greatly impedes the process of invagination and it is never carried very far. Involution occurs extensively in the dorsal

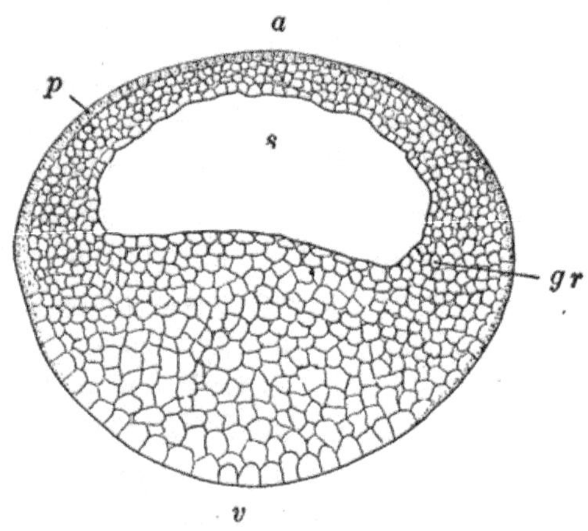

Fig. 154.—Section through late blastula of frog, showing location of germ ring. Later the germ ring is thickened and contracted, forcing the yolk cells upward into the segmentation cavity. After O. Schultze. *a,* animal pole; *gr,* germ ring; *p,* pigment; *s,* blastocœl; *v,* vegetal pole.

region and is accompanied by very pronounced epiboly (Fig. 155); this latter process is very marked in the lateral and ventral regions where little or no invagination occurs. All of these processes are relatively less extensive than in the cœlogastrula of Amphioxus however, probably on account of the larger amount of yolk contained in all of the cells. Their place in development is taken, in a way, by a new process, namely, *delamination.* This is a process of splitting, whereby an extended mass or sheet of cells becomes cleaved apart by the rearrangement of the cells into two more or less distinct layers, separated by a definite space. This process of delamination

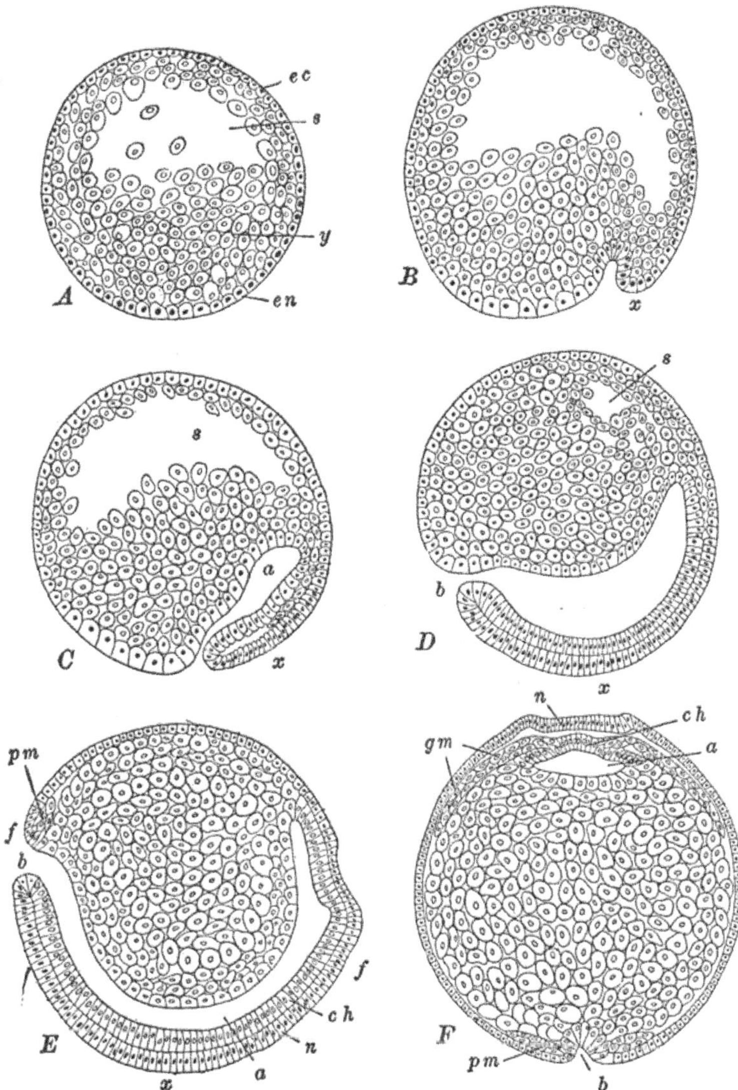

Fig. 155.—Series of diagrammatic drawings of sections showing the process of gastrulation in the Urodele, *Triton*. After Greil and Ruffini. *F*. Transverse section, others sagittal. *A*. Blastula. *B*. Commencement of gastrulation (invagination). *C*. Continued invagination accompanied by epiboly and involution. Formation of archenteron. *D*. Continuation of all three processes of gastrulation. Blastocœl nearly obliterated. *E*. Archenteron completely formed. Rudiments of notochord and neural plate differentiated. *F*. Transverse section through embryo shown in *E*, through the plane marked *ff*. *a*, archenteron (gut cavity in *F*) lined with endoderm; *b*, blastopore; *ch*, rudiment of notochord; *ec*, ectoderm; *en*, endoderm; *ff*, plane of section shown in *F*; *gm*, gastral (axial) mesoderm; *n*, neural plate (in *F*, bounded by neural folds); *pm*, peristomial mesoderm; *s*, blastocœl or segmentation cavity; *x*, marks corresponding points on the surface ectoderm, showing extent of epiboly; *y*, yolk cells.

begins where the processes of invagination and involution leave off, and it is important to recognize that the didermic character of the gastrula of the Amphibian results partly from all three of these processes. The chief result of involution is the formation of the rudiments of the notochord and gastral mesoderm, as in Amphioxus. Figure 155 shows how the blastocœl is finally obliterated by the invaginated and involuted regions. The germ ring finally completes its growth over the yolk cells or endodermal floor of the archenteron, and closes together much as in Amphioxus.

The formation of the mesoderm offers some points of difference when compared with Amphioxus. The peristomial mesoderm forms typically in the margin of the blastopore, out of the undifferentiated cell mass of the germ ring. Sometimes, in the region just within the blastopore dorsally, traces of enterocœlic outgrowths can be seen (Fig. 156), but most of the gastral

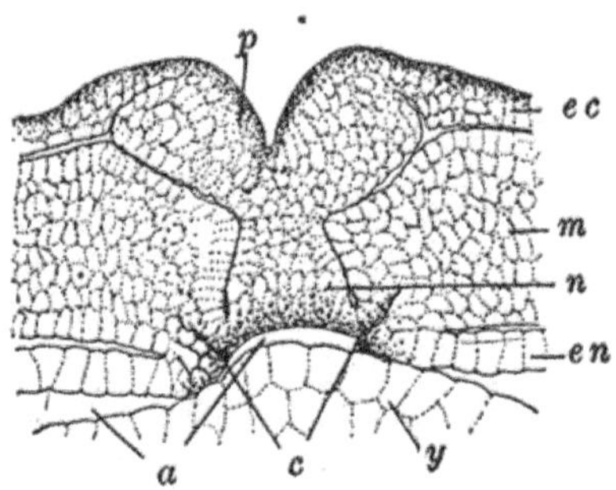

Fig. 156.—Part of a section through the body of an embryo of the frog, *Rana fusca*, showing traces of enterocœl formation. After O. Hertwig. *a*, archenteron; *c*, enterocœls; *ec*, ectoderm; *en*, endoderm; *m*, mesoderm; *n*, notochord; *p*, neural plate; *y*, yolk cells.

mesoderm is formed either from involuted cells derived from the germ ring, or later from the surface of the endoderm by a process of delamination or splitting off of the superficial cells lying next the ectoderm; these come off first as a solid sheet, which much later itself splits into two layers leaving a cœlomic cavity between them.

Thus while invagination occurs to a slight extent, gastrulation in these forms results largely from the processes of involu-

tion and epiboly combined with delamination. The didermic
condition results, to a considerable extent, from the overgrowth
of the animal hemisphere cells (germ ring) which come to enclose
the yolk cells of the vegetal hemisphere. As in Amphioxus,
however, the yolk, although here so much more abundant, is
finally included in the floor of the gut cavity, and the yolk cells
take a direct share in the formation of the structures of the
later embryo. After the mesoderm and chorda have been
formed from the roof of the archenteron, this is left as a thin
layer, only one cell in thickness, quite in contrast with the thick
mass of cells forming its floor (Fig. 155).

Turning now to the third type of gastrula, that formed from
the discoid blastula, we find conditions which vary widely from
the Amphioxus type, but which after all may be interpreted
in the light of the processes just outlined. In the Ganoid or
Amphibian, both the animal and vegetal hemispheres of the
egg share directly in the processes of cleavage, and blastula and
gastrula formation; and the yolk, contained in typical cells, is
carried directly into the wall of the primitive gut. But in
the extremely meroblastic eggs of the Elasmobranchs, Teleosts,
Reptiles, and Birds, the large yolk mass, which is the equivalent

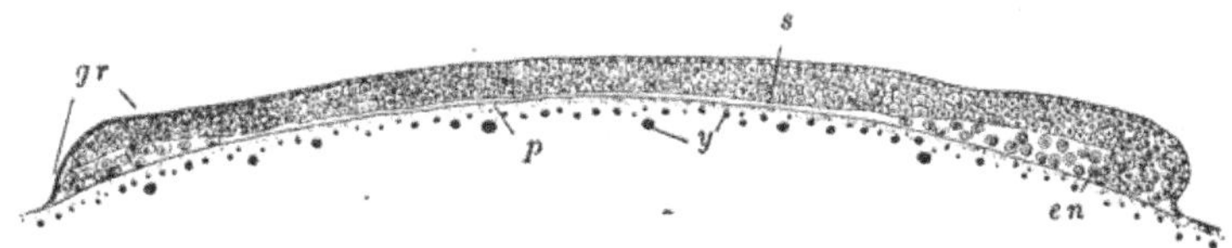

Fig. 157.—Sagittal section through early gastrula of the catfish, *Ameiurus.*
en, endoderm; *gr*, germ ring; *p*, periblast; *s*, segmentation cavity, or sub-germinal cavity; *y*, yolk.

of the vegetal pole of the egg, does not cleave (Figs. 150, 48)
and takes no direct share in the formation of the cellular blastula
and gastrula. For comparative purposes, therefore, we have
already seen that we must recognize the germ disc or "blastula"
of this type as equivalent only to the animal hemisphere of such
a form as the frog or Amphioxus, flattened out and resting
upon the undivided yolk mass. In such a condition as this the
equivalent of the germ ring would be found forming the

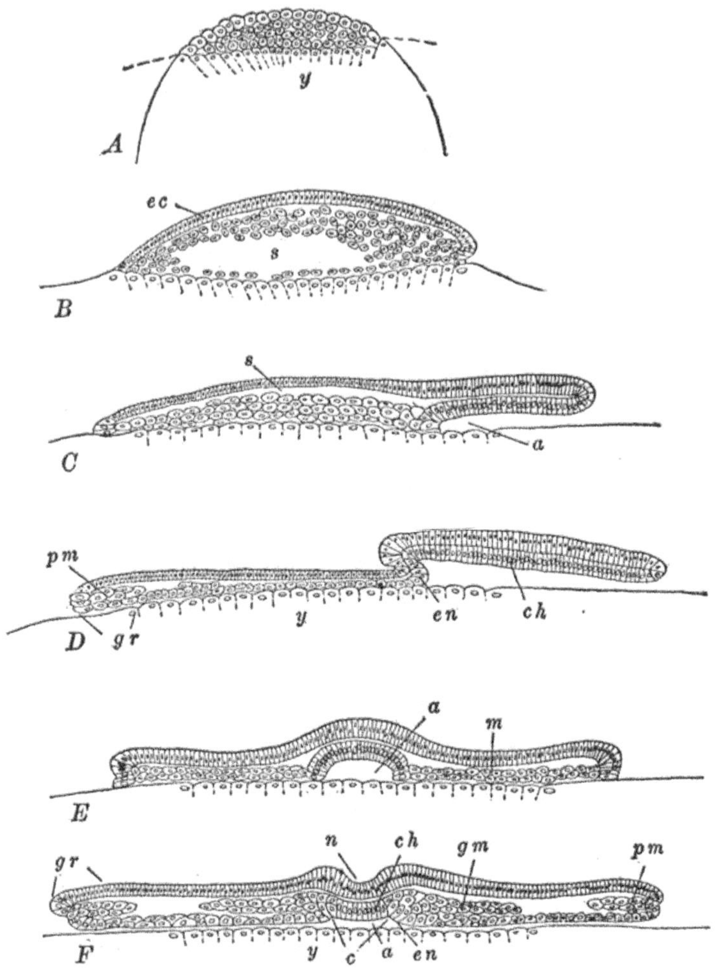

Fig. 158.—Semi-diagrammatic drawings of sections through Elasmobranch embryos. After Greil, after Rückert and Ziegler. *A–D.* Sagittal sections; *E, F,* transverse sections. *A.* Diagram of section to show relations of blastoderm to animal pole of blastula with less yolk. *B.* Commencement of gastrulation by invagination, epiboly, and involution. *C, D.* Continuation of all three processes of gastrulation. *E.* Transverse section showing relation of endoderm to yolk, mesoderm, and the germ ring. *F.* Transverse section farther forward, through stage resembling *D.* Anterior margin of blastoderm toward the left in *A–D. a,* archenteron; *c,* vestiges of enteroccels; *ch,* rudiment of notochord; *ec,* ectoderm; *en,* endoderm; *gm,* gastral (axial) mesoderm; *gr,* germ ring; *m,* mesoderm; *n,* neural groove, bordered laterally by neural folds; *pm,* peristomial mesoderm; *s,* blastoccel or sub-germinal cavity; *y,* yolk.

periphery of the blastodisc. The process of gastrulation concerns only the blastodisc, for the yolk mass takes no more share in this than in the later processes of development; the gastrula forms separately from the yolk, which is left outside the embryo and as we shall see, comes into relation with it only indirectly. The blastula or blastodisc, once formed as a flat plate, many cells in thickness, begins to extend over the surface of the yolk mass. Its central part becomes quite thin in consequence, but

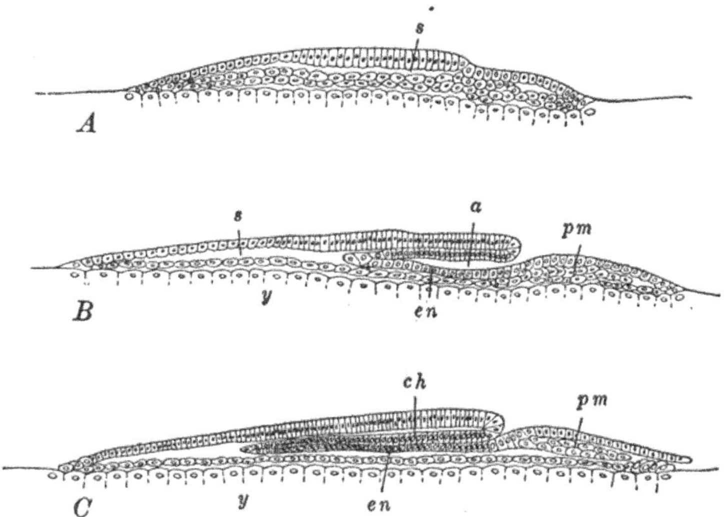

Fig. 159.—Diagrammatic drawings of sagittal sections through embryos of Sauropsids. After Greil, after Will and Schauinsland. *A, B.* Two stages of Reptile. *C.* Bird. *a,* archenteron; *ch,* rudiment of notochord; *en,* gut endoderm; *pm,* peristomial mesoderm; *s,* blastocœl or sub-germinal cavity; *y,* yolk.

the margin of the disc, or germ ring, remains thickened, and as it advances over the yolk its margin becomes slightly involuted, forming a narrow shelf of cells on its inner surface, toward the yolk (Fig. 157). This inner layer is the rudiment of the primary inner layer. Gastrulation is thus primarily accomplished by involution. The margin of the germ disc is clearly the germ ring or rim of the blastopore, although its form and relation to the yolk are quite unlike what we have seen heretofore.

The Elasmobranchs and Reptiles afford important transitional conditions here, in that a definite process of invagination is indicated (Figs. 158, 159). Invagination is here limited to the posterior margin of the blastoderm, where the germ ring becomes elevated above the yolk as the endoderm is folded under the ectoderm. In the Teleosts little or no indication of invagination can be found. In these forms the germ ring extends rapidly over the yolk, reaches and passes an equator of the egg, and then as it continues to advance, gradually narrows, and finally closes completely, having passed over the entire yolk mass (Fig. 160). This overgrowth of the blastoderm occurs more rapidly in the anterior and lateral directions than in the posterior direction, so that the blastopore finally closes in nearly the same relative position as in the frog and in Amphioxus, *i.e.*, postero-ventrally. As the germ ring extends around the yolk, only a single, and very thin, layer of cells is left behind it as a covering layer. In the posterior and postero-lateral regions alone, is the involution of an inner layer well marked. It should be noted that in the Teleosts the endoderm is largely replaced functionally by a specialized protoplasmic region on the surface of the yolk, known as the *periblast*, which contains free nuclei derived originally from those of the margin of the blastodisc (Figs. 150, *C*; 157).

During the later stages of the overgrowth of the germ ring, as it contracts after passing the equator of the egg, its substance is payed into its more slowly advancing posterior region, where it forms a longitudinal median thickening (Fig. 160). This thickened region of the blastoderm is the *primitive streak*, the earliest rudiment of the essential parts of the embryo, which gradually differentiate out of its anterior end.

In such a gastrula as this the endoderm forms a flat median plate of cells lying directly upon the surface of the periblast (yolk), and the archenteron is present only virtually as a narrow space between the endoderm and periblast (Fig. 157). In such a case the formation of a true gut cavity is independent of the formation of the inner layer, and occurs later by a process of folding.

The mesoderm is differentiated at a comparatively early stage, and the distinction between peristomial and gastral mesoderm is very clear. The peristomial mesoderm appears as a small mass of slowly differentiating cells lying in the germ ring, between the superficial ectoderm of the blastoderm and the involuted shelf of endoderm (Fig. 158, *F*). The gastral mesoderm is seen budding off laterally from the primitive streak region, also between ectoderm and endoderm or even beyond the region where the endoderm is found (Fig. 158, *F*). Posteriorly

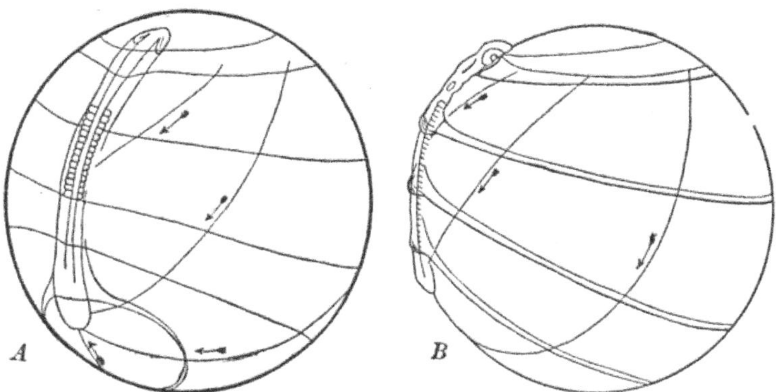

Fig. 160.—Diagrams of the formation of the Teleost embryo by confluence of the germ ring, and the growth of the germ ring around the yolk.   From Kopsch. *A*. In half-profile.   *B*. In profile.

of course the gastral mesoderm of each side becomes continuous with the peristomial mesoderm, and as peripheral portions of the germ ring, where the three layers are slowly differentiating, are continually passing into the posterior end of the primitive streak, it is clear that peristomial mesoderm is constantly becoming gastral, merely through relative change of position, not through any change in the mode of its formation or in its relation to the other germ layers.

The Elasmobranchs and Reptiles are again transitional in that vestiges of enterocœlic grooves may be seen as shallow and narrow longitudinal depressions, either side of the midline, in the region of which the formation of mesoderm is most rapid (Fig. 158, *F*).   In the Teleost, as in the Bird, no traces of enterocœls are to be seen.   As usual the notochord forms from the cell mass lying between the rudiments of the gastral mesoderm, and may be said to have been derived either from the gastral mesoderm, or from the endoderm in the same way that the mesoderm itself is. After the separation of the chorda and mesoderm, the endoderm proper, or *enteroderm*, as it is called, is left as a thin narrow strip of cells spread flat over the periblast (yolk) surface, continuous posteriorly with the diverging limbs of the germ ring.

In the Sauropsids, where the accumulation of the yolk is most pronounced, the blastoderm does not grow entirely around the yolk until long after the gastrula is formed and the embryo established. Correlatively we find no typical germ ring formation in the periphery of the blastoderm, save in that posterior region which is to be concerned in embryo formation. Remembering that in the Reptile both true invagination and enterocœl formation occur, while these processes are not apparent in the birds, we may describe (following Patterson's account) the processes of gastrulation and embryo formation in the pigeon, as illustrating these events in the development of the extremely meroblastic ovum.

The blastoderm first becomes quite thin, particularly toward its posterior side, where, at the same time, the margin thickens forming a segment of a true germ ring (Fig. 161). The extension of this posterior part of the germ ring, however, involves the usual processes of cell multiplication accompanied by involution and epiboly; there is no true invagination here (Fig. 161). The formation of an inner layer is thus limited to the posterior region of the blastoderm. Soon, as this whole region extends posteriorly, this segment of a germ ring begins to contract toward the mid-line, and the result is the formation of a median thickening in the posterior half or third of the germ disc. This thickening is the *primitive streak* (Fig. 161), and as usual it is the seat of the formation of the chief embryonic rudiments. As in the Teleost, the primitive streak, formed by the gradual fusion of the lateral halves of the germ ring, is obviously the equivalent of the blastoporal margin of the frog or of Amphioxus. On its surface is a shallow longitudinal groove marking the separation of the two halves; this is the *primitive groove* (Fig. 161), which may be regarded as representing the blastopore proper. The archenteron, in such a gastrula as this

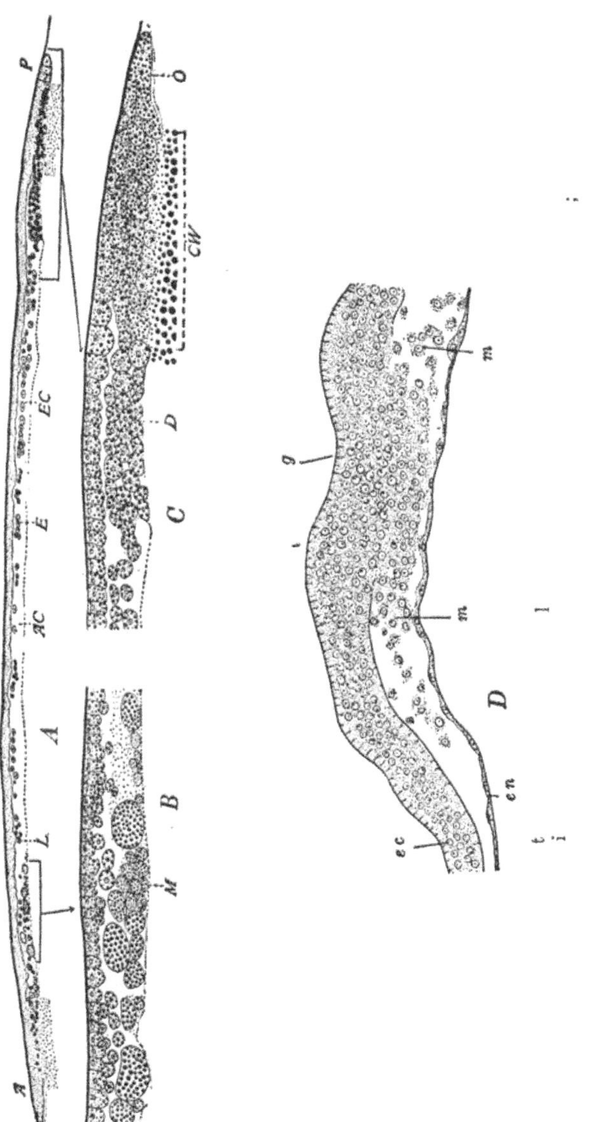

of the chick, may also be said to exist only virtually, for it is represented only by a shallow space left between the endoderm and the yolk.

The mesoderm and chorda are here more closely related with the outer than with the inner layer. In the germ ring there is little indication of separation of the germ layers, other than the distinction of the endoderm, and when the primitive streak is formed, it appears rather as a thickening of the ectoderm. The mesoderm begins to be differentiated along the sides of the primitive streak, and back as far as the region where this is being formed by the fusion of the limbs of the germ ring. Hence the mesoderm is more largely gastral, that is to say, it does not become distinct, as a separate rudiment, until the establishment of the primitive streak has occurred. In the germ ring there is of course a region where ectoderm passes into endoderm, and where the cells may be said to belong to either layer or neither layer. This is the region where the primary "mesoderm" forms and apparently special conditions may determine with which of the primary layers it may seem to have the more intimate relation. Little is gained by attempting to define germ layers in the germ ring.

In the pigeon or chick the rudiment of the notochord appears in the deeper part of the primitive streak after its lateral parts are cut off as mesoderm. The endoderm has therefore the value of an *enteroderm* from the beginning, and has the form of a very thin flat sheet of cells widely spreading over the yolk surface. Ultimately the yolk mass becomes entirely enclosed in a layer of endoderm as well as by the other germ layers, but this does not occur until a comparatively late stage in the development of the embryo.

In Amphioxus and the frog we have seen that the embryo is formed from the entire ovum, that is, the yolk-containing cells become actually included within the wall of the gut. In the Teleosts the yolk mass is so large, and so completely separated from the embryogenic tissues, that the embryo may be said to develop upon the surface of the yolk, which, enclosed within a structure called the *yolk sac*, is only indirectly related to the embryo proper. In forms like the Elasmobranchs and Sauropsids, the accumulation of yolk is still greater and the embryo forms quite apart from the yolk, with which it later acquires a secondary relation. In the Sauropsids, after the rudiments of the embryo are well established, a process of folding begins and a series of infoldings of the cellular blastoderm, anterior, posterior, and lateral, pinch off the embryo from the yolk mass or yolk sac, with which it then remains only indirectly connected by a narrow tube known as the *yolk stalk* which includes a portion of the gut wall and a very abundant blood supply.

In the Sauropsids and Mammals other folds of the blastoderm soon appear, beyond the limits of the embryo proper, which result in the formation of a very special and highly characteristic structure known as the *amnion*. And from the wall of the hind-gut grows out another special and extra-embryonic structure, the *allantois*. The formation and function of these extra-embryonic structures, together called the *embryonic appendages*, cannot be described here. They are of the greatest importance in development and their presence has led to the application of the term *Amniota* to all the forms possessing them (Birds, Reptiles, Mammals) while the other Craniates, without these embryonic appendages (Cyclostomes, Fish, Amphibia) are then known as the *Anamnia*.

On account of the difficulties of comparison it seems wise to omit reference here to the Mammalian gastrula and germ layer formation. For in spite of the nearly alecithal condition of the Mammalian ovum, its development shows marked yolk influence, and the whole course of early development is complicated, not only through the one time presence and the subsequent loss of yolk, but through the very special relations of the early embryo, and particularly the embryonic appendages, with the walls of the maternal cavity in which development proceeds.

CONCRESCENCE

We should consider here, in a particular way, a developmental process which, besides being of great general importance in Chordate development, is of considerable historical interest as well. In the foregoing pages we have seen that where a germ disc is formed, its margin, known as the germ ring, and recognized as the homolog of the lip or margin of the blastopore, is of great importance in the formation of the primary rudiments of the embryo.

The His-Whitman theory of *concrescence* emphasizes the general importance and significance of this relation. First stated fully by His, in 1876, the essential idea of this theory was that each side of the germ ring, not only forms, but really is,

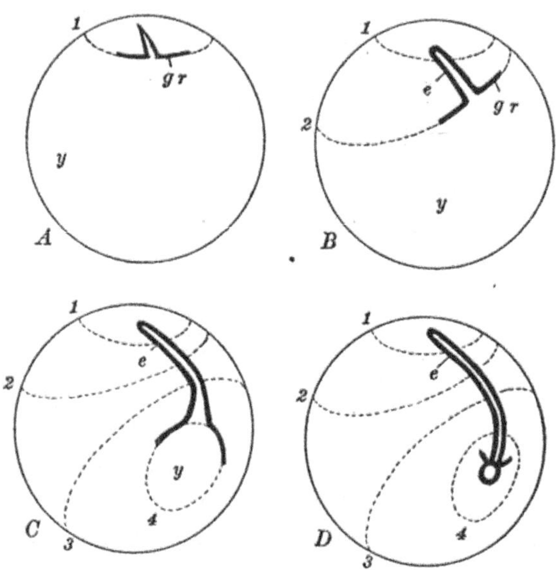

Fig. 162.—Diagrams illustrating four stages in the formation of the Teleost embryo and the growth of the germ ring around the yolk mass. After O. Hertwig. *e*, embryo; *gr*, posterior margin of the germ ring; *y*, yolk mass; 1, 2, 3, 4, successive positions occupied by the germ ring as it advances over the yolk.

the rudiment of the corresponding half of the embryo, which is thus actually formed by the approach and gradual, continued fusion posteriorly of the germ ring. In each half of the ring the essential rudiments of the embryo were thought to be already formed, partly at least, and the process of embryo formation consisted merely or chiefly in the junction or addition of these two originally separate halves. The anterior end of the embryo would thus be formed first, and embryo formation would be complete when the germ ring became fully contracted or closed.

With some modifications of a really fundamental kind, this

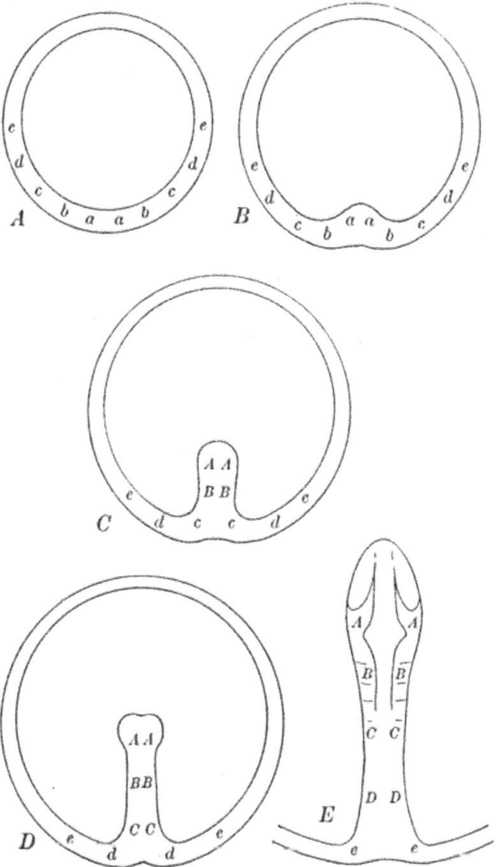

Fɪɢ. 163.—Diagrams of the formation of an embryo by confluence ("concrescence"). *A.* Germ ring before formation of the embryo is indicated. The letters *a–e*, represent symmetrical portions of the germ ring. *B.* Beginning of confluence. *C.* Embryo forming. *AA, BB,* represent regions of the embryo, formed out of the materials of the germ ring at *aa, bb. D, E.* Later stages in the formation of the embryo. The germ ring regions, *cc,* and *dd,* have been differentiated into the embryonic regions, *CC, DD.*

conception is widely adopted to-day. Both observation and experiment have shown, however, that definite halves of an embryo cannot be said to exist preformed in the lateral portions of the germ ring. These regions do contribute to the formation of the median thickening, known as the primitive streak or embryonic rudiment, by a process of gradual fusion posteriorly (Figs. 162, 163). But in this process of coming together, which may better be termed *confluence* (Sumner) than concrescence, the materials from the two sides of the germ ring are fused into a mass which is largely *un*differentiated, and out of this the rudiments of the embryo appear, by a process of differentiation which occurs largely *after* confluence. One side of the germ ring contains not a half of the embryo, but the substance out of which, later, a half of an embryo forms. This process of differentiation is progressive and commences of course in that

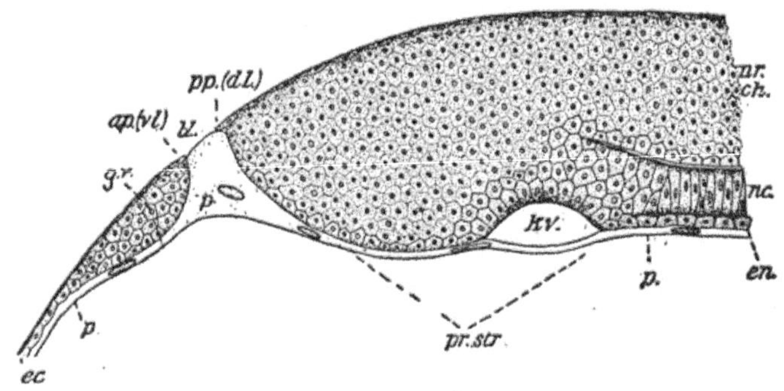

Fig. 164.—Sagittal section through the hinder end of a fish embryo (*Serranus*), showing the undifferentiated primitive streak, anterior to which the structures of the embryo are being differentiated. From H. V. Wilson. *a.p. (v.l.)*, anterior margin of blastoderm or ventral lip of blastopore, after having grown entirely around the yolk mass. *bl.*, blastopore; *ec*, ectoderm; *en.*, endoderm; *g.r.*, germ ring; *k.v.*, Kupffer's vesicle; *nc.*, notochord; *nr. ch.*, nerve cord; *p.*, periblast; *pp. (d.l.)*, posterior margin of blastoderm (dorsal lip of blastopore); *pr. str.*, primitive streak.

part of the primitive streak formed first, *i.e.*, its morphological anterior end (Fig. 163). Then as the primitive streak lengthens posteriorly, the extent of the differentiated region at its anterior end similarly increases posteriorly, roughly keeping pace with the process of elongation (Fig. 164). The primitive streak thus may be regarded as a region which moves backward, receiving posteriorly the diverging limbs of the germ ring, and leaving anteriorly the differentiated rudiments of the embryo. Soon after the germ ring is completely closed or contracted, the primitive streak becomes wholly differentiated, and the rudiments of the embryo may be said to be fully marked out.

The process of concrescence is seen most clearly and typically in those forms with large amounts of yolk and with well-marked germ ring, especially in the Teleosts and Elasmobranchs (Figs. 160, 165).   In the Amphibia, where the amount of yolk is less, and the Sauropsida, where the germ ring is less marked, the process of concrescence, though somewhat modified and slightly obscured, still takes an important part in embryo formation.

THE GERM LAYERS

While this is not the place to give an historical or critical

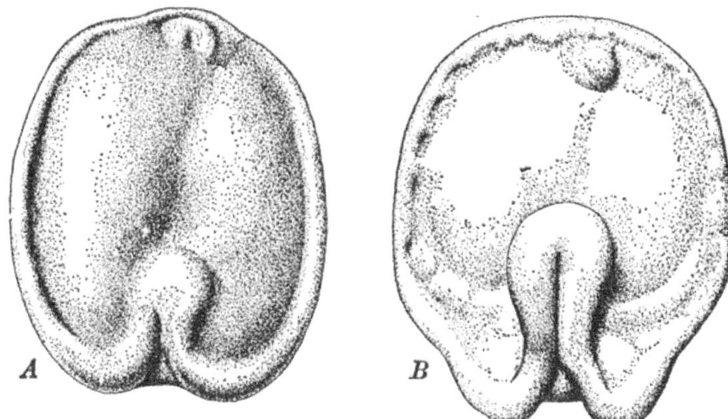

Fig. 165.—Blastoderms of the Elasmobranch, *Torpedo*, showing formation of the embryo. After Ziegler. × 27. *A*. "Stage B." The postero-median thickening is the "embryonic shield," the first indication of the real embryo. *B*. "Stage C." Early embryo; nerve cord rising above the surface of the blastoderm. In both figures the embryonic portion of the blastoderm is directly continuous postero-laterally, with the germ ring, which appears as the thickened margin of the blastoderm.

account of the germ layer theory, it is important that the student should have in mind, before taking up the study of the development of particular organisms, certain fundamental conceptions of the germ layers, and their relation to development, particularly among the Chordata.

When the science of Embryology was itself in a very early stage of its development, the earliest differentiations recognized by the students of animal development, were the sheets or layers of tissue, such as those of the chick, which seemed to give rise to the chief organs of the embryo. These layers of substance were described and their significance recognized, by such pioneers in embryology as C. F. Wolff (1768), Pander (1817), and Von Baer (1828). The whole history of the formation of the systems and organs of the embryo could be read back to these layers, but beyond these, few constant structural features, susceptible of homolgy in different forms, could be made out. The idea became very firmly fixed therefore, that these layers were really the primary differentiations of the embryo, and quite naturally their importance was strongly emphasized.

Subsequent to the statement and establishment of the cell theory, the genesis of the germ layers was traced, the blastula and gastrula fully described in a great variety of forms, and it was found that in spite of the greatest diversity in the earlier processes of development, the general character and structure of the germ layers remained remarkably uniform. And not only were the relations of the germ layers to one another quite constant, but their relation to the tissues and organs of the later embryo were subject to but little variation. In all forms inner and outer layers (endoderm and ectoderm) were present, and in all forms above the Cœlenterates a definite intermediate layer (mesoderm) was to be found. Moreover, in all of these forms the outer layer gave rise to the whole nervous system, central and peripheral, the essential parts of the sense organs, the epidermis and its appendages; from the inner layer came the lining of the digestive tract and its glandular appendages; while the intermediate layer gave rise to the sustentative, vascular, and muscular tissues throughout the body. All of this was finally developed, notably by the Brothers Hertwig (1879–1883), into a carefully and elaborately worked out Germ Layer Theory, the essential points of which were that the three germ layers are entirely homologous throughout the Metazoa, excepting only the Porifera (the Cœlenterates of course lacking a middle layer), and that these layers truly represent the primary and fundamental homologies in the structure of the

Metazoan phyla. Exceptions and contradictions were indeed occasionally noted, but their importance was minimized and they were treated frankly as exceptions, and put down to the account of "cœnogenetic" modifications of "palingenetic" characteristics (see Chapter I).

It is difficult to overestimate the influence of this theory upon the history of Embryology, and upon fundamental embryological ideas. Perhaps no conception, other than the general theory of evolution, has had greater influence in the field of descriptive embryology.

More recently, however, the limitations in the general applicability of this theory have been more fully recognized and the exceptions to the validity of its essential ideas emphasized. At present we must recognize the germ layers as representing a stage in development, just as do the blastula or gastrula, and of no greater or lesser importance than these. The germ layers are descriptive terms of the greatest importance, as such they are indispensable. They are not, however, starting points in any real sense; and to regard them as such is to look forward merely, not backward. Looking both forward and backward we see that the establishment of the germ layers is only one step in the continuous process of development. They represent no more essential homologies than many other features held in common by many developing organisms.

While we cannot consider *in extenso* the facts which have led to this change of opinion regarding the importance of the germ layers, we are bound to state the nature of certain classes of these facts. In the first place are to be noted the difficulties of homologizing the layers of certain groups with their typical condition. For example, in the Porifera that layer which seems entitled to be termed the ectoderm, really gives rise to structures ordinarily derived from endoderm, while the "endoderm" itself forms the covering tissues. In the Mammals the "ectoderm" may contribute little or nothing to the formation of the real embryo and the inner, outer, and middle layers cannot be exactly homologized with those of other Chordates, save by the grace of terminology. In the earlier part of this chapter the varied relations of the mesoderm to the other layers were mentioned; in some cases the middle and inner layers arise from a common rudiment, in others the middle and outer layers. Among the Invertebrates there are many instances of development where even the two primary layers are to be made out only with considerable difficulty, as for example, in the Trematodes, Cestodes, certain of the Bryozoa, *etc.*

In the second place the morphogenetic value of the individual layers is subject to a considerable variation. Thus in the Chordata, leaving aside the Mammals, the mesenchymal connective-tissue cells may be occasionally of "ectodermal" or "endodermal," as well as of "mesodermal" origin. The endothelium of the heart may be "endodermal" or "mesodermal." The notochord may with equal correctness be described as endodermal, mesodermal, or even ectodermal, in various forms.

Single organs like the nephridia may be composites, ectodermal and mesodermal, or, in some cases ectodermal, in others mesodermal.

In the process of regeneration certain contradictions to the germ layer theory become apparent. Organs and tissues normally derived during embryonic development from a certain layer may, during regeneration, be produced from another layer. In certain Oligochætes new mesoderm is of ectodermal origin, and the regenerated pharynx may be lined with endodermal, rather than ectodermal cells.

Especially in the process of budding, as it occurs in a great many groups, do we find abundant exceptions to this theory. In some of the Polyzoa the gut may be of ectodermal origin; the nervous system and pharynx are mesodermal in some of the flatworms. Analogous conditions are very common among the Tunicates; here the pharynx may be endodermal or ectodermal; the atrium and even the nervous system may be ectodermal, mesodermal, or endodermal, in different forms where in egg development the relations of these structures to the germ layers are typical.

Finally, the most important qualifications and limitations of the germ layer theory grow out of the observed facts of normal development prior to the formation of the germ layers. These structures are by no means the earliest constant embryonic differentiations, and as we have seen in the chapter on cleavage, it is just as easy to draw homologies between cell groups in the blastula stage, or in an earlier cleavage stage, as it is between the later appearing germ layers. It is not too much to say that in some cases homologies may be drawn between various formed substances in the undivided egg. Animal and vegetal poles of the ovum, cleavage patterns, cell-groups, micromeres, macromeres, upper and lower poles of the blastula, are all constant and comparable features of development no less than inner, outer and middle germ layers. We may recall that the cell known as 4*d* may be identified and its history and fate compared, in the cleavage of many groups, even in different phyla. This cell whose form, position, and derivation are so constant, may or may not form "mesoderm"; even when it does form mesoderm this may go to form very different parts of the embryonic and adult structure. Often the "mesoderm" may be a *cell*, just as truly as a *layer*.

Summarizing we may say that while the arrangement of the cells of the embryo in the form of definite layers is *almost* universal, at the same time, in the comparison of different groups or of different modes of development, these layers exhibit great inconstancy in their relations to one another, and to the structures forming them and formed from them. The germ layers are valuable, indeed indispensable descriptive units, but they do not represent primary differentiations, and their homologies are no more, though probably no less, fundamental throughout groups larger than phyla, than are many other structures of the developing organism.

MORPHOGENETIC PROCESSES

It remains now to describe some of the more general processes by which the rudiments of the organs and tissues of the embryo may be formed out of the layers or cell masses of the gastrula and post-gastrula stages. We shall not attempt to describe here the actual formation of any specific embryonic structure, but rather shall give a brief classification of the more common and important processes, a few of which have already been mentioned earlier in this chapter.

While the morphogenetic processes within the embryo show the greatest diversity and vary almost infinitely in specific details, yet it is possible to include them all under a few heads, when these matters of detail are omitted. Again it should be recalled that we are limiting our description to the Chordata.

The primary condition of morphogenesis is cell multiplication. After each division the daughter cells increase to practically the size of the parent cell; and numerical increase in cells, together with their growth, *i.e.*, cell proliferation, play either a primary or a secondary part in every morphogenetic process. When the process of cell division is quite general throughout the extent of the germ disc or layer, the result is an increase in the thickness or in the extent of the sheet, respectively, when the plane of the cell divisions is in general parallel with, or perpendicular to, the plane of the whole layer. If there should be little or no regularity in the positions of the division planes, the membrane would increase in all directions (Fig. 166).

Ordinarily, in embryogeny, cell multiplication and growth are more intense in restricted areas of the blastoderm or germ layer. It is convenient then to distinguish between (*a*) those processes in which the multiplying cells tend to remain associated in the same general region, and (*b*) other processes where they become more or less separated, either from one another or from their seat of origin. Under the former head we must again distinguish between the results of increase in thickness and in extent. A localized increase in thickness is frequently termed a *bud;* buds may project either above (limb bud) or below (Teleostean lens) the free surface where they are formed. If the thickening region is elongated the result may be the formation of a strand or plate of cells, again either a ridge-like structure above the surface of the layer (genital ridge), or a keel-like thickening below the surface (Teleostean nerve cord, in part).

Increase in extent of a localized area frequently involves the obstructive action of the region bounding the area. When this form of growth occurs generally, so that the tendency to extension occurs in every direction from the middle of the area concerned, the result is frequently an arching, either outward or inward. This may take the form of a simple arching as in the Teleostean blastula (Fig. 150, *C*), or the same process may be carried farther and followed by a constriction near the base. Such processes are very common indeed and are termed *invagination* and *evagination*, according as the growth is below or above the free surface. Simple illustrations of evagination are afforded by the formation of intestinal villi, the rudiments of lung or thymus, and

the like; typical invaginations are seen in the formation of the optic
cup out of the optic lobe, the auditory sac, *etc.*   When this form of
growth in extent is limited to certain directions instead of occurring
radially, the result is often the formation of a fold which bears somewhat the same relation to the dilation that the strand does to the bud.
The fold also may be above the surface of the membrane, forming a
sort of arch or hollow ridge, usually bounded by lateral depressions
(frog's pronephric duct), or below the surface forming a groove or furrow
bordered by lateral elevations (medullary groove) (Fig. 167).   In some
instances a solid strand may be formed in this way instead of by the
simpler process of direct increase in thickness (Teleostean nerve cord,

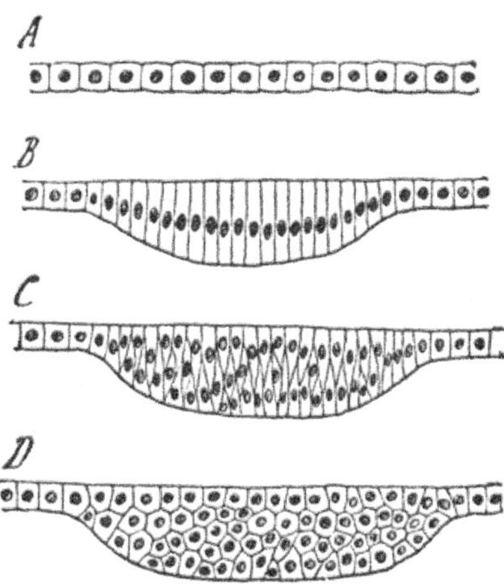

Fig. 166.—Diagrams of four stages in the formation of an epithelial thickening,
several layers of cells deep.   From Korschelt and Heider.

in part).   When the processes leading to the formation of a groove are
continued, the groove may be converted into a closed canal by the
approach, apposition, and fusion of the borders (neural tube) (Fig. 167).

In those conditions where the proliferating cells become, to some
extent, separated either from one another or from the proliferating
region itself, we may note first, instances of actual cell *migration.*
This may be either emigration or immigration, according as to whether
we fix attention upon the source or the destination of the migratory
cells (mesenchyme cells).   Secondary processes of thickening or thinning
may accompany these processes.   In other cases the movement of cells
may be described as rearrangement rather than migration; this may be
illustrated by the formation of mesodermal somites and blood islands
(chick).

One of the common morphogenetic processes is a combination of
increase in thickness and cell rearrangement, such as the usual forma-

tion of the notochord, or the process of, *delamination*, which consists
in the splitting of a single thickened sheet into two separate layers,
either as a whole or in localized areas (formation of mesoderm in the
frog, or division of the mesoderm into somatic and splanchnic layers);
in some instances the initial thickening may not be very apparent.·

Occasionally cells of different layers, or of different rudiments, meet
and *fuse*, forming a continuous rudimentary mass (pituitary body).

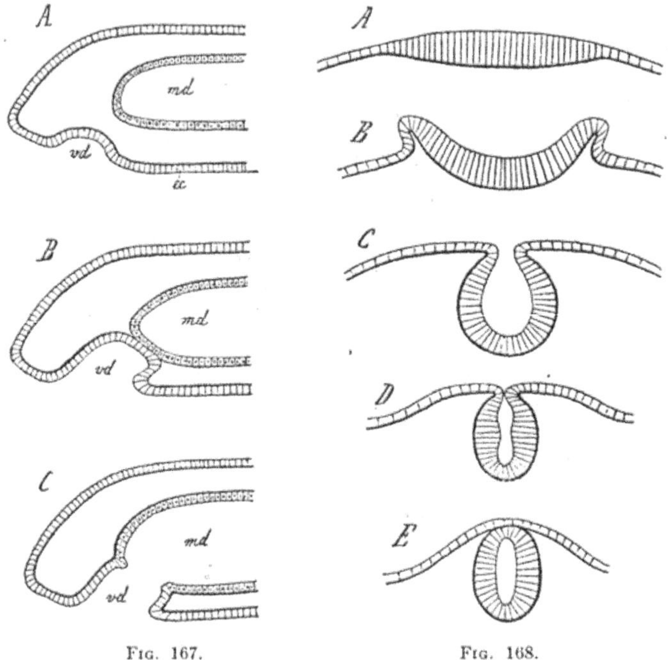

Fig. 167.   Fig. 168.

Fig. 167.—Diagrams of the formation of the medullary canal or neural tube, in
the Vertebrates. From Korschelt and Heider.

Fig. 168.—Diagrams of the formation of the mouth and stomodæum in a
typical form. From Korschelt and Heider. *ec*, ectoderm; *md*, pharynx:
vd, stomodæum.

Finally we may mention certain morphogenetic processes of a wholly
different kind, namely, *resorption*, and changes in the *form* and *size* of
cells. From this point of view merely, the process of resorption may be
regarded as the reverse of proliferation. Definite rudiments may appear
first as spaces thus formed by the gradual dissolution and absorption of
cells in certain areas. Thus the oral and anal openings, gill-clefts, *etc.*,
are usually formed as "perforations" by the resorption of areas where
previously separate and continuous layers became united by a process
of fusion (Fig. 168). In other cases rudiments once established may

gradually disappear, either wholly (tail of tadpole), or in part (pronephros) remaining then as vestigial organs.

Changes in cell form and size chiefly fall under the head of tissue differentiation, or *histogenesis*, but in a few instances such processes are primarily involved in the formation of rudiments. The development of the eye affords illustrations of these processes; the lens forms as a thickening of the cells on one side, and the thinning on the other, of a sac originally formed by invagination, and the optic lobe, at first nearly spherical, flattens and invaginates, one layer becoming thickened as the chief part of the retina (later complicated by cell proliferation) while the other layer forms a thin pigmented layer.

These processes of morphogenetic value are tabulated in the accompanying summary; the arrangement here is merely a convenient one and has no other significance.

It is extremely important to recognize that these morphogenetic processes described above are not merely simple mechanical processes. The arrangement and behavior of the cells in an invaginating or delaminating region are determined by other factors than those of physical resistances, attractions, *etc.* These events are both in fundamentals and in details to be regarded as active phenomena of a living organism. They are frequently also to be understood from the historical or adaptive points of view.

What the precise conditions are which determine the nature of these events, may usually only be conjectured. In some cases they may result from osmotic conditions, absorption of water, *etc.* But for the most part the nature of the specific stimuli, and the conditions within the layer or cell group, which lead to the definite reactions of rudiment formation are unknown. And in this particular field of development we can do little more than to describe what happens from the morphological viewpoint.

SUMMARY OF THE CHIEF MORPHOGENETIC PROCESSES

I. Cell division and growth throughout the layer or disc, resulting in
    (a) Increase in thickness.
    (b) Increase in extent.
    (c) Increase in both thickness and extent.

II. Cell division and growth localized in restricted areas of the layer or disc.

    A. Cells remain related and in continuity
        (a) Increasing in thickness.

            1. Radially—formation of buds $\left\{\begin{array}{l}\text{solid.}\\ \text{hollow.}\end{array}\right.$

            2. Chiefly in one axis—formation of strands $\left\{\begin{array}{l}\text{Ridge.}\\ \text{Keel.}\end{array}\right.$

        (b) Increasing in extent.

            1. Radially $\left\{\begin{array}{l}\text{Dilation.}\\ \text{Invagination.}\\ \text{Evagination.}\end{array}\right.$

            2. Chiefly in one axis $\left\{\begin{array}{l}\text{Groove.}\\ \text{Folds.}\\ \text{Tube (strand secondarily).}\end{array}\right.$

    B. Cells become separated to a varying degree.

        (a) Migration $\left\{\begin{array}{l}\text{Immigration}\\ \text{Emigration}\end{array}\right\}$ with corresponding thickening and thinning.
        (b) Rearrangement.
        (c) Delamination (sometimes preceded by thickening).
        (d) Fusion.

III. Resorption.
    (a) Accompanying a process of perforation.
    (b) Disappearance (degeneration).

IV. Changes in cell form and size.
    (a) Thickening.
    (b) Thinning.

REFERENCES TO LITERATURE

CERFONTAINE, P., Recherches sur le développement de l'Amphioxus. Arch. Biol. **22.** 1906.

DAVENPORT, C. B., Studies in Morphogenesis. IV. A preliminary Catalogue of the Processes concerned in Ontogeny. Bull. Mus. Comp. Zool. Harvard Coll. **27.** 1896.

EYCLESHYMER, A. C., The Formation of the Embryo of *Necturus*, with Remarks on the Theory of Concrescence. Anat. Anz. **21.** 1902.

GREIL, A., Ueber die erste Anlage der Gefässe und des Blutes bei

Holo- und Meroblastiern.   Verh. Anat. Gesell., in Anat. Anz. **32.** 1908.

HERTWIG, O., Die Lehre von den Keimblättern. Handbuch, *etc.* I, **1, 1.** 1903 (1906).

HERTWIG, O., und R., Studies on the Germ Layers. Jena. Zeit. **13-16 (6-9).** 1879-1883.

HIS, W., Untersuchungen über die Entwickelung von Knochenfischen, besonders über diejenige des Salmens. Zeit. Anat. Entw. **1.** 1876. Untersuchungen über die Bildung des Knochenfischembryo. Arch. Anat. Entw. 1878.

JENKINSON, J. W., (Ref. Ch. VII.)

KEIBEL, F., Die Gastrulation und die Keimblattbildung der Wirbeltiere, Ergebnisse Anat. u. Entw. **10.** 1900 (1901).

KOPSCH, F., Untersuchungen über Gastrulation und Embryobildung bei den Chordaten. I. Die Morphologische Bedeutung des Keimhautrandes und die Embryobildung bei der Forelle. Leipzig. 1904.

KORSCHELT UND HEIDER, Lehrbuch, *etc.* I Abschnitt. Experimentelle Entwicklungsgeschichte. Jena. 1902. III Abschnitt. Furchung und Keimblätterbildung. Jena. 1909-1910.

LILLIE, F. R., The Development of the Chick. New York. 1908.

PATTERSON, J. T., On Gastrulation and the Origin of the Primitive Streak in the Pigeon's Egg.—Preliminary Notice. Biol. Bull. **13.** 1907.

PRENANT, A., BOUIN, P., ET MAILLARD, L., Traité d'Histologie. Paris. 1911.

SCHAUINSLAND, H., Beiträge zur Entwickelungsgeschichte und Anatomie der Wirbeltiere. I, II, III. Bibliotheca Zoologica. **39.** Stuttgart. 1901-1903.

SCHULTZE, O., Ueber das erste Auftreten der bilateralen Symmetrie im Verlauf der Entwicklung. Arch. mikr. Anat. **55.** 1900.

SELYS-LONGCHAMPS, M. DE, Gastrulation et formation des feuillets chez *Petromyzon planeri.* Arch. Biol. **25.** 1910.

SUMNER, F. B., Kupffer's Vesicle and its Relation to Gastrulation and Concrescence. Mem. N. Y. Acad. Sci. **2.** 1900. A Study of early Fish Development. Experimental and Morphological. Arch. Entw.-Mech. **17.** 1903.

WHITMAN, C. O., The Embryology of *Clepsine.* Q.J.M.S. **18.** 1878.

WILL, L., Beiträge zur Entwicklungsgeschichte der Reptilien. I. Die Anlage der Keimblätter beim Gecko (*Platydactylus facetanus*). Schreib. Zool. Jahrb. **6.** 1892.

WILSON, H. V., (Ref. Ch. VI.)

ZIEGLER, H. E., (Ref. Ch. III.)

ZIEGLER, H. E., und F., Beiträge zur Entwickelungsgeschichte von *Torpedo.* Arch. mikr. Anat. **39.** 1892.

INDEX

Made in the USA
Monee, IL
07 July 2026

56551649R00225